Ecosystem-Based Management: Applications to Dynamic Environments

Ecosystem-Based Management: Applications to Dynamic Environments

Edited by Leanna Perry

SYRAWOOD
PUBLISHING HOUSE
New York

Published by Syrawood Publishing House,
750 Third Avenue, 9th Floor,
New York, NY 10017, USA
www.syrawoodpublishinghouse.com

Ecosystem-Based Management: Applications to Dynamic Environments
Edited by Leanna Perry

International Standard Book Number: 978-1-64740-377-5 (Hardback)

Cataloging-in-publication Data

Ecosystem-based management : applications to dynamic
environments / edited by Leanna Perry.
p. cm.
Includes bibliographical references and index.
ISBN 978-1-64740-377-5
1. Ecosystem management. 2. Environmental management.
3. Applied ecology. I. Perry, Leanna.
QH75 .E26 2023
333.72--dc23

TABLE OF CONTENTS

Preface VII

Chapter 1 **Approaches for Estimating the Supply of Ecosystem Services: Concepts for Ecosystem-Based Management in Coastal and Marine Environments** 1
Fiona E. Culhane, Leonie A. Robinson and Ana I. Lillebø

Chapter 2 **Prioritizing Stakeholders, Beneficiaries and Environmental Attributes: A Tool for Ecosystem-Based Management** 22
Leah M. Sharpe, Connie L. Hernandez and Chloe A. Jackson

Chapter 3 **Ecosystem-Based Management (EBM) and Ecosystem Services in EU Law, Policy and Governance** 45
Anne Marie O'Hagan

Chapter 4 **Ecosystem-Based Management to Support Conservation and Restoration Efforts** 64
Andrea Funk, Timothy G. O'Higgins, Florian Borgwardt, Daniel Trauner and Thomas Hein

Chapter 5 **Using Stakeholder Engagement, Translational Science and Decision Support Tools for Ecosystem-Based Management** 78
Rebekah Gibble, Lori Miller and Matthew C. Harwell

Chapter 6 **From DPSIR the DAPSI(W) R(M) Emerges... a Butterfly – *'protecting the natural stuff and delivering the human stuff'*** 103
Michael Elliott and Timothy G. O'Higgins

Chapter 7 **Mathematical Modeling for Ecosystem-Based Management (EBM) and Ecosystem Goods and Services (EGS) Assessment** 128
Richard S. Fulford, Sheila J. J. Heymans and Wei Wu

Chapter 8 **Predicting Future Vegetated Landscapes under Climate Change: Application of the Environmental Stratification Methodology to Protected Areas** 143
John M. Johnston, Robert J. Zomer and Ming-cheng Wang

Chapter 9 **Advancing Aquatic Ecosystem-Based Management with Full Consideration of the Social-Ecological System** 163
Gerjan Piet, Gonzalo Delacámara, Marloes Kraan, Christine Röckmann and Manuel Lago

Chapter 10 **Unravelling the Relationship between Ecosystem-Based Management, Integrated Coastal Zone Management and Marine Spatial Planning** 184
Martin Le Tissier

Chapter 11 **Mitigating Negative Unintended Impacts on Biodiversity in the Natura 2000 Vouga Estuary** 195
Ana I. Lillebø, Heliana Teixeira, Javier Martínez-López, Ana Genua-Olmedo, Asya Marhubi, Gonzalo Delacámara, Verena Mattheiß, Pierre Strosser, Timothy G. O'Higgins and António A. J. Nogueira

Chapter 12 **Ecosystem-Based Management and Natural Capital Accounting** 232
Marc Russell, Charles Rhodes, George Van Houtven, Paramita Sinha, Katherine Warnell and Matthew C. Harwell

Permissions

List of Contributors

Index

PREFACE

Every book is initially just a concept; it takes months of research and hard work to give it the final shape in which the readers receive it. In its early stages, this book also went through rigorous reviewing. The notable contributions made by experts from across the globe were first molded into patterned chapters and then arranged in a sensibly sequential manner to bring out the best results.

A dynamic environment refers to an environment where the physical and biological elements are in a state of constant change. Ecosystem-based management (EBM) is an integrated, science-based approach to the management of natural resources with an objective to sustain the health, resilience and diversity of ecosystems while allowing for sustainable use of the goods and services provided by it. One of the areas where ecosystem-based management finds prominent use are marine and coastal environments. The application of EBM to marine and coastal ecosystems seeks to develop decision support tools for evaluating the risks related to potential management actions and facilitates strategic decision making. This book outlines the applications of ecosystem-based management in dynamic environments. A number of latest researches have been included to keep the readers updated with the global concepts in this area of study. The book is appropriate for students seeking detailed information on this topic as well as for experts.

It has been my immense pleasure to be a part of this project and to contribute my years of learning in such a meaningful form. I would like to take this opportunity to thank all the people who have been associated with the completion of this book at any step.

Editor

1

Approaches for Estimating the Supply of Ecosystem Services: Concepts for Ecosystem-Based Management in Coastal and Marine Environments

Fiona E. Culhane, Leonie A. Robinson, and Ana I. Lillebø

Abstract Ecosystem services have emerged as a critical concept for Ecosystem-Based Management (EBM) in aquatic environments, namely in coastal and marine environments. However, despite conceptual advances over the last two decades, major challenges remain in the operationalisation of ecosystem service concepts and practical application. This chapter describes a selection of EBM assessment approaches applied to coastal and marine environments, where the ecosystem services approach is key. These approaches range from qualitative to quantitative, all being transdisciplinary. In the first, (ODEMM project, supported by linkage frameworks) trade-offs in EBM management options can be considered, in terms of their potential to reduce ecological risk, maintain sustainable supply of ecosystem services, and the governance complexity associated with implementing them. In the second, (AQUACROSS project, combining causality links relations and spatial multicriteria analysis) trade-offs are supported by maps with governance boundaries, spatially explicit valuation of ecosystem services and ecological risk. In the third, (MCES project, supported by the GIS-based modelling tool InVest) trade-offs are supported by a developed spatially explicit proxy for the habitats' vulnerability to deliver ecosystem services. Finally, we describe a policy-based operational assessment tool that allows users to assess the current and future capacity of regional seas to supply ecosystem services, based on their current and future ecosystem status reporting. We go on to describe some lessons learnt from our experience in applying these approaches.

F. E. Culhane (✉) · L. A. Robinson
School of Environmental Sciences, University of Liverpool, Liverpool, UK
e-mail: F.Culhane@liverpool.ac.uk; Leonie.Robinson@liverpool.ac.uk

A. I. Lillebø
Department of Biology & CESAM, Campus Universitário de Santiago, University of Aveiro, Aveiro, Portugal
e-mail: lillebo@ua.pt

Lessons Learned

- For marine ecosystems, data availability is often a barrier to operationalising the ecosystem service concept. Ideally, spatial data would be available and, in many cases, it is becoming so. However, assessments are needed now. We show that there is existing information that can be applied to ecosystem service assessments for marine environments, and this should not be a barrier to carrying out assessments now.
- Consider all the ways that nature contributes to human wellbeing. There are criticisms of some approaches to ecosystem service classifications and assessments, because the services cannot be clearly linked to market values and economic assessments. This narrows our perspective on the breadth of ecosystem services. We show that by carrying out a supply side assessment, all the ways that nature contributes to wellbeing can be considered. This can be seen as complementary to demand side assessments and as an end-point in itself. Economic valuation does not need to be the only end point.

Needs to Advance EBM

- Scientists and policy makers need to be open to draw on different approaches including expert judgement and stakeholder knowledge, policy information, as well as, detailed habitat mapping or spatially explicit modelling techniques. These can then be used, together or in isolation, to show how ecosystem integrity can affect human well-being, fulfilling the critical need for balanced (economic, ecological, social) management actions to be taken.

1 Introduction

The concept of nature benefits for humans, is not new, however, it was first in 1983, almost forty years ago, that Ehrlich and Mooney used the term '*Ecosystem Services*' in an International Scientific Indexing (ISI) journal (see Flood et al. (2020) for a discussion of earlier development of the concept). Under the title '*Extinction, Substitution, and Ecosystem Services*' and using different biomes, authors showcased the need for a "*conservative approach to the maintenance of services through minimizing anthropogenic extinctions*" (Ehrlich and Mooney 1983). Regarding the marine biome, the biggest biome in the world, and the accompanying ecosystems services, authors highlighted the impact of fisheries over fish stocks and the role of economy as a driver for extinction and for substitution of target species. Their recommendations followed the need for a "*careful preservation of ecosystems and thus of the populations and species that function within them*". Since then the concept of '*Ecosystem Services*' has evolved in order to become effectively operationalised, but major challenges still remain, namely, how to decide who will win and who will lose, as trade-offs are inherent to the decision making process. This is of paramount importance in the context of global socio-ecological challenges and sustainable development strategies for coastal and marine environments. Such challenges include human indirect drivers like sea level rise, extreme weather events (e.g. floods and storm surges) and invasive species, and human direct drivers (e.g. economic activities). Global strategies tackling these challenges include those based around Blue Growth, Biodiversity and Climate Change. Ecosystem-Based Management (EBM) acknowledges that human well-being and ecological status are linked and integrates multiple drivers and pressures into adaptive management plans (UNEP 2011). In this context, ecosystem services have emerged as a critical concept to operationalise EBM.

1.1 Ecosystem Services Concept

Ecosystem services have been defined in different ways over the years, with the term services often being used interchangeably to mean "*the benefits people obtain from ecosystems*" (MA 2005), to the ecosystem structures, processes or functions that generate the services. More recently, the ecosystem service cascade model has been widely adopted to clearly delineate where a service sits in relation to what generates it (in the ecosystem), and what benefits people get from it (in the socio-economic system) (Potschin and Haines-Young 2011). Following the rationale of the ecosystem services 'cascade' model and Culhane et al. (2019a), we define ecosystem services here as:

> *Ecosystem services represent the flow of ecosystem capital that is realised because of a human active or passive demand (modified from EEA (2015)). They are thus the final outputs from ecosystems that are directly consumed, used (actively or passively) or enjoyed by people.* (Fisher et al. 2009; Haines-Young and Potschin 2013; Maes et al. 2013)

Examples of coastal and marine ecosystem services include nutrition from fish and shellfish, flood and coastal protection from saltmarsh habitats, and artistic inspiration from seascapes and marine animals (Culhane et al. 2019a, and see Fig. 1 for further examples). In order to recognise all the services that ecosystems and marine environments supply, typologies of services have been developed. These categorise services and make their assessment operational. Early international initiatives to develop typologies classified services under four broad categories: provisioning (such as food from fish); regulation and maintenance (such as waste regulation); supporting (such as primary production); and cultural services (such as marine species to observe or to research) (MA 2005; TEEB 2010). The concept of Final Goods and Ecosystem Services (FEGS) also developed (see DeWitt et al. (2020) for a summary of this). These are a subset of ecosystem services, generally not including the supporting services, that can be directly linked to a beneficiary, thus avoiding double-counting when assigning a monetary or market value. Following these, ecosystem service typologies have been further developed and/or adapted for the marine environment (e.g. Böhnke-Henrichs et al. 2013), see Fig. 1 for some examples from this typology, which retains supporting services). CICES, the Common International Classification of Ecosystem Services (Haines-Young and Potschin

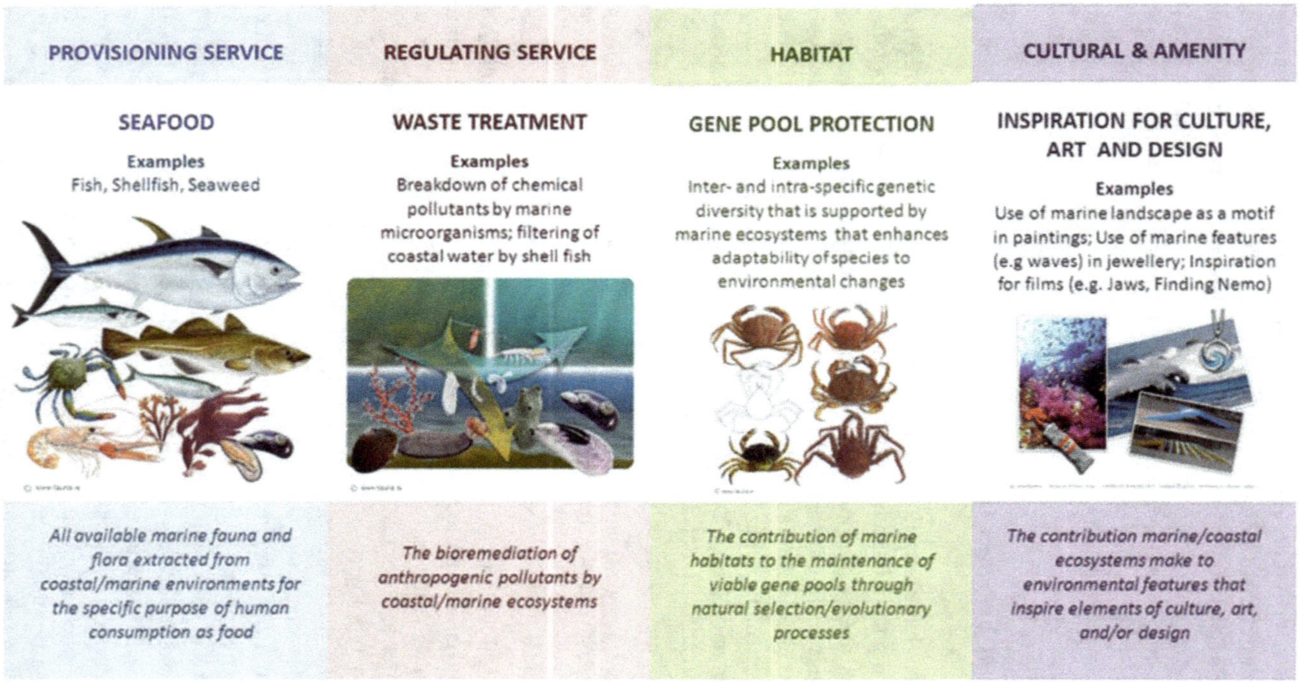

Fig. 1 Examples of marine ecosystem services from the typology of Böhnke-Henrichs et al. 2013, displayed as cards developed in the ODEMM project (Robinson et al. 2014). Habitat services link to those in the TEEB classification and are included under the Supporting Services of other classifications, such as the Millennium Ecosystem Assessment. The ODEMM ecosystem services cards are available to download for use from https://www.odemm.com/content/cost-and-benefits-analyses

2013) is the reference typology for the European Union's assessments linked to the Biodiversity Strategy (EC 2011). It is a hierarchical typology that recognises three categories of ecosystem services: provisioning, regulation and maintenance, and cultural services at the broadest level and further subdivides these categories into more specific services at lower levels (see Flood et al. (2020) for further discussion of ecosystem service classification systems). CICES includes only 'final' services, though one criticism of the typology is that some of the included services are, in fact, supporting (or intermediate). CICES was developed primarily for terrestrial environments but has been adapted for the marine environment (Culhane et al. 2019a), and should be applicable across biomes allowing a Europe wide assessment.

1.2 Policy Background

Around twenty years ago, Costanza and colleagues estimated the value of the world's Ecosystem Services and natural capital, showing that coastal (tidal marsh and mangroves) and marine (open ocean, shelf, estuaries, seagrasses) services accounted for circa 68% of the global value for ecosystem services (Costanza et al. 1997). Just after, in the year 2000, the Millennium Ecosystem Assessment was called for by the United Nations Secretary-General Kofi Annan, as part of the '*The Role of the United Nations in the 21st Century*'. Regarding ecosystems, the objective was "*to assess the consequences of ecosystem change for human well-being and to establish the scientific basis for actions needed to enhance the conservation and sustainable use of ecosystems and their contributions to human well-being*" (MA 2005).

Then ten years ago, the global initiative '*The Economics of Ecosystems and Biodiversity (TEEB)*' focused on '*making nature's values visible*' in order to mainstream the values of biodiversity and Ecosystem Services into decision-making at all levels (TEEB 2010). Within TEEB special attention is given to 'blue growth' and human dependence on healthy ocean ecosystems and on coastal and marine biodiversity. About the same time, the European Union (EU) Biodiversity Strategy for 2020 aimed to halt the loss of biodiversity and Ecosystem Services, reflecting the commitments taken in 2010 within the International Convention on Biological Diversity. As part of the strategy, working groups for mapping and assessment of ecosystem services, including a marine pilot, provided supporting guidance for EU member states (Maes et al. 2016). Just after, in 2012, an independent intergovernmental body of the United Nations '*The Intergovernmental Science-Policy Platform on Biodiversity and Ecosystem Services (IPBES)*' was established "*to strengthen knowledge foundations for better policy through science, for the conservation and sustainable use of biodiversity, long-term human well-being and sustainable development*". This platform stands for nature and biodiversity, in the same way that the '*Intergovernmental Panel on Climate Change (IPCC)*' provides the latest science-based assessments related to climate change, including possible response options.

In the last five years, different but complementary initiatives have taken place acknowledging the need for global actions towards unprecedented changes in nature, climate and human population growth. In 2015, the United Nations, together with governments, businesses and civil society, agreed on the Sustainable Development Agenda for 2030 that integrates seventeen interlinked Sustainable Development Goals (SDG's). Although, SDGs should not be addressed individually, SDG 14 is devoted to '*Conserve and sustainably use the oceans, seas and marine resources for sustainable development*'. In the same year the 'Paris Agreement on climate change' entered into force, being essential for the achievement of the SDGs. The former initiatives, TEEB, IPBES, as well as the EU Strategy for biodiversity beyond 2020 are also framed in the scope of the SDGs.

In this context, Ecosystem-Based Management (EBM), a holistic approach that aims to balance the multiple interrelated dimensions of ecological integrity and human well-being, appears a useful framework to operationalise the concept of ecosystem services (Gómez et al. 2016, 2017). Likewise, one can argue that achieving EBM might be more attainable where the ecosystem services approach is included, since it enables stakeholders and decision makers to see a tangible way in which the integrity of ecosystems directly (actively or passively) affects the well-being of humans. To this end this chapter aims at showcasing selected EBM assessment approaches applied to coastal and marine environments that incorporate understanding and assessment of ecosystem services and to draw lessons learnt from our experience.

2 Operationalising Ecosystem Services in EBM

2.1 Ecosystem Services and Trade-Offs in EBM Management Options

Different decision support tools can support trade-off options regarding the provisioning of ecosystem services. These can be supported by linkage frameworks (e.g., see Robinson and Culhane 2020), by causality links relations (e.g., AquaLinksTool, Nogueira 2018), by spatially-explicit GIS-based modelling tools (e.g., Willaert et al. 2019) or by a combination of the above mentioned decision support tools (e.g., see Lillebø et al. 2020). Three selected examples are now presented.

2.1.1 ODEMM Project: https://www.odemm.com

In the ODEMM project a typology of marine ecosystem services was developed (Böhnke-Henrichs et al. 2013) and an assessment undertaken whereby stakeholders compared management options based on three major criteria: ecological risk, ecosystem service supply and governance complexity (Robinson et al. 2014: Chap. 7).

As illustrated in Fig. 2, management options could be applied to reduce impacts through a number of different pathways, and the reduction in risk to ecological components was considered in terms of any resultant change in the supply of ecosystem services (a framework that aligns with the DPSIR, and later DAPSI(W)R(M) concepts, see Elliott and O'Higgins (2020)). This was considered against the complexity of governance required to implement the compared management options, and the effect of ecological risk reduction on potential achievement of ecological goals as set out by the Marine Strategy Framework Directive.

It was not always the case that management options delivered benefits in terms of all three criteria considered in the same way (Fig. 3). In terms of the examples explored, stakeholders generally found that the management option that delivered the best reduction in risk to achievement of the MSFD good environmental status objectives, was also least complex in terms of governance required to instigate it, but was not the most promising in terms of the benefits to sustainable supply of services. This helps to illustrate the trade-offs experienced in EBM. Participants involved in the ODEMM trade off exercises found the approach to be very useful in terms of helping to visualise "how the sea works", providing a "practical approach to link management options with potential changes in the provision of ecosystem services" (ODEMM Deliverable 19).

2.1.2 AQUACROSS Project: https://aquacross.eu

The EBM approach (see Piet et al. 2020) was applied to the Vouga river coastal watershed, along a continuum of freshwater to marine habitats under the classification of the Natura 2000 network (Lillebø et al. 2018). This case study also aimed to showcase causality links in a linkage chain relating activities, pressures and habitats/highly mobile biotic groups and ecosystem services (Teixeira et al. 2018, 2019; Culhane et al. 2019b) and to assess the vulnerability of ecosystem components regarding the provisioning of ecosystem services (Lillebø et al. 2018).

In this social-ecological system and as part of the EBM framework, stakeholders were invited to value ecosystem services through a spatial multi-criteria analysis that took place at two different spatial scales (Lillebø et al. 2019; Martínez-López et al. 2019). From the resulting prioritization maps representing stakeholder's perceptions and beliefs regarding ecosystem services, valuations were generated and supported the discussion of the areas for potential interventions (ecosystems restoration) and associated trade-offs. As part of the EBM trans-disciplinary approach, stakeholders were also invited to express their concerns regarding the foreseen management options. Simultaneously, causality links and risk assessment were undertaken though a tool that establishes a linkage chain relating activities, pressures and habitats/highly mobile biotic groups and ecosystem services, to assess the vulnerability of ecosystem components regarding the provisioning of ecosystem services (AquaLinksTool, Nogueira (2018)) as shown in Fig. 4. The full linkage matrices dataset is freely available for download https://zenodo.org/record/1101159#.XbKsfS3MyqA.

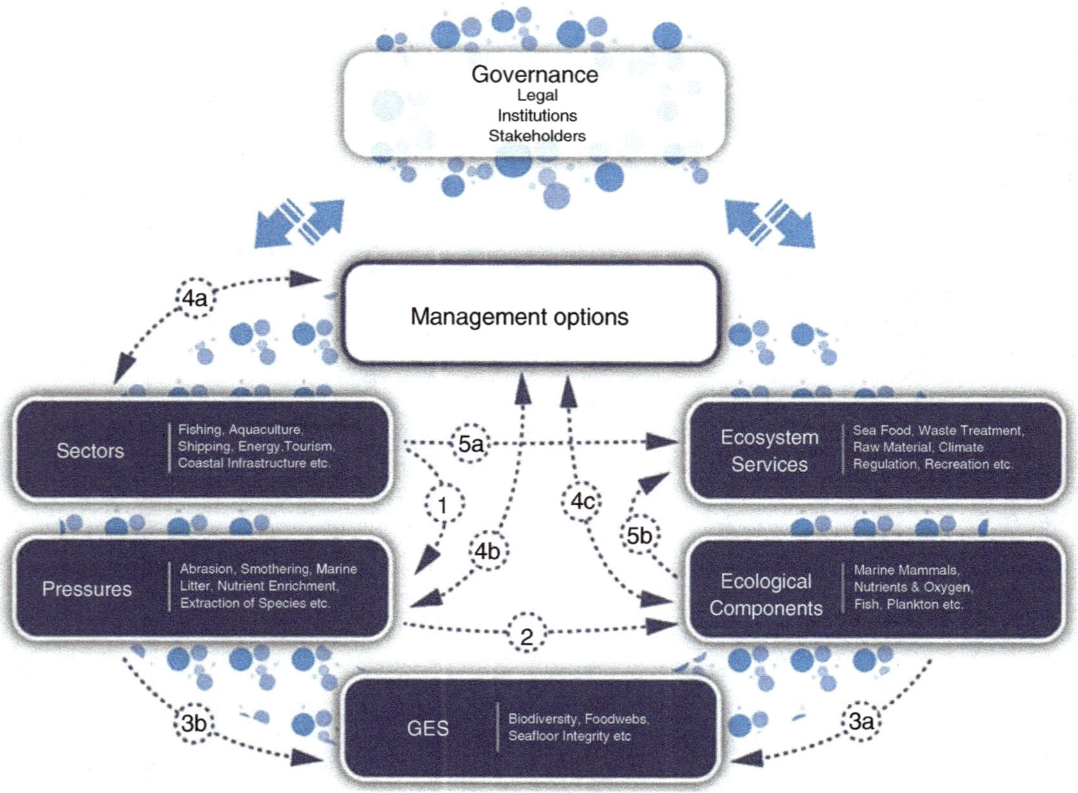

Fig. 2 The ODEMM Cost and Benefits analyses consider how the appraisal of management options can take into consideration both associated costs and benefits (where benefits are described by the supply of ecosystem services). Here, the supply of ecosystem services can be altered when management options are instigated (4 a–c), either though controls on sectors that supply key ecosystem services (5a) (e.g. instigating catch control on fisheries can alter supply of Seafood directly), or though the alteration in state of ecological components that contribute to the supply of ESs (5b). [State of ecological components may change due to management options acting directly on ecological components (4c) or acting on pressures (4b) and/or sectors (4a) that impact ecological components (2)]. Costs associated with management options may arise at both the level of the sector and/or at institutes involved in the governance of those management options. (From Robinson et al. 2014)

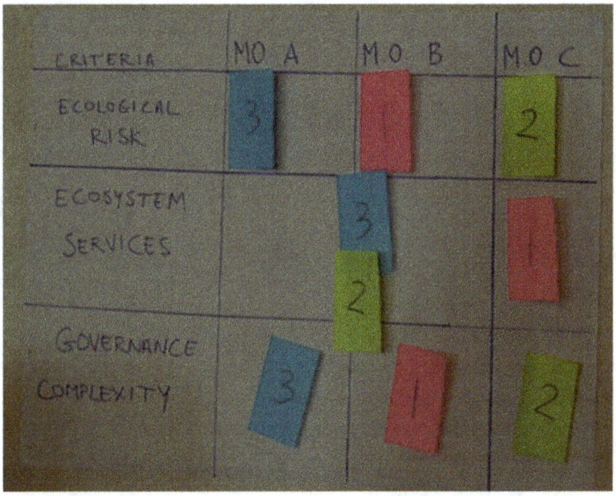

Fig. 3 A final outcome table following completion of all three exercises, where numbers represent the order in which the Management Options (MO A, B, C) work best, in terms of outcomes for the different criteria considered

The vulnerable habitats selected through the AquaLinksTool clearly matched stakeholders' concerns, as well as their ecosystem services prioritization maps (Lillebø et al. 2019; Martínez-López et al. 2019). The combined approach contributed to the effectiveness (hitting the environmental target), the efficiency (making the most for human wellbeing), and to equity and fairness (sharing the benefits) of the proposed EBM responses. Lillebø et al. (2020) detail the co-development process of an EBM plan foreseen to mitigate unintended impacts on biodiversity in Vouga estuary and to its end support the development of the Vouga estuary management plan.

2.1.3 MCES Project

In the MCES project, a vulnerability index of the potential of marine and coastal habitats to deliver ecosystem services was developed. This is an example of an approach to implement ecosystem service assessments in EBM using spatially explicit modelling tools, i.e., by generating maps, GIS-based models enable decision makers to assess quantified trade-offs associated with alternative management options and to identify areas where these can take place. Relevant examples of open source models are ARIES—Artificial Intelligence for Ecosystem Services, already applied for machine learning for ecosystem services (Willcock et al. 2018); MARXAN with Zones enabling to '*develop multiple-use zoning plans for natural resource management*' (Watts et al. 2009; Jumin et al. 2018) and InVest—Integrated Valuation of Ecosystem Services and Trade-offs, already used for calculating vulnerability of marine habitats to deliver ecosystem services (Willaert et al. 2019; Cabral et al. 2015). For detailed consideration of EBM modelling tools see Fulford et al. (2020) and Lewis et al. (2020).

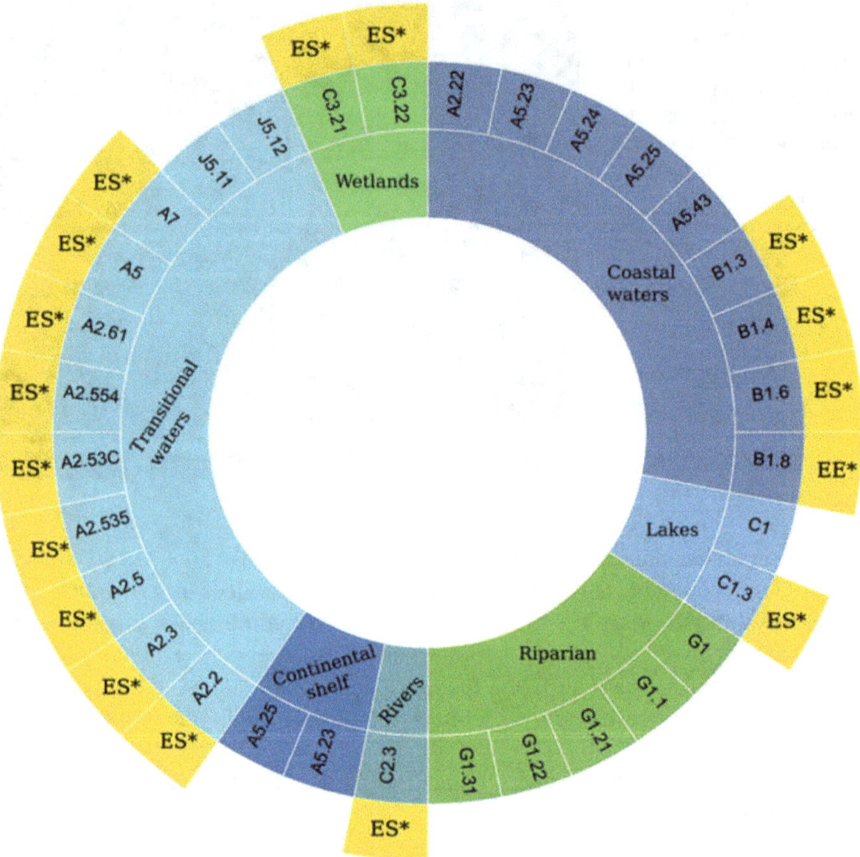

Fig. 4 Vulnerability (ES∗) of the Vouga river coastal watershed habitats under classification of Natura 2000 network to provide ecosystem services defined with AquaLinksTool. (Image source: Adapted from Lillebø et al. (2018), plotted with Mauri et al. (2017)). EUNIS habitats codes: A5.23 Infralittoral fine sand; A5.25 Circalittoral fine sand; A2.22 Barren or amphipod-dominated mobile sand shores; A5.23 Infralittoral fine sand; A5.24 Infralittoral muddy sand; A5.25 Circalittoral fine sand; A5.43 Infralittoral mixed sediments; B1.3 Shifting coastal dunes; B1.4 Coastal stable dune grassland (grey dunes); B1.6 Coastal dune scrub; B1.8 Moist and wet dune slacks; A2.2 Littoral sand and muddy sand; A2.3 Littoral mud; A2.5 Coastal saltmarshes and saline reedbeds; A2.535 *Juncus maritimus* mid-upper saltmarshes; A2.53C Marine saline beds of Phragmites australis; A2.554 Flat-leaved Spartina swards; A2.61 Seagrass beds on littoral sediments; A5 Sublittoral sediment; A7 Pelagic water column; J5.11 Saline and brackish industrial lagoons and canals; J5.12 Saltworks; C1 Surface standing waters; C1.3 Permanent eutrophic lakes ponds and pools; C3.21 Common reed (Phragmites) beds; C3.22 Common clubrush (Scirpus) beds; C2.3 Permanent non-tidal smooth flowing watercourses; G1 Broadleaved deciduous woodland; G1.1 Riparian and gallery woodland (Alnus Betula Populus or Salix); G1.21 Riverine Fraxinus—Alnus woodland; G1.22 Mixed Quercus—Ulmus—Fraxinus woodland of great rivers; G1.31 Mediterranean riparian Populus forests)

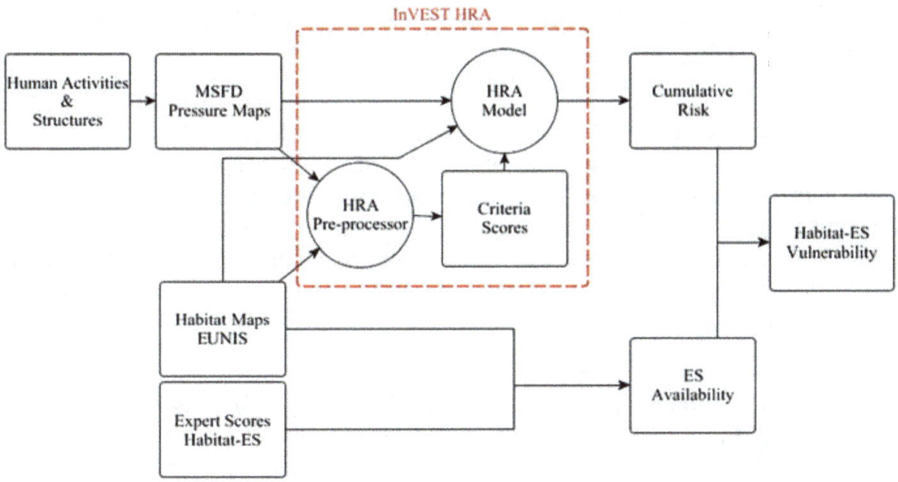

Fig. 5 Schematic representation of the workflow to assess the vulnerability of marine and coastal habitats' potential to deliver ecosystem services. (Image source: Willaert et al. 2019. Note: MSFD—Marine Strategy Framework Directive; HRA—InVEST Habitat Risk Assessment tool; ES—Ecosystem Services; EUNIS—habitat classification system)

The MCES project example showcases a vulnerability-based approach to capture Ecosystem Services in EBM. To this end, the considered approach combined the InVEST habitat risk assessment tool with the identified ecosystem services to create a proxy for the habitats' vulnerability to deliver ecosystem services in support of coastal and marine EBM. Figure 5 illustrates the framework combining the supply and demand for coastal and marine ecosystem services followed by Willaert et al. (2019). The case study was the western-Atlantic coast of Portugal that included the Nazaré Canyon (> 3000 m depth within the study region), Óbidos Lagoon (transitional waters), São Martinho do Porto bay (marine inlet), and Berlengas Archipelago (UNESCO world biosphere reserve).

As concluded by Willaert et al. (2019) "*The mapping of benthic habitats has opened new avenues, contributing to improve not only marine spatial plans, but also the EBM approach by facilitating the combination of spatial explicit GIS tools with supply and demand of marine ES, human activities and their related impacts, as well as with other natural impacts (e.g. climate change) to forecast scenarios (including marine ES trade-offs) and to open the floor to discussion (namely in stakeholders participatory processes) and to sustainable decision making processes in a 'Blue Growth' context by maximizing the net benefits provided by marine environments over time*".

The diversity of habitats in the selected area, the proposed approach to capture vulnerability, and the generated maps (EBM base line and EBM management options prospective scenarios) showcase maritime spatial planning and 'Blue Growth' in the context of the SDGs for 2030.

3 A Policy-Based Regional Seas Assessment of the Capacity to Supply Ecosystem Services

In this section we describe an assessment approach (MECSA: Marine ecosystem capacity for service supply assessment) developed with the European Environment Agency that utilises policy-reported ecosystem status information (from the Marine Strategy Framework Directive and other relevant reporting) to assess current and future capacity to supply ecosystem services (Culhane et al. 2020; Culhane et al 2019a). The key steps and the types of output that can be generated are summarised.

3.1 Using Ecosystem State Information in Ecosystem Service Assessments

The state of the ecosystem can tell us something about the capacity of the ecosystem to supply services, and this is the underlying assumption of this assessment approach. Across multiple services it can be assumed that good ecosystem state underpins good capacity for service supply, since the ecosystem service supply relies on the integrity of the ecosystem (Burkhard et al. 2012; Culhane et al. 2019b). However, at an individual service level, this is not necessarily the case, and we cannot assume that good ecosystem state will necessarily equate to good ecosystem services supply. Thus, the second underlying assumption of this assessment is that we understand something about the ecosystem state-service supply relationship, and we can use this to interpret state information in relation to what it means for the supply of services (Box 1).

Box 1 Understanding the State-Service Relationship
Different services will have different types of relationship with the state of the parts of the ecosystem that supplies them.

Examples:

- Flood protection capacity is better in the presence of a greater area of saltmarsh habitat (King and Lester 1995).
- Heavily-grazed seagrass beds, even though in worse state than ungrazed areas, still support up to three times more coastal protection through their root system that unvegetated areas (Christianen et al. 2013).
- Macroalgae habitat quality is more important than habitat area for service supply e.g. for maintaining fish habitat (van Lier et al. 2018).

It is important to understand this relationship in order to use state information to truly represent the capacity of the ecosystem to supply services. Different services rely on different aspects of the ecosystem. These need to be identified.

3.2 *Steps to Implementing the Method*

This approach entails three main steps, all set in the context of understanding the state-service relationship (Box 1).

- Firstly, the parts of the ecosystem that supply the service being assessed are identified. We described an approach to identify service providing units (SPUs) (see linkage framework chapter and Culhane et al. 2018). The relative contribution of these SPUs can be determined, and only the critical ones of these taken forward for a full assessment.
- The second step involves determining the relative contribution of SPUs to the service supply and identifying which are critical.
- In the third step, the specific state service relationship for the critical SPUs is described and the most appropriate indicators of these identified. State information is then used to fulfil these indicators, though in practice, this is often determined by what is available rather than what is most appropriate. This is interpreted in relation to ecosystem service supply capacity, using understanding of the state-service relationship.

3.3 *Example Case Study: North East Atlantic*

The three steps of the method are illustrated through an example of a cultural ecosystem service—recreation and leisure from whale watching in the North East Atlantic.

As well as whale watching occurring from the shore and other non-commercial routes, it is a growing enterprise in many regions (e.g. Elejabeitia et al. 2012; IUCN-ACCOBAMS 2016), often representing important economic benefits to rural communities (Ryan et al. 2018). In the North East Atlantic, commercial whale watching tours operate around the shores of Ireland, Scotland, England, Portugal and Northern Spain. In the first step of identifying the SPUs, we define these for this service as the whales that are watched by people, and the habitats that support those whale populations. Since whales are highly mobile and may make use of many habitats, we focus here on the whales themselves, as their populations are likely to reflect the condition of their habitats. We identified the relevant SPUs by checking whale tour operator websites and the species advertised as being seen on tours (Table 1).

To identify the critical SPUs for step two, we consider whether some species are more important than others. We find that, while common species are more reliably seen, rare species may constitute a special and equally important experience. Thus, in this case, we take all species forward as being critical. We define the state-service relationship as being that a greater whale population will result in a greater likelihood of seeing whales. Thus, in this case, good ecosystem state of whale populations

Table 1 Summary of results for the current state and direction of change of the metrics relevant to assess the 'whale watching' service in the North East Atlantic

Indicator	Classification	% Whale species assigned	Service supply capacity
State	Pass	56	Good
	Fail	17	
	Insufficient information	28	
Direction of change	Increasing	22	Unknown
	Stable	0	
	Decreasing	0	
	Insufficient information	78	

would correspond with a good capacity for the supply of this service. We can then look to policy assessments to find information on the status of each of these whale populations in the North East Atlantic.

To interpret this state information in relationship to the service supply capacity, we go back to our state-service relationship, where we had determined that more whales would lead to greater likelihood of spotting whales, and that each whale species is as important as any other. Therefore, we assess what is happening overall to whale populations in the North East Atlantic. From the policy status assessments, we see that most (56%) are in a good state, passing their policy objectives, but there is insufficient information about the direction of change of their populations (Fig. 6). Thus, the overall capacity for the ecosystem to supply this service is good but we do not know how this is changing.

3.4 Conclusions

This assessment approach uses available ecosystem state information to assess the capacity of the ecosystem to supply services. In doing so, a number of assumptions need to be made that relate to understanding the state-service relationship, and to the suitability of the state information available. In this example, we could assess the current capacity of the service but we could not assess the direction of change. While policy information allows a source of status assessments that can be used for this purpose, there are many unknowns and uncertainties. Nevertheless, this approach allows for assessments at broad regional scales, using existing information where it exists and does not rely on the availability of spatial or other data that may not be available.

It is also important to note that, although this service was found to have a good capacity overall, not every species is meeting its ecological objectives. There is still a

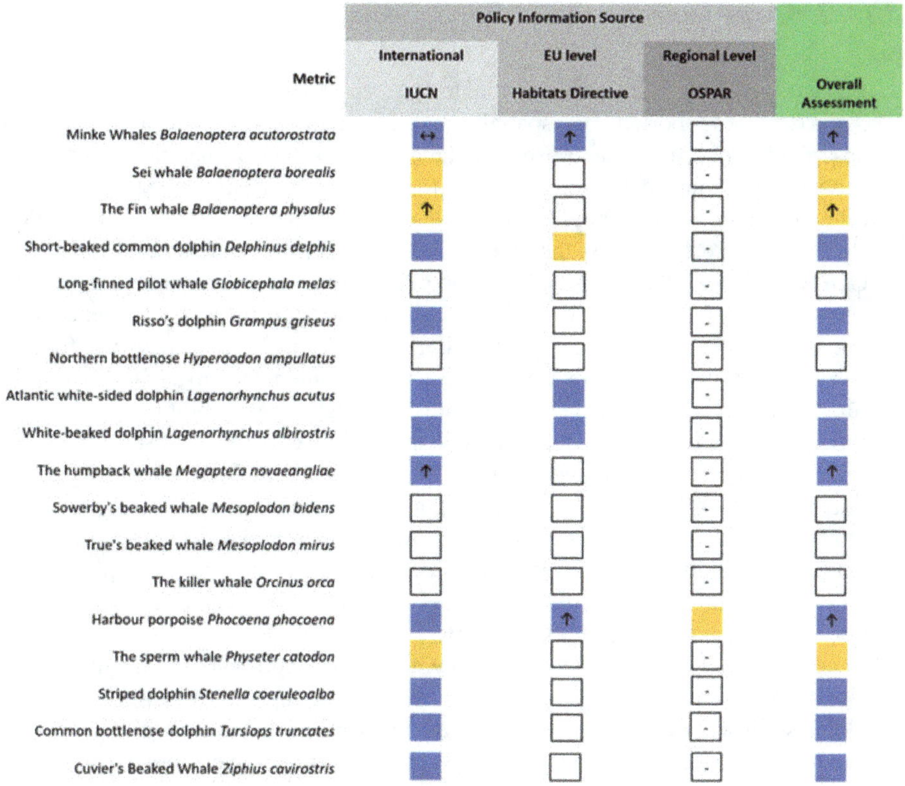

Fig. 6 List of commonly, occasionally or rarely spotted cetacean species advertised by whale watch tour operators (nine tour operators consulted from Ireland, Scotland, England, Portugal and Northern Spain) in the North East Atlantic region. The state of each whale species metric reported under each policy for the North East Atlantic is given. Legend: blue box = policy objectives achieved (here meaning that the whale population is in a good state); yellow box = fail to meet policy objectives; white box = unknown state/no assessment could be made; upward/downward arrow = Trend towards/away from achieving policy objectives; arrow on both sides = no trend; no arrow = unknown trend, 'Minus' symbol means the metric is not relevant for a particular policy)

need for conservation measures to protect species, and an ecosystem service assessment does not replace a biodiversity assessment.

4 Lessons Learnt and Next Steps

We draw a number of conclusions and lessons from our experience in trying to make the ecosystem services concept operational in Ecosystem-Based Management (adapted from Culhane et al 2019a).

4.1 Lessons Learnt

1. Make use of what is currently available
   Approaches supported by spatially explicit data (AQUACROSS, MCES) described above, can make use of detailed spatial data on habitats, ecosystem service supply and demand. However, for marine environments, spatial data is often scarce, in particular at the large spatial scales that may be relevant for policy assessments. Alternatively, the other approaches that we demonstrated (MECSA, ODEMM) show that it is still possible to carry out assessments of marine ecosystem services using what is available. Although these assessments may not be ideal in terms of the information underpinning them, they can form a baseline and indicate where there are potential problems in sustainability, while future assessments can make use of better information availability.
2. Small scale studies are needed to complement high level regional sea studies
   We demonstrated assessments that are high level, at broad ecological (regional sea) scales (ODEMM, MECSA), as well as more local (AQUACROSS, MCES). However, both these types of studies are underpinned by knowledge found from studies on specific habitats, species and ecosystem services, and how people use them. We continue to need these studies to improve understanding and confidence in ecosystem service assessments.
3. Ecosystem service assessments are not equivalent to ecosystem assessments
   Good ecosystem state does not always equal good capacity to supply services. Good service capacity may be satisfied by a broad habitat or group of taxa supplying it, for example in the whale watching example given above. But within this broad taxa group, several species may be failing their ecological assessments and be vulnerable. Ecosystem service assessments do not replace assessments of biodiversity and ecosystem condition—it is essential that these are seen as complementary.
4. Consider all of the contributions of nature to human wellbeing
   There are a number of ecosystem service typologies that can be used today. Although each have their critics, typologies like CICES (v5.1), TEEB and that used in the IPBES (International Panel on biodiversity and Ecosystem Services, (IPBES 2019), which include services that cannot be given a monetary or market value (see discussion on FEGS, DeWitt et al. (2020)), have a broad and encompassing set of ecosystem services that recognises all of the ways that nature contributes to our wellbeing. It is important to recognise services even where they cannot be given a market value, because ultimately, they are contributing to the long-term sustainability of society. It is possible to measure changes to these contributions by using a supply-side approach demonstrated here (e.g. MECSA, MCES), rather than a demand side approach.
5. Establish who are the beneficiaries of ecosystem services
   In order to recognise all the contributions of nature to human wellbeing (see 4 above) and to understand how the marine ecosystem contributes to our wellbeing, we need to explicitly identify all of the beneficiaries of coastal and

marine ecosystem services and how these benefits are perceived. This would help to establish how services are 'final' (sensu FEGS, see DeWitt et al. (2020)) and to establish how best we might measure specific contributions to different parts of society. This may not always be through measuring an economic value, and different, complementary approaches may need to be used side by side to fully capture how nature contributes to human wellbeing.

6. Need for adaptive response to ecosystem service assessment
   Ecosystem services are embedded in the current social and ecological context. As this changes, ecosystem service assessments also need to change to keep up. These assessments can inform management about the need to change and adapt in response to changing ecosystem service supply or demand. Furthermore, as mentioned above, a critical point in the decision-making process is to decide who will win and who will lose over time and space. Therefore, adaptive responses are crucial.
7. Link human activities and pressures to ecosystem service supply
   Increasing activities and pressures in the marine environment, alongside global climate change, requires urgent assessment of how these activities and pressures are affecting ecosystem service supply and what this will mean for long-term sustainability. The ODEMM approach described above demonstrates how trade-offs can be explored for different management options, whilst the MCES vulnerability index allows prioritisation of the most vulnerable or most important habitats needed for service supply. These can help to consider where human activities will impact service supply, and where management can act to be most effective, efficient and equitable.

4.2 Next Steps

The demand side of ecosystem services is socio-economically driven, while the supply side is dependent on ecosystems capacity to provide the required ecosystem services underpinning maritime activities. Europe's Biodiversity Strategy, aims to conserve and restore the supply side of ecosystem services, by halting biodiversity loss and deterioration of services by 2020 (EC 2011). On the other hand, Climate change strategies in the EU are aiming to significantly cut greenhouse gas emissions reaching net zero by 2050 (EC 2018). This will be achieved, at least partly, by increasing the share of renewable energy, including offshore energy. The Blue Growth strategy brings the supply and demand sides together for marine and coastal environments. It aims at supporting an effective implementation of maritime, marine and coastal-related policies, and at "*realising the potential of our seas and oceans for jobs and growth*" following the principles of conservation and of sustainable development (EC 2012). The strategy also foresees approaches to restore marine and coastal habitats, biodiversity and ecosystem services, being in line with the United Nations SDGs for 2030.

Therefore, the operationalization of any of these strategies requires balanced trade-offs between economic, social and environmental aspects supported by coastal and marine ecosystem services (Lillebø et al. 2017). To this end, Ecosystem-Based Management incorporating adaptive management is likely to have a critical role (e.g. Lillebø et al. 2020).

Balancing the demands we put on our ecosystems, with what they can sustainably supply, is the challenge of EBM. Ecosystem service assessments are one tool that can help us to do this, by allowing recognition of all the ways that the ecosystem contributes to human wellbeing and all the ways that human activities can impact the ecosystem and the supply of services. While the original ethos of ecosystem services was based in conservation (Ehrlich and Mooney 1983), it is clear that ecosystem service assessments alone do not replace nature conservation, in particular when individual services of interest are assessed, as opposed to multiple ecosystem services (MA 2005; Culhane et al. 2018; Teixeira et al. 2019).

Using the ecosystem service approach to benefit EBM is still challenging. The challenges lie in our knowledge and understanding of how the ecosystem works, in our understanding of human behaviour and the demands of society and in the resources that we have available, such as data or the means to collect data. Nevertheless, we have presented a number of ways showing that ecosystem service assessments can be integrated into EBM decisions. These assessments draw on different approaches in data-limited situations, including expert judgement and stakeholder knowledge, policy information, as well as, detailed habitat mapping or spatially explicit modelling techniques. These are all ways that can be used to achieve a key tenet of EBM, namely that it allows decision makers and stakeholders to see how ecosystem integrity can affect human well-being, thus allowing more balanced actions to be taken.

Acknowledgements ODEMM (EU FP7 Programme 'Options for Delivering Ecosystem Based Marine Management' (ODEMM); Grant number 244273). AQUACROSS (Grant Agreement no. 642317) collaborative research project was supported by the European Commission under the Horizon 2020 Programme for Research, Technological Development and Demonstration. MCES research project was supported by Calouste Gulbenkian Foundation, Portugal, in the context of the 'Gulbenkian Oceans´ Initiative. Thanks, are also due to the Portuguese Foundation for Science and Technology (FCT) for the financial support to CESAM (UID/AMB/50017/2019). MECSA was produced under an European Topic Centre grant agreement (Negotiated Procedure EEA/NSV/14/002) with the European Environment Agency (EEA). Opinions expressed are those of the authors and do not necessarily reflect the official opinion of the EEA or other European Community bodies and institutions.

References

Böhnke-Henrichs, A., Baulcomb, C., Koss, R., Hussain, S. S., & de Groot, R. S. (2013). Typology and indicators of ecosystem services for marine spatial planning and management. *Journal of Environmental Management, 130*, 135–145. https://doi.org/10.1016/j.jenvman.2013.08.027.

Burkhard, B., Kroll, F., Nedkov, S., & Müller, F. (2012). Mapping ecosystem service supply, demand and budgets. *Ecological Indicators, 21*, 17–29. https://doi.org/10.1016/j.ecolind.2011.06.019.

Cabral, P., Levrel, H., Schoenn, J., Thiébaut, E., Le Mao, P., Mongruel, R., et al. (2015). Marine habitats ecosystem service potential: A vulnerability approach in the Normand-Breton (Saint Malo) Gulf, France. *Ecosystem Services, 16*, 306–318. https://doi.org/10.1016/j.ecoser.2014.09.007.

Christianen, M. J. A., van Belzen, J., Herman, P. M. J., van Katwijk, M. M., Lamers, L. P. M., van Leent, P. J. M., et al. (2013). Low-canopy seagrass beds still provide important coastal protection services. *PLoS One, 8*(5), e62413–e62413. https://doi.org/10.1371/journal.pone.0062413.

Costanza, R., d'Arge, R., de Groot, R., Farber, S., Grasso, M., Hannon, B., et al. (1997). The value of the world's ecosystem services and natural capital. *Nature, 387*(6630), 253–260. https://doi.org/10.1038/387253a0.

Culhane, F. E., Frid, C. L. J., Royo-Gelabert, E., White, L. J., & Robinson, L. A. (2018). Linking marine ecosystems with the services they supply: What are the relevant service providing units? *Ecological Applications, 28*(7), 1740–1751. https://doi.org/10.1002/eap.1779.

Culhane, F., Frid, C. L. J., Royo-Gelabert, E., & Robinson, L. A. (2019a). EU policy-based assessment of the capacity of marine ecosystems to supply ecosystem services. ETC/ICM Technical Report 2/2019: European Topic Centre on Inland, Coastal and Marine Waters, 269p.

Culhane, F., Teixeira, H., Nogueira, A. J. A., Borgwardt, F., Trauner, D., Lillebø, A., et al. (2019b). Risk to the supply of ecosystem services across aquatic ecosystems. *Science of the Total Environment, 660*, 611–621. https://doi.org/10.1016/j.scitotenv.2018.12.346.

Culhane, F., Frid, C. L. J., Royo-Gelabert, E., Piet, G., White, L. J., & Robinson, L. A. (2020). Assessing the capacity of European regional seas to supply ecosystem services using marine status assessments. *Ocean and Coastal Management*, 190, 105154

DeWitt, T. H., Berry, W. J., Canfield, T. J., Fulford, R. S., Harwell, M. C., Hoffman, J. C., Johnston, J. M., Newcomer-Johnson, T. A., Ringold, P. L., Russel, M. J., Sharpe, L. A., & Yee, S. J. H. (2020). The final ecosystem goods and services (FEGS) approach: A beneficiary-centric method to support ecosystem-based management. In T. O'Higgins, M. Lago and & T. H. DeWitt (Eds.), *Ecosystem-based management, ecosystem services and aquatic biodiversity: Theory, tools and applications* (pp. 127–148). Amsterdam: Springer.

EC. (2011). Communication from the commission to the European parliament, the council, the economic and social committee and the committee of the regions. Our life insurance, our natural capital: An EU biodiversity strategy to 2020, Brussels, 3.5.2011 COM (2011) 244 Final, 17p.

EC. (2012). Communication from the commission to the European Parliament, the Council, the European Economic and Social Committee and the Committee of the Regions. Blue Growth opportunities for marine and maritime sustainable growth. Brussels, 13.9.2012, COM (2012) 494 final.

EC. (2018). Communication from the commission to the European Parliament, the European Council, the Council, the European Economic and Social Committee, the Committee of the Regions and the European Investment Bank. A Clean Planet for all. A European strategic long-term vision for a prosperous, modern, competitive and climate neutral economy. Brussels, 28.11.2018. COM(2018) 773 final.

EEA. (2015). State of Europe's Seas. EEA Report No 2/2015. Copenhagen, Denmark: European Environment Agency.

Ehrlich, P. R., & Mooney, H. A. (1983). Extinction, substitution, and ecosystem services. *Bioscience, 33*(4), 248–254. https://doi.org/10.2307/1309037.

Elejabeitia, C., Urquiola, E., Verborgh, P., & de Stephanis, R. (2012). Towards a sustainable whale-watching industry in the Mediterranean Sea. In L. M. Rosalino, A. Silva, & A. Abreu (Eds.), *New trends towards Mediterranean tourism sustainability*. Nova Science Publishers.

Elliott, M., & O'Higgins, T.G. (2020). From the DPSIR, the D(A)PSI(W)R(M) emerges... a butterfly-'protecting the natural stuff and delivering the human stuff' In T. O'Higgins, M. Lago, & T. H. DeWitt (Eds.), *Ecosystem-based management, ecosystem services and aquatic biodiversity: Theory, tools and applications* (pp. 61–86). Amsterdam: Springer.

Fisher, B., Turner, R. K., & Morling, P. (2009). Defining and classifying ecosystem services for decision making. *Ecological Economics, 68*(3), 643–653. https://doi.org/10.1016/j.ecolecon.2008.09.014.

Flood, S., O'Higgins, T. G. and Lago, M. (2020). The promise and pitfalls of ecosystem services classification and valuation. In T. O'Higgins, M. Lago, & T. H. DeWitt (Eds.), *Ecosystem-based management, ecosystem services and aquatic biodiverstiy: Theory, tools and practice* (pp. 87–104). Amsterdam: Springer.

Fulford, R. S., Heymans, S. J. J., & Wu, W. (2020). Mathematical modelling for ecosystem-based management (EBM) and ecosystem goods and services (EGS) assessment. In T. O'Higgins, M. Lago, & T. H. DeWitt (Eds.), *Ecosystem-based management, ecosystem services and aquatic biodiversity: Theory, tools and applications* (pp. 275–290). Amsterdam: Springer.

Gómez, C., Delacámara, G., Arévalo-Torres, J., Barbière, J., Barbosa, A., Boteler, B., et al. (2016). The AQUACROSS innovative concept-deliverable 3.1.

Gómez, C., Delacámara, G., Jähnig, S., Langhans, S. D., Domisch, S., Hermoso, V., et al. (2017). Developing the AQUACROSS assessment framework deliverable 3.2.

Haines-Young, R., & Potschin, M. (2013). Common International Classification of Ecosystem Services (CICES): Consultation on version 4, August–December 2012. EEA Framework Contract No EEA/IEA/09/003.

IPBES. (2019). *Summary for policymakers of the global assessment report on biodiversity and ecosystem services of the intergovernmental science-policy platform on biodiversity and ecosystem services.* Bonn, Germany: IPBES Secretariat.

IUCN-ACCOBAMS. (2016). Assessment of whale watching activities in the Gibraltar Strait. By Cazalla, E., Casado, J., Catala, T., Tilot, V., Bernal, C. p. 66.

Jumin, R., Binson, A., McGowan, J., Magupin, S., Beger, M., Brown, C. J., et al. (2018). From Marxan to management: Ocean zoning with stakeholders for Tun Mustapha Park in Sabah, Malaysia. *Oryx, 52*(4), 775–786. https://doi.org/10.1017/S0030605316001514.

King, S. E., & Lester, J. N. (1995). The value of salt marsh as a sea defence. *Marine Pollution Bulletin, 30*(3), 180–189. https://doi.org/10.1016/0025-326X(94)00173-7.

Lewis, N. S., Marois, D. E., Littles, C. J., & Fulford, R. S. (2020). Projecting changes to coastal and estuarine ecosystem goods and services—models and tools. In T. O'Higgins, M. Lago, & T. H. DeWitt (Eds.), *Ecosystem-based management, ecosystem services and aquatic biodiversity: Theory, tools and applications* (pp. 235–254). Amsterdam: Springer.

Lillebø, A. I., Pita, C., Garcia Rodrigues, J., Ramos, S., & Villasante, S. (2017). How can marine ecosystem services support the Blue Growth agenda? *Marine Policy, 81*, 132–142. https://doi.org/10.1016/j.marpol.2017.03.008.

Lillebø, A. I., Teixeira, H., Morgado, M., Genua-Olmedo, A., Nogueira, A., Delacámara, G., et al. (2018). Case study 5 report: Improving integrated management of Natura 2000 sites in the Ria de Aveiro Natura 2000 site, from catchment to coast, Portugal. Deliverable 9.2, European Union's Horizon 2020 Framework Programme for Research and Innovation Grant Agreement No. 642317. Retrieved from https://aquacross.eu.

Lillebø, A. I., Teixeira, H., Morgado, M., Martínez-López, J., Marhubi, A., Delacámara, G., et al. (2019). Ecosystem-based management planning across aquatic realms at the Ria de Aveiro Natura 2000 territory. *Science of the Total Environment, 650*, 1898–1912. https://doi.org/10.1016/j.scitotenv.2018.09.317.

Lillebø, A. I., Teixeira, H., Martinez-Lopez, J., Genua-Olmedo, A., Marhubi, A., Delacámara, G., et al. (2020). Mitigating negative unintended impacts on biodiversity in the Natura 2000 Vouga estuary (Ria de Aveiro, Portugal). In T. O'Higgins, M. Lago, & T. H. DeWitt (Eds.), *Ecosystem-based management, ecosystem services and aquatic biodiversity: Theory, tools and applications*. (pp. 461–498). Amsterdam: Springer.

MA. (2005). Millennium ecosystem assessment. Washington, DC.

Maes, J., Teller, A., Erhard, M., Liquete, C., Braat, L., Berry, P., et al. (2013). Mapping and assessment of ecosystems and their services. An analytical framework for ecosystem assessments under action 5 of the EU biodiversity strategy to 2020. Luxembourg.

Maes, J., Liquete, C., Teller, A., Erhard, M., Paracchini, M. L., Barredo, J. I., et al. (2016). An indicator framework for assessing ecosystem services in support of the EU biodiversity strategy to 2020. *Ecosystem Services, 17*, 14–23. https://doi.org/10.1016/j.ecoser.2015.10.023.

Martínez-López, J., Teixeira, H., Morgado, M., Almagro, M., Sousa, A. I., Villa, F., et al. (2019). Participatory coastal management through elicitation of ecosystem service preferences and modelling driven by 'coastal squeeze'. *Science of the Total Environment, 652*, 1113–1128. https://doi.org/10.1016/j.scitotenv.2018.10.309.

Mauri, M., Elli, T., Caviglia, G., Uboldi, G., & Azzi, M. (2017). RAWGraphs: A visualisation platform to create open outputs. In Proceedings of the 12th biannual conference on Italian SIGCHI Chapter. ACM, p. 28.

Nogueira, A. (2018). *AquaLinks Tool. Zenodo Dataset*. https://doi.org/10.5281/zenodo.1101159.

Piet, G., Delacámara, G., Kraan, M., Rockman, C., & Lago, M. (2020). Advancing aquatic ecosystem-based management with full consideration of the social-ecological system. In T. O'Higgins, M. Lago, & T. H. DeWitt (Eds.), *Ecosystem-based management, ecosystem services and aquatic biodiversity: Theory, tools and applications* (pp. 17–38). Amsterdam: Springer.

Potschin, M. B., & Haines-Young, R. H. (2011). Ecosystem services: Exploring a geographical perspective. *Progress in Physical Geography: Earth and Environment, 35*(5), 575–594. https://doi.org/10.1177/0309133311423172.

Robinson, L., & Culhane, F. (2020). Linkage frameworks: An exploration tool for complex systems. In T. O'Higgins, M. Lago, & T. H. DeWitt (Eds.), *Ecosystem-based management, ecosystem services and aquatic biodiversity: Theory, tools and applications* (pp. 213–234). Amsterdam: Springer.

Robinson, L. A., Culhane, F. E., Baulcomb, C., Bloomfield, H., Boehnke-Henrichs, A., Breen, P., et al. (2014). Towards delivering ecosystem-based marine management: The ODEMM approach. Deliverable 17, EC FP7 Project (244273) 'Options for Delivering Ecosystem-based Marine Management'. University of Liverpool. ISBN: 978-0-906370-89-6: 96p.

Ryan, C., Bolin, V., Shirra, L., Garrard, P., Putsey, J., Vines, J., et al. (2018). The development and value of whale-watch tourism in the west of Scotland. *Tourism in Marine Environments, 13*(1), 17–24.

TEEB. (2010). The economics of ecosystems and biodiversity (TEEB). Retrieved from http://www.teebweb.org/.

Teixeira, H., Lillebø, A., Culhane, F., Robinson, L., Trauner, D., Borgwardt, F., et al. (2018). Assessment of causalities, highlighting results from the application of meta-ecosystem analysis in the case studies—Synthesis report. Deliverable 5.2, European Union's Horizon 2020 Frame work Programme for Research and Innovation Grant Agreement No. 642317. Retrieved from https://aquacross.eu.

Teixeira, H., Lillebø, A. I., Culhane, F., Robinson, L., Trauner, D., Borgwardt, F., et al. (2019). Linking biodiversity to ecosystem services supply: Patterns across aquatic ecosystems. *Science of the Total Environment, 657*, 517–534. https://doi.org/10.1016/j.scitotenv.2018.11.440.

UNEP. (2011). Taking steps toward marine and coastal ecosystem-based management - An introductory guide, by Agardy, T., Davis, T., Sherwood, K., Vestergaard, O. UNEP Regional Seas Reports and Studies No. 189. ISBN: 9789280731736.

van Lier, J. R., Wilson, S. K., Depczynski, M., Wenger, L. N., & Fulton, C. J. (2018). Habitat connectivity and complexity underpin fish community structure across a seascape of tropical macroalgae meadows. *Landscape Ecology, 33*(8), 1287–1300. https://doi.org/10.1007/s10980-018-0682-4.

Watts, M. E., Ball, I. R., Stewart, R. S., Klein, C. J., Wilson, K., Steinback, C., et al. (2009). Marxan with zones: Software for optimal conservation based land- and sea-use zoning. *Environmental Modelling and Software, 24*(12), 1513–1521. https://doi.org/10.1016/j.envsoft.2009.06.005.

Willaert, T., García-Alegre, A., Queiroga, H., Cunha-e-Sá, M. A., & Lillebø, A. I. (2019). Measuring vulnerability of marine and coastal habitats' potential to deliver ecosystem services: Complex Atlantic region as case study (original research). *Frontiers in Marine Science, 6*(199). https://doi.org/10.3389/fmars.2019.00199.

Willcock, S., Martínez-López, J., Hooftman, D. A. P., Bagstad, K. J., Balbi, S., Marzo, A., et al. (2018). Machine learning for ecosystem services. *Ecosystem Services, 33*, 165–174. https://doi.org/10.1016/j.ecoser.2018.04.004.

2

Prioritizing Stakeholders, Beneficiaries and Environmental Attributes: A Tool for Ecosystem-Based Management

Leah M. Sharpe, Connie L. Hernandez, and Chloe A. Jackson

Abstract Successful Ecosystem-Based Management (EBM) approaches have advanced both a socio-ecological approach to systems thinking and the application of principles of structured decision making. This chapter presents a scoping tool designed to help decision makers in the early stages of their efforts by providing a transparent, repeatable, defendable approach for identifying and prioritizing stakeholders, the ways in which they use the environment (their beneficiary roles), and the most relevant environmental attributes for those uses as part of a set of decision criteria within a larger decision context. This scoping tool is a multi-criteria decision analysis approach that uses formalized criteria in stakeholder prioritization, along with the theoretical framework of the Final Ecosystem Goods and Services (FEGS) Classification System, to translate those prioritized stakeholders into the language of ecosystem services. The FEGS Scoping Tool is predicated on the idea that the decisions being made in a community can be complex, and that relevant and meaningful environmental decision criteria, let alone ecosystem services decision criteria, can be hard to identify and incorporate into the decision-making process.

Lessons Learned

- The EBM field lacks a clear pathway to prioritize stakeholders, develop a beneficiary profile, and focus management decisions on the environmental attributes most meaningful to a community.
- Transparent stakeholder prioritization provides clarity over who is included and why, facilitating a decision process that connects more directly to shared values,

---

L. M. Sharpe (✉)
U.S. EPA, Gulf Ecosystem Measurement and Modeling Division, Gulf Breeze, FL, USA
e-mail: Sharpe.leah@epa.gov

C. L. Hernandez · C. A. Jackson
Oak Ridge Institute for Science and Education (ORISE) at U.S. EPA, Pacific Ecological Systems Division, Newport, OR, USA
e-mail: Hernandez.conniel@epa.gov; Jackson.chloe@epa.gov

uses, and experiences, ultimately leading to increased legitimacy of the final decision.
- Development of a beneficiary profile allows managers to directly connect the ecosystem to the community's array of benefits, creating a holistic view of people's interactions to identify commonalities.
- Identification of key environmental attributes on beneficiary uses allows managers to focus their decision objectives on the most relevant metrics when evaluating tradeoffs.
- The FEGS Scoping Tool elucidates which attributes of the environment are highly valued based on the intersecting and overlapping interests of stakeholders and the beneficiaries they represent, which may lead to improved EMB design and buy-in.

Needs to Advance EBM

- There is a need to advance EBM practices through a transparent, repeatable, defendable approach for identifying and prioritizing stakeholders, the ways in which they experience the environment, and the most relevant environmental attributes for those uses.
- There is a need to identify how the FST users feel the tool influences their decision-making process and leads to improved EBM.

1 Introduction

Structured Decision Making (SDM) is a method of approaching a decision-making process in a formalized way that allows for more transparency and deliberation. Use of SDM is particularly valuable when the decisions being made are difficult ones—touching on a variety of issues, impacting a wide range of stakeholders, surrounded by uncertainty, or involving competing values (Gregory et al. 2012). Particularly difficult decisions are known as "wicked" problems. Some of the characteristics of what makes a problem "wicked" include problems that are essentially unique every time, with no clear stopping rule, with solutions that are neither "right" nor "wrong," with no clearly defined set of existing solutions, and where every solution is essentially a "one shot operation" (Conklin 2006). Ecosystem-Based management (EBM) decisions, with their need to include social, economic, and political considerations as well as complex ecological issues, are a perfect example of the type of "wicked" problem that could benefit from the use of an SDM approach (Van Bueren et al. 2003).

There are a wide range of SDM approaches (Gregory et al. 2012), but most contain some version of the same generic steps (Fig. 1). These steps are as follows:

1. Clarify the decision context—understand the context for the decision and why you are making it;

Fig 1 The decision steps in a generic structured decision-making process

2. Define objectives—clearly identify what decision makers or stakeholders value in the context of this decision and metrics for assessing how well alternatives meet those objectives;
3. Develop alternatives—identify possible alternatives for consideration;
4. Estimate consequences—estimate how well each alternative meets the decision objectives;
5. Evaluate trade-offs—examine trade-offs in how well the alternatives meet the decision objectives relative to one another;
6. Select preferred alternative—select an alternative; and
7. Implement, monitor, and review—monitor how well that alternative meets those objectives after implementation and whether any information from its real-world performance could be used to inform future decisions.

Although working through these steps can seem like an increased burden on decision makers, essentially, SDM is a formalization of the steps already being used in an *ad hoc* or unconscious fashion when making any decision (Gregory et al. 2012). For example, when deciding what to have for breakfast, explicitly clarifying the context, developing performance measures, and carrying out the remaining steps is unnecessarily burdensome. If, however, you were to explain to someone else how you decided upon your breakfast choice, you would see how you rapidly went through each step in the SDM process:

- I wanted to eat breakfast before I left for work (clarifying the context);
- I wanted something that was quick and filling (defining objectives and metrics);
- I had cereal and eggs in the kitchen (developing alternatives);
- I didn't have time to make eggs and I am trying to eat more fiber (evaluating trade-offs);
- So, I had cereal (selecting alternatives).

For more complex decisions, such as in EBM, explicitly working through the steps of a decision-analysis process has many benefits, but these steps will be worked through even without a formal process. The benefits of a formal process include improved guidance for information gathering activities, improved communication with stakeholders, increased opportunities for stakeholder engagement and involvement, improved documentation with a clear and transparent record of what happened during a decision process, and increased creativity in alternative development (Yee et al. 2017).

Decision Support Tools (DSTs) are tools that support decision-making processes. They can be powerfully effective methods for incorporating an SDM approach and increasing the transparency and repeatability of the decision-making process. The concept of decision support systems was developed in the 1970s and came out of the fields of organizational decision making and interactive computing systems (Keen and Morton 1978). By providing a mechanism for conducting one or more of the SDM steps and guiding the user through them, DSTs can facilitate incorporation of SDM thinking into a decision-making process. The tools themselves also provide users with clear and impactful ways to communicate with stakeholders (Fedra 1995).

Both the SDM approach and the use of DSTs support increased engagement with stakeholders and help decision makers identify multiple points in the decision-making process where stakeholder involvement could take place. Stakeholder involvement in making these difficult decisions is important for a number of reasons. Not only do they have the right to participate in making the decisions that impact their lives, stakeholders can also contribute valuable local knowledge that may otherwise be overlooked, and their involvement can lead to a better informed, and ideally, more legitimate (i.e., a "fair") decision process that considers representative perspectives (Cash et al. 2003; Fiorino 1990; NRC 1996). Despite its benefits, stakeholder engagement can be challenging. The process can be time consuming and expensive; it can be difficult to identify the complete set of stakeholders, and even if all stakeholders can be identified, it may not be feasible to include all of them in the decision-making process (Reed et al. 2009; Luyet et al. 2012), for one of a number of potential reasons (e.g., regulatory constraints, temporal constraints, or willingness or ability to engage). An example from this volume where not all stakeholders were fully engaged in an EBM effort is presented in O'Higgins et al. (2020).

These challenges are heightened in the complex context of EBM decisions. The effects of different management decisions on the environment can be highly uncertain and the ways in which stakeholders benefit from the environment can be easy to overlook. The concept of ecosystem services was developed to better quantify those benefits (MEA 2005). This concept was further refined as Final Ecosystem Goods and Services (FEGS), the "components of nature, directly enjoyed, consumed, or used to yield human well-being" (Boyd and Banzhaf 2007). Although the language of FEGS provides stakeholders with a way of identifying and articulating the benefits they receive and value, the concepts are ones that require introduction and education (DeWitt et al. 2020).

The concept of ecosystem services is one that can be viewed as having a "supply" side—the goods and services the ecosystem is capable of producing—and a "demand" side—the goods and services that humans are interested in enjoying, consuming, or using (Culhane et al. 2020). Many approaches to ecosystem services focus on the question of supply (e.g., from an ecological standpoint, how large a fish population can be sustained). These questions rely on answers from the biological and ecological realms. In order to incorporate FEGS into decision making as effectively as possible, however, it is valuable to have a clear understanding of the "demand" side of that equation. This includes a clear understanding of who is benefiting from the ecosystem services (Culhane et al. 2020) as well as the specific aspects of the environment necessary to realize those benefits (DeWitt et al. 2020). These questions require answers from society and require engagement with stakeholder groups and community decision makers.

2 Stakeholders as Beneficiaries

The suite of FEGS are the attributes of the environment from which humans directly benefit, such as fish for food, property protection (i.e., protection from storm surge or wave action) provided by coastal habitats, or drinking water from a stream. The concept of FEGS is useful for decision making because it serves as a foundation for defining, classifying, and measuring ecosystem services (Landers and Nahlik 2013). The FEGS concept helps avoid ambiguity, minimizes double counting of a good or service (from a valuation perspective), bridges natural and social sciences to facilitate communication and collaboration, and is beneficiary-specific so it is directly connected to what people value (Landers and Nahlik 2013; Russell et al. 2020). Beneficiary roles are the ways in which an individual or group enjoys, uses, consumes, or interacts with some aspect of the environment. Beneficiaries are those who directly benefit from a FEGS (e.g., people who eat the fish, who own property that is protected by coastal habitats, or who drink the water from the stream) (Landers and Nahlik 2013). Defining a beneficiary helps identify the specific FEGS and connects them to human well-being (Landers and Nahlik 2013; DeWitt et al. 2020). This helps decision makers involved in EBM projects make decisions based on what matters, is directly valued, and directly benefits community members, ultimately improving human health and well-being.

Stakeholders are interested and affected parties. Stakeholder groups result from the roles the individuals within them play in society and the community. For example, a sporting club representing recreational anglers and boaters could be considered a stakeholder group. There is a lack of data on how people use the environment, and the ways in which individuals use a particular part of the environment is highly variable (e.g., one individual might choose a particular fishing spot because they went with their family as a child, while another individual might choose the same fishing spot because it is close to a car park). By identifying the beneficiary groups within stakeholder groups, decision makers are better able to identify and articulate the ways in which those in the community benefit from the environment. The individual members of the sporting club act as representatives of the club, but the members of the sporting club may benefit from the environment in many different ways and therefore would be composed of several beneficiary groups covering the different aspects of their interaction with the environment (i.e., fishing, boating, swimming, appreciation of views, etc.).

Both stakeholder and beneficiary concepts are valuable for community-level decision making and EBM applications because stakeholder groups make up the groups that should be consulted in the decision-making process and may include those most affected by the decision action. They are also the groups that community decision makers are used to thinking of and considering when making management decisions. Their roles as stakeholders, however, do not necessarily explicitly connect to how they are engaging with and benefiting from the environment. Beneficiary categories, on the other hand, are not necessarily a useful place to start when engaging in community consultation activities as groups and individuals are not used to thinking of themselves in these roles. By using both concepts, decision makers can connect how community members identify themselves within the community to how they benefit from the environment.

3 FEGS Scoping Tool

3.1 Identifying and Prioritizing Attributes Relevant to Stakeholders

Although there are repeated calls for inclusion of ecosystem services in environmental decision making (NRC 2005; Ruckelshaus et al. 2015), they are often less influential in the decision-making process than other social or economic considerations (MEA 2005; NRC 2005; Ruckelshaus et al. 2015; Yee et al. 2017). Oftentimes this is because the ecosystem services metrics that are used in decision making are ones that are easy to measure or commonly thought of (Yee et al. 2017) but may not be relevant or meaningful to decision makers or the communities they serve (Wasson et al. 2015; Yee et al. 2017). DeWitt et al. (2020) presents an argument on how the FEGS approach allows a decision maker to focus in on those ecosystem services most relevant to stakeholders and Nahlik et al. (2012) states that FEGS are more easily understood by the general public because the FEGS are determined by beneficiaries. By identifying more relevant ecosystem services, the services can be more influential in the decision-making process. Ecosystem services thinking, however, may not be intuitive for many decision makers, and long lists of potential ecosystem services (e.g., MEA 2005; Haines-Young and Potschin 2012) alone may provide insufficient guidance. Leaving ecosystem services out of the discussion entirely or selecting less relevant services can lead to decisions that omit commonly shared benefits (i.e. derived from recreational, cultural, and existence values), disconnect from what matters to people, and undermine biodiversity, human well-being, and social goals (Chan et al. 2012). For example, conservation and economic development programs in Papua New Guinea that neglected to incorporate culturally-based ecosystem services in restoration design undermined triple bottom line goals, leading to undesirable changes to impacted communities and the cultural values attached to the forest (Chan et al. 2012).

The FEGS Scoping Tool (FST) was designed to help decision makers identify and prioritize stakeholders, beneficiaries, and environmental attributes in a structured, transparent, repeatable process. The relevant and meaningful attributes can then be used to evaluate decision alternatives. The FST uses an SDM approach and the FEGS framework to identify the environmental attributes most relevant to the decision and valued by stakeholders in a transparent and structured fashion. The goal of the tool is to identify the most relevant environmental attributes for inclusion in the larger decision process so that valued FEGS are represented alongside other decision criteria. The level of stakeholder involvement in this or any other part of the decision-making process is entirely in the decision makers' hands and beyond the scope of this tool.

The FST uses a specific type of SDM, known as Multi-Criteria Decision Analysis (MCDA), with steps that mirror the generic steps of SDM. A MCDA is a formal decision-making framework that aims to represent decision goals in terms of explicitly evaluated criteria (Stewart 1992). This framework informs decision making in a transparent fashion by formalizing key criteria, explicitly stating priorities, and supporting easy replication and justification of results.

There are a range of MCDA approaches. The FST uses the method of ranking the alternatives on the sum of weighted criteria, which has been used in a variety of participatory environmental decision-making contexts (Ralls and Starfield 1995; OST n.d.). There are two main elements of an MCDA: (i) the decision alternatives that decision makers are considering, and (ii) the decision criteria used to prioritize those alternatives. In the FST, the decision criteria are a set of stakeholder prioritization criteria that were developed from the literature across a range of fields (Sharpe et al. under review) and the decision alternatives come from decision maker inputs and the National Ecosystem Services Classification System Plus (NESCS Plus) (DeWitt et al. 2020).

The benefit of using an MCDA approach for prioritizing amongst decision alternatives is that it focuses the discussion on the importance of stakeholder values rather than just the components of individual options. It does this by beginning the conversation with what decision makers are trying to achieve in meeting their goals (i.e., identifying the objective (MCDA Step 1 in Table 1)) and then identify what

Table 1 The MCDA steps in each of the three parts of the FEGS Scoping Tool. The colored boxes indicate how the output of one tier is used as an input in the next. The output from Tier 1 (red box) is used as an input in Tier 2 (red box). The output from Tier 2 (black dashed box) is used as an input in Tier 3 (black dashed box). (*modified from* Sharpe and Jenkins 2018)

MCDA Steps	Tier 1: Stakeholders	Tier 2: Beneficiaries	Tier 3: Attributes
1. Objective	Prioritize stakeholders	Prioritize beneficiaries	Prioritize environmental attributes
2. Decision criteria	Used when prioritizing stakeholder groups – supplied by the tool	Used when prioritizing beneficiaries – the stakeholder groups themselves are used as these criteria (i.e., which beneficiary groups are relevant to stakeholders)	Used when prioritizing attributes – the beneficiary groups are used as these criteria (i.e., which attributes are relevant to beneficiaries)
3. Metrics & Value functions	Used to score each stakeholder group for each criterion – supplied by the tool	Used to score each beneficiary group for each stakeholder group – this is done in the Beneficiary step of the tool when users are asked to identify those beneficiary categories found within each stakeholder group	Used to score each attribute for each beneficiary group – this is done in the Attribute step of the tool when users are asked to identify attributes of concern for each beneficiary group
4. Alternatives	Stakeholders identified by tool users	Beneficiary list from the FEGS Classification System and NESCS Plus	Attribute list from the FEGS Classification System and NESCS Plus
5. Weighting	Done by users in the first step of the tool	Stakeholder MCDA values from the output of Tier 1 are used as weights in this step	Beneficiary MCDA values from the output of Tier 2 are used as weights in this step
6. Score alternatives	User input at the Stakeholder step	User input at the Beneficiary step	User input at the Attribute step
7.Calculate value	Output at Stakeholder step	Output at Beneficiary step	Output at Attribute step

criteria are necessary for meeting them (Step 2). It then uses metrics (Step 3) to determine as objectively as possible how well each decision alternative (Step 4) meets those objectives (Step 6). This is separate from the question of the relative importance of those criteria (Step 5) to the decision makers. When choosing a car, for example, two people may disagree on whether storage capacity is more or less important than fuel efficiency, but they can easily agree on how each of those options score on those criteria. Together, how well each alternative scores on the criteria and the relative importance of each criterion to the decision makers are used to determine the MCDA value of each alternative and their overall prioritization (Step 7).

3.2 *Tiers of the FEGS Scoping Tool*

The FST has three tiers: (1) a stakeholder prioritization; (2) development of a beneficiary profile; and (3) identification of key environmental attributes, with each tier feeding into the next (Fig. 2a). Because the FST has three objectives—prioritizing the stakeholders, the beneficiaries, and the key attributes—each of the MCDA steps is run through three times in succession in a tiered MCDA approach,

Fig. 2 Overview of the FEGS Scoping Tool. The FST has three tiers (a) —stakeholder prioritization, development of a beneficiary profile, and identification of key environmental attributes. Each tier feeds into the next. MCDA weights and scores decision alternatives on a range of criteria and uses the combination of weights and scores to prioritize those alternatives. In this schematic (b), circles are the user inputs and squares are the tool outputs. In the first tier, users weight and score the prioritization criteria to rank the decision alternatives (the stakeholder groups). In the second tier, the stakeholder prioritization (first tier output) is used as the weights and users score each of the stakeholder groups as to their beneficiary roles. These weights and scores are used to prioritize the decision alternatives (the beneficiary groups). In the third tier, the beneficiary profile (second tier output) is used as the weights and users score each of the beneficiary groups as to key attributes needed to realize those benefits. These weights and scores are used to prioritize the decision alternatives (the environmental attributes)

with the output from Tier 1 feeding into Tier 2 and the output from Tier 2 feeding into Tier 3 (Fig. 2b and Table 1). Specifically, the output from the stakeholder prioritization (Tier 1, Step 7) is used to weight the influence of those stakeholder groups in the development of the beneficiary profile (Tier 2, Step 5). Continuing that approach, the output from the beneficiary profile (Tier 2, Step 7) is used to weight the influence of those beneficiary groups in the identification of key environmental attributes (Tier 3, Step 5).

The FST was designed to be relatively simple for decision makers to incorporate in their existing decision-making processes. It does not require decision makers to collect specific data about the stakeholder groups beyond knowledge of the community and familiarity with the stakeholder groups. The more iterative and participatory a process is structured, the more likely the process will lead to results that are defensible and well-founded. The intent of the tool is to provide a simple, transparent process for prioritizing stakeholder groups and letting that inform a prioritization of the benefits they receive from the environment and the elements of the environment necessary for receiving those benefits. It is anticipated that tool users will often be the decision makers, but this is not necessarily the case. Non-decision makers, such as those interested in evaluating or analyzing a decision made by others, might find the FST valuable as well.

3.2.1 Stakeholder Prioritization

During the stakeholder prioritization, decision makers are asked to review and weight the stakeholder prioritization criteria found in Table 2. This is a question of values. In this step, decision makers must ask themselves which of these decision criteria are most meaningful to them when looking at the stakeholder groups they are prioritizing. Criteria weighting is a key element of MCDA and the FST. Priorities

Table 2 Stakeholder prioritization criteria used in the FST

Stakeholder prioritization criteria	Definition
Level of interest	The amount of interest a stakeholder group has in the decision-making process or the decision outcome
Level of influence	The amount of influence a stakeholder group has over the decision-making process
Magnitude of impact	The degree of potential impact to the stakeholder group as a result of the decision
Probability of impact	The likelihood of potential impact to the stakeholder group as a result of the decision
Urgency/temporal immediacy	The degree to which a stakeholder group would like to see a decision made or an action taken
Proximity	How frequently a stakeholder group comes into contact with the environment for which a decision is being made
Economic interest	Whether a stakeholder group's livelihoods or assets could be impacted by the decision outcome
Rights	Whether a stakeholder group has legal, property, consumer, or user rights associated with the decision-making process, the decision outcome, or the environment for which the decision is being made
Fairness	Whether the exclusion of a stakeholder group from the decision-making process would lead to the process being viewed as unfair by the community
Underrepresented/under-served populations	Whether a stakeholder group includes any underrepresented or underserved populations

will differ from community to community as well as among groups within a community. For example, if a business development group is using this tool to incorporate FEGS into their decision making, they are likely to weigh the criterion of economic interest substantially higher than a non-profit group focused on social justice. In this step, decision makers are asked to identify the criterion most important to them and give that criterion a weight of 100. After that most valued criterion has been identified, all other criteria are weighted relative to that criterion on a 0–100 scale. Those criteria that are not considered by decision makers should be given a weight of 0. The FST provides a visual aid to allow decision makers to see the relative impact each criterion will have on the prioritization process.

Current methods of stakeholder analysis in environment decision making focus on stakeholder identification (who is/should be considered a stakeholder), categorization (often focused on distinguishing groups based on level of engagement), and relationship analysis (using tools like social network analysis to understand how the different groups relate to and influence one another). Stakeholder prioritization is discussed in the fields of business, management, and public relations, but the concerns of researchers in those fields are imperfectly analogous to environmental decision making. Therefore, a proposed set of ten stakeholder prioritization criteria was developed specifically for the field of environmental decision making and this tool (Table 2) (Sharpe et al. under review).

These criteria are not entirely independent from one another; however, each captures some element that has been found to be useful or important in stakeholder analyses and it is critical to include all criteria that could be relevant for decision makers (Sharpe et al. under review).

Once the criteria have been weighted, the decision makers are asked to identify all stakeholder groups relevant to the decision context. There is no step or process embedded within the tool itself that ensures that all possible stakeholder groups are being included in the decision process. That is, this tool, just like any stakeholder engagement effort, relies upon the good faith efforts of decision makers to cast a wide net when it comes to stakeholder identification and inclusion. However, by having a record of which stakeholder groups were considered and by using the FST in a transparent and iterative fashion, there will be opportunities for decision makers and their constituents to identify missing stakeholder groups and include them in the process. After the stakeholder groups have been identified, users will then score them on each of the decision criteria. The tool itself lays out specific scoring metrics for each criterion with the goal of making the scoring as objective as possible. Although different decision makers could disagree on how important economic interest is in making a stakeholder group a priority, it should be clear whether (or not) a given group has an economic interest in the decision.

Once the decision criteria have been weighted and the decision alternatives (e.g., the stakeholder groups) have been scored on those criteria, the FST calculates a value for each alternative by summing the weighted scores for each alternative. The value, $y(i)$, of an alternative, i, is calculated as:

$$y(i) = \sum\nolimits_{m=1}^{M} w_m z_i$$

where M is the number of possible metrics for which i can be scored, w_m is the weight given to each criterion, and z_i is the score of alternative i on metric m. The value, $y(i)$, is then normalized, $n(i)$, to a 0—100 scale by dividing $y(i)$ by the sum of all weights:

$$n(i) = \frac{y(i)}{\sum\nolimits_{m=1}^{M} w_m}$$

This results in an output of a prioritized list of stakeholders with each group given a "value." This value is only meaningful in describing the relative differences in priority for a collection of stakeholders that have been evaluated in a single exercise. It provides comparative information about the relative priority of different groups but has no meaning beyond that.

3.2.2 Beneficiary Profile

In the second part of the tool, users develop a beneficiary profile of the decision context to better understand the ways in which the community benefits from the ecosystem under consideration. It helps decision makers take a more holistic view of various groups' interactions with the environment and identify commonalities among them. In this step, users are asked to segment each stakeholder group into its component beneficiary groups by percentage, for a total of 100%. Once this has been completed, the FST will once again calculate a "value" for each beneficiary group using the same calculations as in the stakeholder prioritization. In this calculation, however, the output values, $n(i)$, from the stakeholder prioritization are used as the weights, w_m, for the beneficiary profile.

The beneficiary categories in this step come directly from the Final Ecosystem Goods and Services Classification System (Landers and Nahlik 2013) and the National Ecosystem Services Classification System Plus (NESCS Plus) (DeWitt et al. 2020). At this step, tool users should ask themselves how each stakeholder group benefits from, uses, or values the ecosystem under consideration. For example, a stakeholder group consisting of a fishing club could benefit from the ecosystem through a waterbody that can be navigated by their fishing boats (the "Boaters" beneficiary), fish that can be caught (the "Anglers" beneficiary), and a pleasing view when traveling to and from the fishing site (the "Experiencers/Viewers" beneficiary).

3.2.3 Key Attribute Identification

In the final portion of the FST, users build upon the previous steps to identify the key environmental attributes of the decision context. These key environmental attributes are those attributes that are necessary for the stakeholders to receive the benefits that they value. In this step, users are asked to identify, by percentage, the environmental attributes that are necessary for each beneficiary group to succeed in using and benefiting from the environment. After the environmental attributes of concern have been identified for each beneficiary group, the FST will calculate a "value" for each attribute using the same calculations as in the stakeholder prioritization and beneficiary profile. In this calculation, however, the output values, $n(i)$, from the beneficiary profile are used as the weights, w_m, for key attribute identification. This step allows the user to see what environmental attributes are relevant for evaluation of decision alternatives and provides a clear explanation of why those attributes are relevant.

The list of attributes used in the FST was developed for NESCS Plus (DeWitt et al. 2020). To continue the earlier example, the "Anglers" beneficiary would likely care a great deal about "Charismatic fauna" (i.e., fish that are of interest to anglers) and "Edible fauna" (i.e., fish that are safe to eat). Attributes related to the fuel, fiber materials, or fungal communities, for example, would likely not have any impact on their ability to realize this benefit.

The prioritized set of environmental attributes are the set that should be considered when evaluating different management options. The combination of prioritized beneficiaries and attributes provides the decision makers with guidance on the appropriate metrics to use when evaluating those options. The metric(s) used to assess the management options should be ones relevant to the beneficiaries that care about that attribute. The NESCS Plus has released a report on the development of national metrics and indicators for a number of ecosystem/beneficiary/attribute combinations and would provide useful guidance for developing sets of metrics for local-scale decisions (DeWitt et al. 2020; US EPA under review).

3.3 Using the FEGS Scoping Tool

The FST was designed to be used at an early stage of decision making, when decision makers are aware a decision needs to be made, but before any actions are taken (e.g., step 1 of the SDM process—clarify the decision context). Once those key environmental attributes have been identified, they can be included as objectives for the decision under discussion (e.g., step 2 of the SDM process—define objectives). These FEGS-related objectives can be used later in the decision-making process alongside other, non-environmental attributes, such as cost of the alternatives or job creation associated with each alternative, to estimate the overall consequences for each alternative (e.g., step 4 of the SDM process—estimate consequences). The FST itself is not designed to work through the entire SDM process for a given management decision. Rather, it is designed for use in step 1 and for its outputs to be applied in subsequent steps. Beyond this, the FST provides no additional guidance in conducting those steps, or in making a final decision.

The FST was designed to be used during the scoping phase for any decision with an environmental context by community-level decision makers who are involved in articulating the overall decision objectives and choosing amongst various decision options. Ideally, the FST is used in a participatory, iterative fashion with input from stakeholders, but can be applied in a variety of ways depending on the community's existing decision-making processes and requires no technical expertise or data collection beyond familiarity with the community and its stakeholder groups.

Decision makers prioritize stakeholders, either consciously or subconsciously. This tool formalizes and records the stakeholder prioritization process and makes those priorities transparent. The stakeholder prioritization results are then used to systematically identify environmental attributes most relevant to the prioritized stakeholders. The FST is predicated on the idea that the decisions being made in these communities are complex and that relevant and meaningful environmental decision criteria, let alone ecosystem services decision criteria, can be hard to identify. Thus, the use of a structured stakeholder prioritization approach results in the ability to provide a transparent, repeatable, and defendable approach for selecting the more relevant environmental attributes for use as decision criteria in that larger decision.

4 FEGS Scoping Tool Applications for Ecosystem-Based Management

Along the coast of Oregon (Fig. 3), communities are inextricably interconnected with estuarine wetland ecosystems and upland watersheds. The region provides highly viable industries (e.g., dairy, agriculture, timber, and fishing) and recreation centered around natural resources, and communities affect ecosystem functions through recreation, resource use, urban development, and dredging for shipping and port access. Many small communities along the Oregon coast have experienced an increase in retirement age migration and are seeing the tourist service industry, residential and resort developments, charter fishing, and whale watching increase in importance to the local economy while extractive industries in timber and fishing have declined (Radtke and Davis 1994; Ackerman et al. 2016). Additionally, historic natural disasters and climate change are straining the ability for communities to solely depend on these resources and seek support in mitigating adverse impacts and restoring ecological functions.

Because the health of Oregon's coastal habitats is vital to the safety and well-being of its human and wildlife communities, federal, state, non-governmental organizations, academic, and private institutions in Oregon have invested heavily to research and restore impacted ecosystems. We are going to walk through a hypothetical example of using the FST at the beginning of a wetland restoration project to more clearly illustrate the FST process. This example is a simplified version of the prioritization that might happen at a site that is being restored back

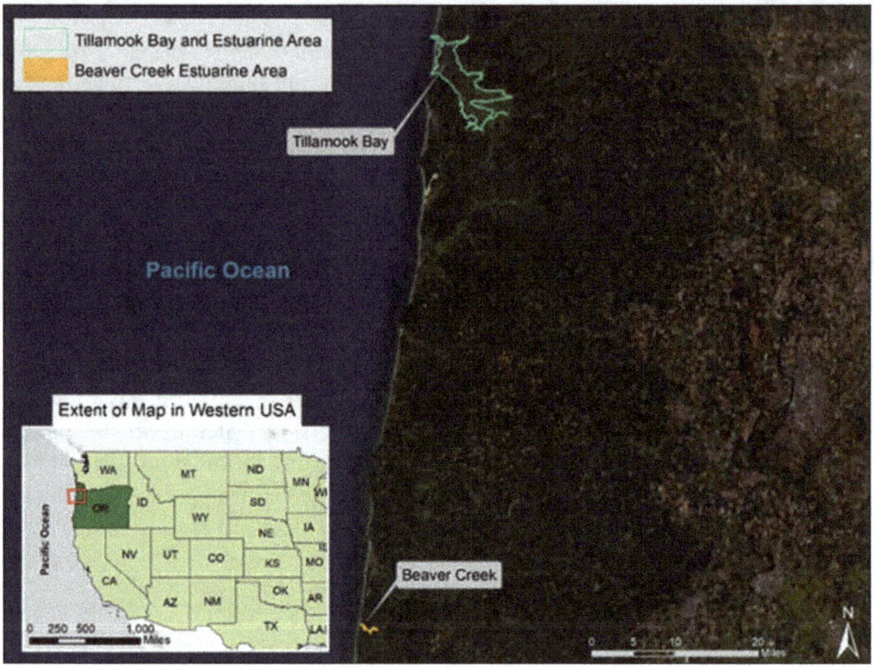

Fig. 3 Location of Tillamook Bay and Beaver Creek and their tidally influenced wetland areas. Extent area data layer was retrieved from Oregon Spatial Data Library on 11/14/2019, the estuaries data set from Adamus Resource Assessment Inc. (2004)

to a wetland, and does not include all likely stakeholders, beneficiaries, or attributes that would be included in a real ongoing project.

In the first tier of the tool, Stakeholder Prioritization, decision makers must assign weights to each criterion and then score the stakeholder group for those weights. When assigning weights, the decision makers might highly weight criteria such as Level of Influence and Level of Interest, if they are interested in making sure that influential and interested groups were prioritized and they might highly weight Underrepresented and Underserved Representation if they are interested in environmental justice concerns. Decision makers might also give Proximity moderate weight if they are interested in prioritizing those groups most likely to come into contact with the area and give Economic Interest moderate weight if they are interested in prioritizing those who may see an economic impact, either positive or negative, as a result of the project. For the purposes of this example, all other criteria are being considered unimportant to the decision makers and given a weight of zero. Figure 4 shows the user input form for the weights as well as the visual representation of the weights relative to one another.

After the weights are input into the tool, users identify the relevant stakeholder groups and score them, using provided metrics, on those criteria. In this case, some of the stakeholders sitting at the decision-making table might include Funding

Color	Criterion	Weight
	Magnitude & Probability of Impact	0
	Level of Influence	100
	Level of Interest	100
	Urgency & Temporal Immediacy	0
	Proximity	50
	Economic Interest	50
	Rights	0
	Fairness	0
	Underrepresented & Underserved Representation	100

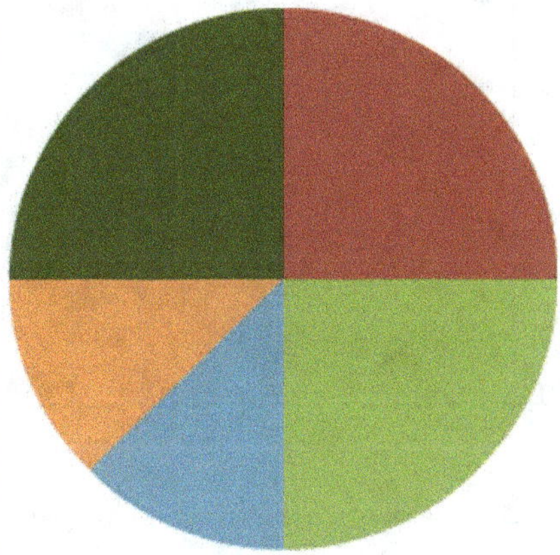

Fig. 4 The FST user input form for the weights (left) and the visual representation of the weights relative to one another (right)

Organizations, an NGO Conservation Trust, County Government Agencies, and Neighbors & Landowners. The metrics vary from criterion to criterion. In the case of Underrepresented and Underserved Representation, users are asked a yes or no question as to whether the stakeholder group contains any underrepresented or underserved groups. In this hypothetical example, only the Neighbors & Landowners group does. For the criterion of Proximity, however, users select from a range of scores based on how frequently the stakeholder group is in contact with the area in question or adjacent spaces. Once users have input the weights and the scores, they are combined to produce the stakeholder prioritization (Fig. 5). In this example, the Neighbors & Landowners are likely to include underrepresented or underserved groups and therefore have a higher prioritization given that this criterion was weighted higher than Economic Interest, which the other three stakeholders have. The County Government is prioritized the lowest because this stakeholder has less of an influence over the decisions being made than the other three stakeholders. This

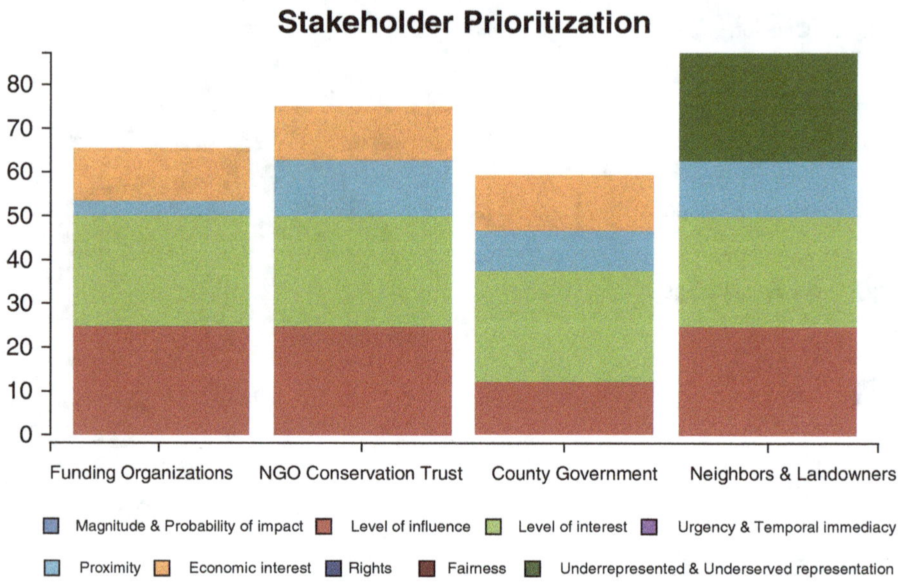

Fig. 5 Tier 1 output of the FST: Stakeholder Prioritization

would be a good time to communicate with the stakeholders and ensure that a key stakeholder group was not left out in the process.

In the second tier of the tool, Beneficiary Profile, users are asked to identify the ways in which each stakeholder group is benefiting from the area. In this example, decision makers might find that the Neighbors & Landowners group, while principally benefiting through owning the area or land adjacent to it, also benefit through recreational opportunities such as fishing (angler beneficiary category) and hiking (experiencer/viewer beneficiary category). These beneficiary scores are then combined with weights arising from the stakeholder prioritization to lead to a beneficiary profile (Fig. 6). The beneficiary profile allows decision makers to identify commonalities among stakeholder groups. In this case, we can see that all four stakeholder groups have common ground in that they all care about the continued existence of a healthy wetland ecosystem.

In the third tier of the tool, Key Attribute Identification, users are asked to identify the ecosystem attributes necessary for each beneficiary group to receive their benefit. In this case, a recreational angler beneficiary would likely care about water quality, water quantity, flora community, fauna community, edible fauna, viewscapes, and the ecological condition of the site. These attribute scores are then combined with weights arising from the beneficiary profile to lead to identification of key environmental attributes (Fig. 7). This final output allows decision makers to identify the key environmental attributes that should be considered and measured when contemplating restoration options for the site. It is unlikely that the entire suite of attributes can be considered, but if the user focuses on the attributes related most directly to sustaining the availability of the FEGS that anglers care about, it is more likely

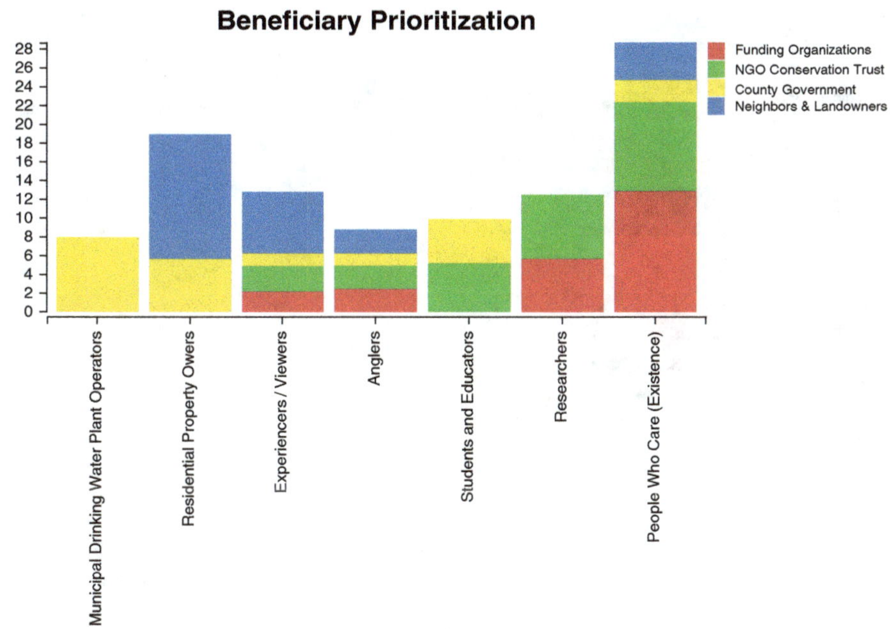

Fig. 6 Tier 2 output of the FST: Beneficiary Profile. Depicted as a bar chart showing the influence of each stakeholder group on the beneficiary group

that the ultimate decision will include ecosystem service metrics relevant to the stakeholders and the ways in which they benefit from the site.

While this example is hypothetical, there are numerous restoration efforts that have taken place in estuarine systems along Oregon's Northern Coast, such as the Southern Flow Corridor (SFC) project and the Beaver Creek watershed restoration project (Fig. 3). The SFC project was implemented by the Tillamook Estuary Partnership (TEP), one of EPA's 28 National Estuaries Programs that was established to conserve and restore estuaries and watersheds in Tillamook County (TEP 2019). The Tillamook Estuary Partnership first implemented a Comprehensive Conservation and Management Plan in 1999, and work was done to encourage broader community and stakeholder participation to identify values relevant to the estuary, prioritize conservation goals, and specify resource management actions (Gregory and Wellman 2001). The focus of the SFC project was restoring tidal wetland habitats and ecological function at the deltas of the Wilson, Trask, and Tillamook rivers (SFC 2019). The SFC produced a Project Effectiveness Monitoring Plan in which stakeholders and decision makers outlined four flood mitigation and restoration goals and expected ecological and economic benefits (Brophy and van de Wetering 2014). Interventions included a conditions assessment and various infrastructure changes to ditches, levees, dikes, floodgates, and buildings (Brophy and van de Wetering 2014). The project is currently in the monitoring phase to determine if the flood attenuation and ecological function goals were met (Brown et al. 2016;

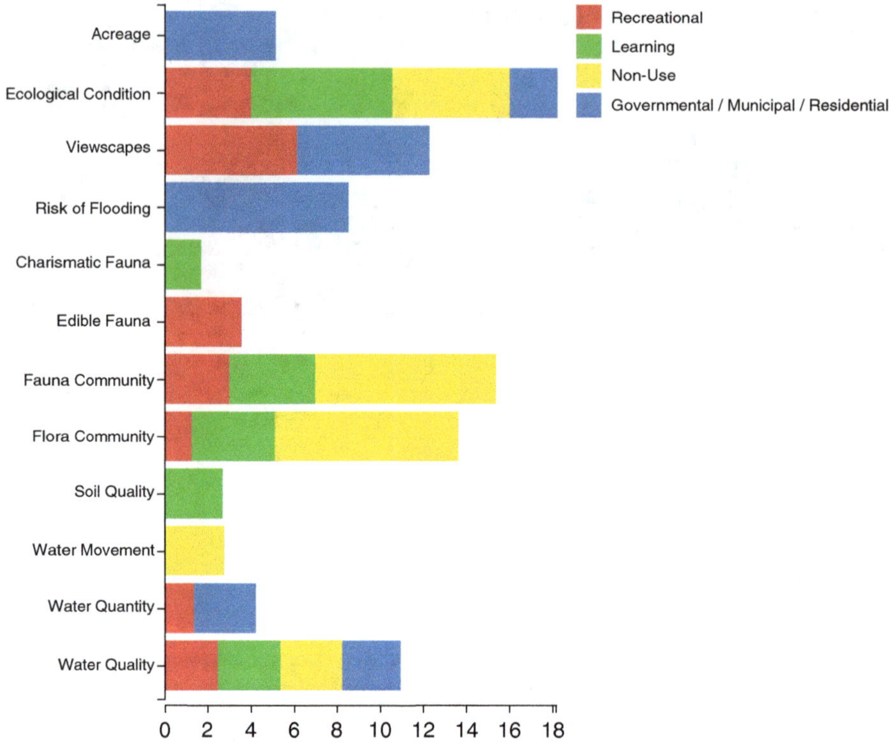

Fig. 7 Tier 3 output of the FST: Environmental Attribute Prioritization

SFC 2019). The conservation and restoration being done in the Beaver Creek watershed addresses critical watershed restoration issues (The Wetlands Conservancy 2018) and aims to simplify instream habitat, move roads that are too close to the stream, and increase the number of riparian trees and shrubs (TEP 2019).

Given that restoration is so important to the state of Oregon, it is important to assess the effectiveness of restoration relative to the achievement of a project's restoration goals that are meaningful to adjacent communities. By using locally-relevant ecosystem metrics and indicators derived from stakeholder and beneficiary goals, the local ecology, and the NESCS Plus, restoration effectiveness can be evaluated by assessing a restoration site based on its capacity to produce and deliver priority nature-derived benefits (i.e., FEGS). The set of FEGS will vary with, and be dependent on, community values and priorities. The FST is important to this process because it can be used in the initial stages of the restoration to identify priority FEGS, beneficiaries, and benefits for a particular restoration site. For example, the FST could be a useful approach to use at a nascent project site along the Tillamook River where a parcel of land that is currently farmland has been acquired and designated for restoration and enhancement. Dairy farms, forest, and rural homes surround the site, along with a road that often floods from the river. The driving interests of stakeholders at the site include the construction of flood mitigation structures, habitat restoration to support salmonid growth, waterfowl hunting, and a designated shooting range, while aligning with the county's policy of no-net farmland loss.

Prioritizing stakeholders in restoration projects such as this is necessary because of potential conflicting stakeholder interests. Although there is no mechanism in the FST to ensure that all stakeholder groups are considered and represented, the SDM approach it builds on supports the inclusion of as wide a range of stakeholder groups and interests as possible. All engagement efforts rely on the good faith of the decision makers. This tool is one way for them to be transparent about who is considered, who is most relevant to the decision and the community impacted by it, and why ultimately not everyone's interests may be feasibly incorporated into decisions and goals. Additionally, an iterative and participatory approach can be taken in the stakeholder characterization step to allow for additional stakeholders to be identified and incorporated into the process. Using the FST to identify a broad range of potential stakeholders functions as a documented structure to prioritize stakeholders, beneficiaries, and ecological attributes and can help site planners decide on which type of site to target, on which FEGS to focus, and evaluate available options towards achieving restoration goals—whether that be salmon for fishing, waterfowl for hunting, or habitat for flood protection. From there, it can be determined how local FEGS metrics and indicators may be used to assess progress towards achieving those desired benefits.

5 Conclusion

Ecosystem-based management is a field requiring complex tradeoffs for its decision makers. In addition to the difficulty and uncertainty surrounding ecosystem forecasting, weighing socioeconomic concerns against environmental ones can be challenging. The concept of FEGS helps provide managers with language that more directly connects environmental concerns to the community's values. Using an SDM approach like the FST provides a clear pathway to prioritize stakeholders, develop a beneficiary profile, and focus management decisions on the environmental attributes most meaningful to the community, all of which help facilitate effective communication of the value of proposed work. It is rarely feasible to include all possible stakeholders in a decision-making process. Stakeholder prioritization provides clarity over who is included and why, enabling managers to be transparent about the perspectives being given weight in the decision process and, ultimately, leading to increased legitimacy (as defined as a "fair" decision process that considers representative perspectives; Cash et al. 2003) of the final decision. Development of a beneficiary profile allows managers to directly connect their community to the array of uses it has for the area under discussion. This allows managers to find common goals across beneficiary groups as well as potential points of contention. The holistic view of these uses also lowers the chance that valued uses are overlooked and then impacted by the decision. Identification of key environmental attributes of concern based on these uses then allows managers to focus their decision objectives on the most relevant metrics when evaluating tradeoffs. Ultimately, inclusion of the FST in an EBM process could lead to identification and consideration of more relevant ecosystem service metrics as a result of a more deliberate approach to stakeholder engagement and an improved understanding of community priorities.

Disclaimer This chapter has been subjected to Agency review and has been approved for publication. The views expressed in this paper are those of the author(s) and do not necessarily reflect the views or policies of the U.S. Environmental Protection Agency.

References

Ackerman, R., Neuenfeldt, R., Eggermont, T., Burbidge, M., Lehrman, J., Wells, N., & Chen, X. (2016). *Resilience of Oregon coastal communities in response to external stressors.* M.S. Thesis, University of Michigan.

Adamus Resource Assessment Inc. (2004). Oregon estuaries GIS data. Retrieved November 14, 2019, from https://spatialdata.oregonexplorer.info/geoportal/.

Boyd, J., & Banzhaf, S. (2007). What are ecosystem services? The need for standardized environmental accounting units. *Ecological Economics, 63*, 616–626.

Brophy, L. S., & van de Wetering, S. (2014). *Southern Flow Corridor project effectiveness monitoring plan.* Corvallis, Oregon: Institute for Applied Ecology and the Confederated Tribes of Siletz Indians.

Brown, L. A., Ewald, M. J., Brophy, L. S., & van de Wetering, S. (2016). *Southern Flow Corridor baseline effectiveness monitoring: 2014.* Corvallis, Oregon: Estuary Technical Group, Institute for Applied Ecology.

Cash, D., Clark, W. C., Alcock, F., Dickson, N. M., Eckley, N., & Jäger, J. (2003). Salience, credibility, legitimacy and boundaries: Linking research, assessment and decision making. KSG Working Papers Series. Retrieved October 28, 2019, from http://nrs.harvard.edu/urn-3:HUL.InstRepos:32067415.

Chan, K. M., Guerry, A. D., Balvanera, P., Klain, S., Satterfield, T., Basurto, X., et al. (2012). Where are cultural and social in ecosystem services? A framework for constructive engagement. *BioScience, 62*(8), 744–756.

Conklin, J. (2006). *Dialogue mapping: Building shared understanding of wicked problems.* Wiley Publishing, Chichester, England. ISBN: 978-0-470-01768-5.

Culhane, F. E., Robinson, L. A., & Lillebø, A. I. (2020). Approaches for estimating the supply of ecosystem services for ecosystem-based management in coastal and marine environments. In T. O'Higgins, M. Lago, & T. H. DeWitt (Eds.), *Ecosystem-based management, ecosystem services and aquatic biodiversity: Theory, tools and applications* (pp. 105–126). Amsterdam: Springer.

DeWitt, T. H., Berry, W. J., Canfield, T. J., Fulford, R. S., Harwell, M. C., Hoffman, J. C., Johnston, J. M., Newcomer-Johnson, T. A., Ringold, P. J., Russel, M. J., Sharpe, L. A., & Yee, S. J. H. (2020). The final ecosystem goods and services (FEGS) approach: A beneficiary-centric method to support. In T. O'Higgins, M. Lago, & T. H. DeWitt (Eds.), *Ecosystem-based management, ecosystem services and aquatic biodiversity: Theory, tools and applications* (pp. 127–148). Amsterdam: Springer.

Fedra, K. (1995). Decision support tools for natural resources management: Models, GIS and expert systems. *AI Applications, 9*(3), 3–19.

Fiorino, D. J. (1990). Citizen participation and environmental risk: A survey of institutional mechanisms. *Science, Technology, & Human Values, 12*, 226–243.

Gregory, R., & Wellman, K. (2001). Bringing stakeholder values into environmental policy choices: A community-based estuary case study. *Ecological Economics, 39*(1), 37–52.

Gregory, R., Failing, L., Harstone, M., Long, G., McDaniels, T., & Ohlson, D. (2012). *Structured decision making: A practical guide to environmental management choices*. Chichester, UK: Wiley-Blackwell.

Haines-Young, R., & Potschin, M. (2012). Common international classification of ecosystem services (CICES, Version 4.1). *European Environment Agency, 33*, 107.

Keen, P. G. W., & Morton, M. S. S. (1978). *Decision support systems: An organizational perspective*. Reading, MA: Addison Wesley.

Landers, D., & Nahlik, A. (2013). *Final ecosystem goods and services classification system (FEGS-CS)*. EPA/600/R-13/ORD-004914. Washington, DC: U.S. Environmental Protection Agency.

Luyet, V., Schlaepfer, R., Parlange, M. B., & Buttler, A. (2012). A framework to implement stakeholder participation in environmental projects. *Journal of Environmental Management, 111*, 213–219.

Millennium Ecosystem Assessment (MEA). (2005). *Ecosystems and human well-being: Synthesis*. Washington, DC: Island Press.

Nahlik, A. M., Kentula, M. E., Fennessy, M. S., & Landers, D. H. (2012). Where is the consensus? A proposed foundation for moving ecosystem service concepts into practice. *Ecological Economics, 77*, 27–35.

National Research Council (NRC). (1996). *Understanding risk: Informing in a democratic society*. Washington, DC: National Academies Press.

National Research Council (NRC). (2005). *Valuing ecosystem services: Toward better environmental decision-making*. Washington, DC: National Academies Press.

O'Higgins, T. G., Culhane, F., O'Dwyer, B., Robinson, L., & Lago, M. (2020). Combining methods to establish potential management measures for invasive species Elodea nutallii in Lough Erne Northern Ireland. In T. O'Higgins, M. Lago, & T. H. DeWitt (Eds.), *Ecosystem-based management, ecosystem services and aquatic biodiversity: Theory, tools and applications* (pp. 445–460). Amsterdam: Springer.

Office of Science and Technology (OST). (n.d.). Fish discard and release mortality science. National Marine Fisheries Service, National Oceanic and Atmospheric Administration. Retrieved October 28, 2019, from https://www.st.nmfs.noaa.gov/ecosystems/bycatch/discard-and-release-mortality.

Radtke, H., & Davis, S. (1994). *A demographic and economic description of the Oregon Coast*. Prepared for the Oregon Coastal Zone Management Association. Corvallis, Oregon: The Research Group.

Ralls, K., & Starfield, A. M. (1995). Choosing a management strategy: Two structured decision-making methods for evaluating the predictions of stochastic simulation models. *Conservation Biology, 9*(1), 175–181.

Reed, M. S., Graves, A., Dandy, N., Posthumus, H., Hubacek, K., Morris, J., Prell, C., Quinn, C. H., & Stringer, L. C. (2009). Who's in and why? A typology of stakeholder analysis methods for natural resource management. *Journal of Environmental Management, 90*(5), 1933–1949.

Ruckelshaus, M., McKenzie, E., Tallis, H., Guerry, A., Daily, G., Kareiva, P., Polasky, S., Ricketts, T., Bhagabati, N., Wood, S. A., & Bernhardt, J. (2015). Notes from the field: lessons learned from using ecosystem service approaches to inform real-world decisions. *Ecological Economics, 115*, 11–21.

Russell, M. J., Rhodes, C., Sinha, R. K., Van Houtven, G., Warnell, G., & Harwell, M. C. (2020). Ecosystem-based management and natural capital accounting. In T. O'Higgins, M. Lago, & T. H. DeWitt (Eds.), *Ecosystem-based management, ecosystem services and aquatic biodiversity: Theory, tools and applications* (pp. 149–164). Amsterdam: Springer.

Sharpe, L., & Jenkins, S. (2018). *FEGS scoping tool user manual. EPA/600/R-18/288*. Gulf Breeze, FL: U.S. Environmental Protection Agency.

Sharpe, L. M., Harwell, M. C., & Jackson, C. under review. Stakeholder prioritization for environmental management.

Southern Flow Corridor (SFC). (2019). Southern Flow Corridor—Landowner preferred alternative. Retrieved October 28, 2019, from https://tillamookoregonsolutions.com/.

Stewart, T. J. (1992). A critical survey of the status of multiple criteria decision making theory and practice. *Omega, 20*, 569–586.

The Wetlands Conservancy. (2018). *Beaver Creek marsh*. Retrieved October 28, 2019, from https://wetlandsconservancy.org/beaver-creek-marsh/.

Tillamook Estuaries Partnership (TEP). (2019). *Tillamook estuaries partnership—A National Estuary Project*. Retrieved October 28, 2019, from https://www.tbnep.org/.

U.S. Environmental Protection Agency (US EPA). under review. *Metrics for national and regional assessment of aquatic and terrestrial final ecosystem goods and services*. Washington, DC: U.S. Environmental Protection Agency.

Van Bueren, E. M., Klijn, E.-H., & Koppenjan, J. F. M. (2003). Dealing with wicked problems in networks: Analyzing an environmental debate from a network perspective. *Journal of Public Administration Research and Theory, 13*(2), 193–212.

Wasson, K., Suarez, B., Akhavan, A., McCarthy, E., Kildow, J., Johnson, K. S., Fountain, M. C., Woolfolk, A., Silberstein, M., Pendleton, L., & Feliz, D. (2015). Lessons learned from an ecosystem-based management approach to restoration of a California estuary. *Marine Policy, 58*, 60–70.

Yee, S., Bousquin, J., Bruins, R., Canfield, T. J., DeWitt, T. H., de Jesús-Crespo, R., Dyson, B., Fulford, R., Harwell, M., Hoffman, J., Littles, C. J., Johnston, J. M., McKane, R. B., Green, L., Russell, M., Sharpe, L., Seeteram, N., Tashie, A., & Williams, K. (2017). *Practical strategies for integrating final ecosystem goods and services into community decision-making. EPA/600/R-17/266*. Gulf Breeze, FL: U.S. Environmental Protection Agency.

3

Ecosystem-Based Management (EBM) and Ecosystem Services in EU Law, Policy and Governance

Anne Marie O'Hagan

Abstract Ecosystem-Based Management has become the dominant desired paradigm for environmental management globally yet what it entails and how it can be implemented presents many challenges. This chapter seeks to set out the legal bases for Ecosystem-Based Management (EBM) and Ecosystem Services in EU law and policy frameworks. It traces both concepts with a view to establishing how their legal status internationally and regionally has influenced their uptake within national governance frameworks. The rationale for EBM is to manage resources in a way that maintains the health of the ecosystem alongside appropriate human use of the marine environment, for the benefit of current and future generations, and accordingly is intrinsic to achieving sustainable development. EBM therefore represents two keys challenges for existing governance systems: firstly, the need to move away from sectoral based management and towards more integrated approaches and secondly, to embed the notion of 'healthy' ecosystems into all law and policy instruments. The chapter begins with the international and regional levels of governance and how both approaches have been incorporated. It then proceeds to a brief overview of the EU legal system, examines the key elements of biodiversity policy and conclude with a discussion and conclusions on whether these have enabled the implementation of EBM and ecosystem services.

Lessons Learned

- There is no definition of EBM or ecosystem services in EU law.
- EBM necessitates a move away from traditional sectoral focussed management as policy-makers need to manage for multiple ecosystem services that cannot be achieved if a single sectoral or policy 'lens' is taken.
- Regional Conventions have been instrumental in developing understanding of EBM and ES but can be limited by their individual remits.

A. M. O'Hagan (✉)
MaREI: The SFI Research Centre for Energy, Climate and Marine, Environmental Research Institute: Beaufort Building, University College Cork, Cork, Ireland
e-mail: a.ohagan@ucc.ie

- In the EU, EBM is partially implemented and in a top-down manner giving Member States significant discretion and often resulting in extensive differences between countries.
- Many existing EU policies in their current form contradict the commitment to implementing EBM and preserving ecosystem services, such as measures and activities under Common Agricultural Policy and Common Fisheries Policy.

Needs to Advance EBM

- There is a need for clear definition and agreement on what EBM is and how it can be implemented in practice at EU and national levels.
- EBM also necessitates adaptive management in order to deal with dynamic ecosystems and the absence of complete knowledge or understanding of their functioning.
- There is a critical need for clarity on what EBM actually requires, on how implementation progress can be measured and sharing of successful 'better' EBM practices.

1 Introduction to Ecosystem-Based Management (EBM) and Ecosystem Services (ES) in International Law and Policy

Somewhat surprisingly, there is no universally accepted definition of Ecosystem-Based Management (EBM) or the ecosystem approach in international or EU law, yet this has not limited implementation efforts to date. Aspects of the ecosystem approach can be traced back to the Stockholm Declaration on the Human Environment in 1972, which called for cooperation on conservation, protection and restoration of the Earth's ecosystem (United Nations 1972). The two key international legal instruments of relevance to EBM and ES are the UN Convention on Biological Diversity and the UN Convention on the Law of the Sea.

1.1 UN Convention on Biological Diversity (CBD)

The Ecosystem Approach was adopted as the primary framework for the conservation and sustainable use of biodiversity under the Convention on Biological Diversity (CBD) in 1995 hence it is from the CBD that the most widely used definition is derived. The fifth Conference of the Parties describes the Ecosystem Approach as "a strategy for the integrated management of land, water and living resources that promotes conservation and sustainable use in an equitable way" (CBD 2000). It went further to state that it "requires adaptive management to deal with the complex and dynamic nature of ecosystems and the absence of complete knowledge or understanding of their functioning". Implementation of the Ecosystem Approach is facilitated through 12 interlinked principles, known as the Malawi Principles (see Enright and Boetler 2020). Whilst the term 'ecosystem services' was not in use when the CBD was endorsed, the concept is implicit in the Convention text. Principle 5 goes some way towards explaining how the Ecosystem Approach and ecosystem services are connected: "conservation of ecosystem structure and functioning, in order to maintain ecosystem services, should be a priority target of the ecosystem approach", with a focus on ecosystem functioning.

The Malawi Principles provide the context for further action and, at the tenth Conference of the Parties, a global Strategic Plan for Biodiversity 2011–2020 was adopted. The plan's mission is "to take effective and urgent action to halt the loss of biodiversity in order to ensure that by 2020 ecosystems are resilient *and continue to provide essential services*, thereby securing the planet's variety of life, and contributing to human well-being, and poverty eradication". This means that the CBD provides the global framework for regional and national action to protect biodiversity but also the services that biodiversity provides. The Strategic Plan includes the Aichi Biodiversity Targets on the management of protected areas and the conservation of all ecosystems through the application of the precautionary approach and the Ecosystem Approach. The Aichi Biodiversity Targets consist of five Strategic Goals each of which are accompanied by specific targets. The goal on improving the status of biodiversity by safeguarding ecosystems, species and genetic diversity, for example, is the basis for Target 11 on conservation designations: "by 2020, at least 17 per cent of terrestrial and inland water, and 10 per cent of coastal and marine areas, especially areas of particular importance for biodiversity and ecosystem services, are conserved through effectively and equitably managed, ecologically representative and well-connected systems of protected areas and other effective area-based conservation measures, and integrated into the wider landscapes and seascapes" (CBD 2010). Ecosystem services are also addressed directly by Strategic Goal D: "Enhance the benefits to all from biodiversity and ecosystem services" which includes Target 14: "By 2020, ecosystems that provide essential services, including services related to water, and contribute to health, livelihoods and well-being, are restored and safeguarded, taking into account the needs of women, indigenous and local communities, and the poor and vulnerable."

1.2 United Nations Convention on the Law of the Sea (UNCLOS)

The Law of the Sea Convention defines the rights and responsibilities of nations with respect to their use of the world's oceans, protection of the marine environment, and the management of marine natural resources. There is no mention of the Ecosystem Approach in the Convention though debatably, the approach is included implicitly as it contains an absolute obligation to protect and preserve the marine environment (Article 192) and to adopt measures to protect and preserve rare or fragile ecosystems as well as the habitat of depleted, threatened or endangered species and other forms of marine life (Article 194(5)). Both these provisions, however, require further implementation mechanisms at national level in order to be achieved. The 1995 Fish Stocks Agreement[1] establishes an obligation for States to protect marine biodiversity via the protection of target and non-target species as well as the ecosystems associated with those species. The Agreement also includes a duty to cooperate in Regional Fisheries Management Organisations (RFMOs) or similar arrangements in an effort to recognise the transboundary nature of aquatic biodiversity and need for concerted coordinated action. Alas this remains a key challenge as the existing RFMOs do not cover all ocean space and organise their work according to geographic area and specific fish species meaning that many areas and species remain ineffectively managed. The links between the CBD and UNCLOS are recognised under CBD's Article 22(2) which provides that Parties will implement the CBD with respect to the marine environment "consistently with the rights and obligations of States under the Law of the Sea".[2]

1.3 OSPAR Convention

At the regional level, application of the Ecosystem Approach can be found in many of the Regional Seas Conventions and their associated mechanisms. The OSPAR Convention, for example, provides a framework for the regulation of almost all human activities[3] that have an adverse effect on marine ecosystems and biodiversity in the North-East Atlantic. The Helsinki and OSPAR Commissions adopted a statement on how the EA could be implemented under their respective legal instruments. This states that the Ecosystem Approach is "the comprehensive integrated management of human activities based on the best available scientific knowledge about the ecosystem and its dynamics, in order to identify and take action on influences which are critical to the health of marine ecosystems, thereby achieving sustainable use of ecosystem goods and services and maintenance of ecosystem integrity" (Helsinki and OSPAR Commissions 2003). The OSPAR Commission's work is guided by the Ecosystem Approach and it is implemented in the North-East Atlantic by means of the programmes and measures developed under OSPAR's six

[1] Agreement for the Implementation of the Provisions of the UN Convention on the Law of the Sea of 10 December 1982 relating to the Conservation and Management of Straddling Fish Stocks and Highly Migratory Fish Stocks, 4 August 1995 (into force 11 December 2001) 2167 UNTS 3.

[2] An exception to this is provided in Article 22(1) whereby "CBD provisions shall not affect the rights and obligations of any Party deriving from any international agreement except where the exercise of those rights and obligations would cause a serious damage or threat to biological diversity".

[3] Except fisheries and pollution from ships.

thematic Strategies.[4] The International Council for the Exploration of the Seas (ICES), working with OSPAR and HELCOM, has been instrumental in developing a better understanding of the EA and providing guidance and recommendations on how it can be implemented (e.g. ICES 2005). This includes the identification of practical steps in applying the approach by those tasked with implementing marine policy in the EU and informed the design of the EU's Marine Strategy Framework Directive. Unfortunately, by design, the Regional Seas Convention bodies are limited in the actions they can take in implementing the EA due their pre-defined geographical area and remit for actions.

Essentially the rationale for EBM is that whilst the ecosystem itself may not be managed, the human uses and activities that interact and impact upon the ecosystem may be managed so as to conserve biodiversity and ensure sustainable development (Long 2012). Ultimately the aim is to preserve ecosystem structure and functioning so as to ensure the ongoing provision of products and services. Therefore, management of the impacts of human activities must focus on the entire ecological system and not its component parts. This necessitates a move away from traditional sectoral management approaches towards those that are integrated, adaptive and coherent across policy domains so as to take account of social, economic and environmental aspects. EBM recognises that new forms of valuation and assessment are needed, and that different sectors of society will view ecosystems from their own environmental, economic and societal needs. The role of ecosystem services therefore is to provide information on the values and services that flow from ecosystems to humans. Internationally, the Millennium Assessment (MA) and The Economics of Ecosystems and Biodiversity (TEEB) initiatives have sought to capture information on the value of and benefits from ecosystems in a format that can be used by policy and decision-makers.

2 EBM and ES in EU Law and Policy

Whilst EBM is not expressly mentioned in the European treaties, under Article 11 of the Treaty on the Functioning of the EU, there is a duty to integrate environmental protection into the definition and implementation of EU policies "in particular with a view to promoting sustainable development". Article 191 TFEU states that EU environmental policy should promote measures at international level to deal with regional or worldwide environmental problems, which could be interpreted to include EBM. Essentially EBM (or the Ecosystem Approach) is introduced and implemented in the EU in a top-down fashion through a wide range of Directives and policy documents, resulting in much national autonomy in terms of implementation. The EU Biodiversity Baseline found that only 17% of habitats and species and 11%

[4]Biodiversity, eutrophication, radioactive substances, hazardous substances, offshore industry and assessment/monitoring.

of key ecosystems protected under EU legislation were classified as being in a favourable state (EEA 2010). In an attempt to address reverse this loss and assist in becoming a more resource efficient and green economy, the EU adopted a biodiversity strategy in 2011.

2.1 Biodiversity Law and Policy

The **EU Biodiversity Strategy 2020** aims to implement the CBD's Strategic Plan for Biodiversity 2011–2020 and the Aichi Targets (European Commission 2011). It recognises the role that biodiversity plays in underpinning the economy and the services it provides. The strategy sets out six targets to achieve the over-arching 2020 target of "halting the loss of biodiversity and the degradation of ecosystem services in the EU by 2020, and restoring them in so far as feasible, while stepping up the EU contribution to averting global biodiversity loss". The targets relate to nature (target 1), ecosystems and their restoration (target 2), the sustainable use of Europe's nature, land and sea resources via agriculture, forestry and fisheries (targets 3 and 4), alien species (target 5) and the EU's global impacts (target 6). A mid-term review of the Biodiversity Strategy was conducted in 2015 and found that biodiversity loss and degradation of ecosystem services in the EU has continued and, whilst some progress has been made at the policy level, this has not yet halted the trend of degradation of ecosystems and services (European Commission 2015a). The review concludes by stating that the 2020 biodiversity targets will only be achieved if implementation and enforcement efforts become "considerably bolder and more ambitious." An associated Fitness Check of the Birds and Habitats Directive, as the cornerstones of biodiversity policy in the EU, evaluated these instruments in terms of their effectiveness, efficiency, relevance, coherence and EU added value (European Commission 2016a). In terms of effectiveness, the evaluation found that the general objectives of the Directives have not yet been met and that it was not possible to predict when the objectives would be fully achieved but acknowledged that improvements in the status of species and habitats occur where there are targeted actions at a sufficient scale.

The Biodiversity Strategy is complemented by a wide range of legal instruments including the **Birds and Habitats Directives**, the Water Framework Directive, the Marine Strategy Framework Directive and the EU Regulation on Invasive Alien Species (No. 1143/2014). These four Directives, and one Regulation, provide the legal basis to protect aquatic biodiversity across the freshwater—marine continuum and hence seek to enable implementation of EBM. Bastmeijer (2019) states that the Birds and Habitats Directive are not based on the ecosystem approach and Rouillard et al. (2018) state neither Directive explicitly mentions ecosystem services or takes them into account implicitly. Despite this they are essential in protecting certain types of biodiversity. The Birds Directive provides for the designation of sites for the protection of bird species listed in Annex I, along with the designation of sites designated for "regularly occurring migratory species not included in Annex I" under Article 3(2). This is complemented by the Habitats Directive which, under Article 3(1), requires Member States to select and designate "sites hosting the natural habitat types listed in Annex I and habitats of the species listed in Annex II". Together the protected sites designated under each Directive form the Natura 2000 network, the largest global network of protected areas, which seeks to achieve the objective of maintaining or restoring natural habitats and species of Community interest at Favourable Conservation Status. The Appropriate Assessment procedure under Article 6, determines whether a plan or project can be implemented without damaging a Natura 2000 site, through an examination of the implications of a proposed development for the Natura 2000 site and its conservation objectives. Though this can add to regulatory and consenting requirements, the majority of plans and programmes subjected to this assessment are permitted to proceed (European Commission 2016a).

The **Invasive Alien Species Regulation** (IAS, No. 1143/2014) is mentioned here as it is the only EU legal instrument to explicitly contain a definition of ecosystems services as "the direct and indirect contributions of ecosystems to human wellbeing" in Article 3(6), perhaps attributable to its relatively recent entry into force (2015) when compared to other biodiversity instruments. It consists of three types of measures: prevention; detection and eradication; and management measures, based on the list of IAS of Union Concern contained within the Regulation. This list is updated regularly commencing with a proposal from a Member State or the EC, supported by a risk assessment; followed by an expert evaluation of the available evidence and inputs from a range of stakeholders and the Member States. This proposal must then be approved by a Committee comprised of Member States' representatives before final adoption by the Commission. There are no timelines associated with this process. Bouwma et al. (2018) state the IAS Regulation requires that its effectiveness on biodiversity, ecosystem services and, human health and the economy, if applicable, is monitored though the development of concrete measures and comprehensive action plans to prevent the "unintentional introduction and spread of invasive alien species" is left up to Member States. There is no dedicated funding instrument associated with the IAS though actions have been supported by the Commission through Horizon 2020, LIFE+, the rural development programme and other funding programmes such as Interreg.[5]

The European Commission (2007) has stated that the implementation of the Natura 2000 network forms one of the legal components to implementing the ecosystem approach in the marine environment. The mid-term review of the Biodiversity Strategy, states that the Natura 2000 network now covers approximately 18% of land but marine coverage is much lower at 6%, well below the 10% Aichi target. This could be attributed to the fact that there are much less marine species and

[5]The Interreg IVA programme, for example, funded the INVEXO project to support joint management efforts on four priority invasive alien species in Flanders and southern part of the Netherlands.

habitats listed in the Annexes of both Directives.[6] Marine species are also more difficult to protect in the same manner as terrestrial species as the marine species concerned may have a wider geographic range, perhaps taking in multiple jurisdictions. The EEA (2015a) concluded that only 21% of the habitat assessments and 23% of the non-bird species assessments were at favourable conservation status and 52% of the bird species assessed were secure, based on the reporting required under both Directives between 2007 and 2012. The same report (p. 8) states that the most frequently cited pressures and threats for marine ecosystems are fishing, modification of natural conditions and pollution. This brings into focus the need for greater interplay between the nature conservation Directives and other thematic legislation, not only to better protect the environment, but to implement the ecosystem approach. This is echoed by the EC's mid-term review of the EU's Biodiversity Strategy (2015a) which notes that "a lot remains to be done to halt the loss of ordinary biodiversity outside the Natura 2000 network." A key consideration in this context is the costs associated with implementation. The Fitness Check (EC 2016a, p. 5) states that as Member States do not have to report on the costs and benefits of the nature conservation legislation, there is limited quantitative information available at the EU scale to underpin assessments on efficiency and that "compliance costs of designating, protecting and managing Natura 2000 sites have been estimated to be at least € 5.8 billion annually across the EU."

2.2 *Water Law and Policy*

Alongside the nature conservation Directives, the **Water Framework Directive (WFD)** seeks to prevent the deterioration of freshwater ecosystems and restore their good ecological status. The Marine Strategy Framework Directive (MSFD), adopted slightly later in 2008, aims to achieve good environmental status of EU marine waters by 2020. The Directives therefore provide an ecological continuum from a river basin to the sea and, by taking a cyclical approach to implementation, should enable an adaptive management approach, one of the fundamental principles of EBM. The WFD was the first EU legal instrument to adopt a holistic approach to aquatic regulation via a move away from management on the basis of administrative boundaries. Evaluation of water quality under Annex V of the WFD requires consideration of the quality of the structure and functioning of aquatic ecosystems (surface waters), the physical-chemical nature of the water and sediment, the flow characteristics of the water, and the physical structure of water bodies. Article 4 of the Directive contains exemptions from the ecological goals, either by way of *force majeure*, reasons of overriding public interest or lack of ability of the Member State to achieve the goals due to external factors (e.g. impacts from other States). Whilst

[6]There are nine marine habitat types and 16 species listed in the Habitats Directive, and 60 bird species listed in the Birds Directive.

elements of the Directive are based on physical, chemical and biological parameters, the River Basin Management Plans, required under the Directive must also consider the socio-economic environment of the region and all activities that might impact on the status of a water body. A fitness check of the WFD (EC 2012) recommended that greater consideration be given to ecosystem services in the Directive, its Common Implementation Strategy and other policies "so that they can be better reflected in the implementation on the ground."

The **Marine Strategy Framework Directive (MSFD)** primarily aims to prevent any further deterioration of the marine environment, recognising it is the basis of the blue economy or, in other words, recognising the ecosystem services the marine environment provides. The Directive tries to better integrate the concepts of environmental protection and sustainable use. The MSFD was also viewed by the European Commission (2005) as a way of addressing sectoral fragmentation in marine governance and attaining the international obligations the EU had in terms of biodiversity under the CBD and Regional Seas Conventions to which the EU is a party. The MSFD requires in Article 1(1) that EU Member States to "take the necessary measures to achieve or maintain good environmental status in the marine environment by the year 2020 at the latest." According to Article 3(5), this objective is to be achieved by applying "adaptive management on the basis of the ecosystem approach." It is clear, from the 11 descriptors used in the Directive, that the concept of "good environmental status" includes the conservation of biodiversity and the maintenance of ecosystem health and integrity. The descriptors have since been supplemented in an European Commission Decision (2010) which subdivided them into 29 criteria and 56 associated 'indicators' so as to determine more precisely what attributes of the ecosystem features should be considered when assessing environmental status, ultimately with the aim of achieving more uniform assessment. Ecosystem services are only indirectly included in the descriptors. Berg et al. (2015) explain that Descriptor 3, on commercially exploited fish and shellfish, defines criteria on fish demographics aiming to ensure that fish populations are able to be caught (ecosystem service: food provision) and still be viable and productive. They conclude that although the MSFD includes sustainable use of the marine environment as part of the definition of GES, the Decision does not include criteria targeting ecosystem services that can be used to inform on the aspect of sustainable use. Like under the WFD, Member States are required to produce Marine Strategies for marine areas under their sovereignty and jurisdiction which must contain a comprehensive assessment of the state of the marine environment, a definition of "GES" at the regional level, clear environmental targets and monitoring programmes.

Subsequent to the initial assessment, setting of targets and monitoring plan, Member States must design a programme of measures (POM) to deliver the targets, each measure being supported by a cost-benefit analysis. In designing their POM, Member States are obliged to consult competent authorities in the field of water and nature conservation policy but the involvement of other sectoral authorities is at the discretion of the Member State. Perhaps in recognition of the need to take a broader perspective to implement the Ecosystem Approach, the MSFD advocates that operational and implementation measures are adopted through the Regional Seas Conventions. The Programmes of Measures "shall include spatial protection measures, contributing to coherent and representative networks of marine protected areas (MPAs) adequately covering the diversity of the constituent ecosystems" such as SACs, SPAs and other forms of MPAs under the RSCs or other international agreements. Sites in the Natura 2000 network, with marine qualifying interests, are the single largest contributor to European MPAs in terms of coverage, though it is acknowledged that geographic coverage lessens further offshore: Natura 2000 sites covered 33.3% of nearshore waters, 11.3% of coastal waters and only 1.7% of offshore waters (EEA 2015b). The EEA differentiates between three types of MPAs in the EU: marine Natura 2000 sites, MPAs designated under Regional Sea Conventions, and individual national MPAs. The Commission (2015b) has stated that in order for MPAs to fully deliver their potential, they must include management measures and require effective monitoring and enforcement. MPAs will be an integral part of Maritime Spatial Plans, which are legally required by 2021. As the MSFD is intended to make marine regulation and decision-making more integrated in form and content it will influence both cross-cutting management approaches such as Maritime Spatial Planning and Integrated Coastal Management, as well as sectoral policies, such as the Common Fisheries Policy.

2.3 *Sectoral Law and Policy*

2.3.1 Common Fisheries Policy (CFP)

As outlined above, to achieve the over-arching objectives of the MSFD, particular marine activities will have to reduce their environmental impacts or plan to mitigate these. One example of this is the requirement to consider the effects of the Common Fisheries Policy (CFP) on GES.[7] Article 2 of the Basic Fishery Management Regulation (2371/2002) states that one aim of the CFP is "to minimise the impact of fishing activities on marine ecosystems and to ensure the progressive implementation of an ecosystem-based approach to fisheries management." The European Parliament and Council's 7th Environmental Action Programme (2013), in relation to the exploitation of marine resources, recognises that "care needs to be taken to ensure their exploitation is compatible with the conservation and sustainable management of marine and coastal ecosystems." For GES, Descriptor 3 requires that populations of all commercially exploited fish and shellfish are within safe biological limits, exhibiting a population age and size distribution that is indicative of a healthy stock. It is left up to Member States to decide how to achieve this. Fisheries policy with regard to resource exploitation is an exclusive competence of the EU. The MSFD sets out objectives to be achieved but is unable to subject the CFP to its terms.

[7]Recital 40, MSFD and Article 2(5)(j) of Regulation 1380/2013.

As such, the MSFD merely notes in Recital 39 that "measures regulating fisheries management can be taken in the context of the Common Fisheries Policy" . . . "with a view to supporting the achievement of the objectives addressed by this Directive". A significant weakness however is that there is no requirement for the CFP to conform to or harmonise with the environmental objectives of the MSFD, and no specific measures to ensure coherence, thereby limiting the potential for implementation of an Ecosystem Approach and consideration of ecosystem services in fisheries management.

Article 8 of the revised Fisheries Regulation (No.1380/2013) provides for the creation of fish stock recovery areas, "due to their biological sensitivity, including areas where there is clear evidence of heavy concentrations of fish below minimum conservation reference size and of spawning grounds".[8] In these areas fishing activities may be restricted or prohibited in order to contribute to the conservation of living aquatic resources and marine ecosystems.[9] These areas could also contribute to a coherent network of protected areas, as envisaged under the Biodiversity Strategy and CBD. Article 11(1) of the Fisheries Regulation (No.1380/2013) provides that Member States can adopt conservation measures within waters under their jurisdiction in order to comply with their obligations under the MSFD and other EU environmental law provided that such measures do not affect the fishing vessels of other Member States and are compatible with the objectives of the CFP. Where there is a serious threat to the conservation of marine biological resources, or to the marine ecosystem, based on evidence, Article 12 (No.1380/2013) provides that the Commission may adopt restrictive measures for environmental protection that apply to all vessels but, by definition, these are time bound in that they cannot apply for longer than six months and may only be renewed once. This 'supremacy' of fishing can best be explained by its prominence in the EU Treaties, under the agriculture provisions, and in practice means that all Member States would have to agree to any measures likely to impact on or modify fishing activity. The revised governance structure in the 2013 Reform Package aims to address this somewhat by giving Member States more of a role in customising regional conservation measures, specifically on recommendations for achieving the objectives of conservation measures, provided the Commission is of the opinion such recommendations are compatible with the relevant conservation measure and/or applicable multiannual plan.[10]

---

[8]Regulation (EU) No 1380/2013 of the European Parliament and of the Council of 11 December 2013 on the Common Fisheries Policy, amending Council Regulations (EC) No 1954/2003 and (EC) No 1224/2009 and repealing Council Regulations (EC) No 2371/2002 and (EC) No 639/2004 and Council Decision 2004/585/EC. Official Journal of the European Union, L 354, pp. 22–61.

[9]Article 8(1), Fisheries Regulation (No.1380/2013).

[10]Article 18(3), Fisheries Regulation (No.1380/2013).

2.3.2 Common Agricultural Policy (CAP)

The Common Fisheries Policy is largely modelled on the Common Agricultural Policy, one of the common policy areas of the EU, in operation since 1962. As a common policy area the aim is to ensure there is a level playing field and fair competition between farmers. The objectives of CAP are to improve agricultural productivity, ensure a fair standard of living for those involved in farming, stabilise markets, assure the availability of supplies and ensure that produce reaches consumers at reasonable prices.[11] Reforms to the CAP have expanded it to encourage farmers to provide public goods, enhance biodiversity and help address climate change but not in the official constitutionalised version from the late 1950s. The CAP is financed centrally by the EC through two funds: the European Agricultural Guarantee Fund (EAGF), which provides direct support and funds market measures; and the European Agricultural Fund for Rural Development (EAFRD) which finances rural development programmes. The budget for CAP is €362.8 billion, almost 40% of the total EU budget, for the period 2014–2020, of which €277.9 billion is foreseen for Direct Payments and market-related expenditure (Pillar 1) and €84.9 billion is for Rural Development (Pillar 2) (European Commission 2013a). Agricultural nutrient sources are a major contributor to the status of water quality across the EU and needs to be considered as part of an ecosystem approach to aquatic management. The Nitrates Directive, together with the WFD, requires Member States to monitor water quality and, in particular, to identify areas that are polluted or at risk of pollution due to agricultural activities. These areas are known as "Nitrate Vulnerable Zones" and Member States are required to create Nitrate Action Programmes in order to reduce and prevent water pollution. Measures under these programmes include limits in when fertilizers can be applied, requirements for storage of manure, conditions for fertilizer applications, and limits on the amounts of fertilizer that can be used.

The Nitrates Directive does not refer to the Ecosystem Approach or ecosystem services directly though the definition of "pollution" in Article 2(j) acknowledges that pollution can cause "... harm to living resources and to aquatic ecosystems, damage to amenities or interference with other legitimate uses of water", thereby implicitly recognising ecosystem services. Reform of the CAP in 2013 resulted in a new policy instrument of the first pillar (greening) and covers the provision of environmental public goods. The Green Direct Payment equates to 30% of the national direct payment envelope and recompenses farmers for maintaining permanent grassland, creating ecological focus areas and diversifying crops. A portion of the rural development programme budget (30%) must also be used for measures that are beneficial for the environment and climate change, such as agri–environmental-climate measures, organic farming, Areas of Natural Constraints (ANC), Natura 2000 areas and forestry measures. Whilst well-intentioned, these measures have been largely unsuccessful, in terms of biodiversity conservation (Pe'er et al. 2017),

[11] Article 39(1) Article 39 TFEU (ex Article 33 TEC).

mitigation of climate change (European Commission 2019), and wider public opinion on whether CAP actually does enough to address environmental degradation and climate change (European Commission 2016b). Pe'er et al. (2019) found that highest investments are made into the least effective greening (€789.9/ha), compared to a third as many payments for the more effective agri-environment climate measures (€247.2/ha) and direct payments continue to be 'coupled' to the production of certain crops and livestock including input-intensive systems such as beef fattening and vegetable production which undermines overall sustainability goals. To achieve these, significant reforms are required not only to the substantive provisions of CAP but also its overall governance and integration with other key policy areas.

2.4 Cross-cutting Management Approaches

Integrated marine governance has been a focus area of Commission work since the early 1990s in terms of Integrated Coastal Zone Management (ICZM) and more recently Maritime Spatial Planning (MSP). MSP is seen as essential to delivering the jobs potential provided by maritime sectors, protection of the marine environment and optimisation of the use of marine space. ICZM has a policy basis in the associated Recommendation dating from 2002.[12] This recommends that Member States protect their coastal environment based on an ecosystem approach "preserving its integrity and functioning". Recognising the need for more action on integrated management approaches, the EC proposed a draft Directive, aimed at creating a framework for both Integrated Coastal Management (ICM) and Maritime Spatial Planning, with a view to improving planning and management of the land-sea interface (European Commission 2013b). During the negotiation phase, however, ICM was dropped from its contents. No official explanation is available but significant concerns were expressed by the Committee of the Regions (2013), for example, as it was perceived that ICM impinged substantially on existing Member State competences relating to spatial planning policy and practice at regional and/or local levels. The MSP Directive (2014/89/EU) was adopted in July 2014 and necessitates Member States to establish their first Maritime Spatial Plans by 31 March 2021. Article 5(1) requires Member States to apply an ecosystem-based approach when establishing and implementing MSP. The substantive provisions of the Directive say nothing further on the ecosystem-based approach though Recital 14 could be said to add some clarity saying the aim of applying an ecosystem approach is to ensure that the collective pressure of all activities is kept within levels compatible with the achievement of GES under MSFD and "that the capacity of marine ecosystems to respond to human-induced changes is not compromised, while

[12]Recommendation of the European Parliament and of the Council of 30 May 2002 concerning the implementation of Integrated Coastal Zone Management in Europe. OJ L148, 6 June 2002, pp.24–27. http://data.europa.eu/eli/reco/2002/413/oj

contributing to the sustainable use of marine goods and services by present and future generations". The MSFD identifies MSP as a tool to support the ecosystem-based management of human activities in order to achieve GES, which is a point reiterated in Recital 22 (Preamble) of the MSP Directive.

The principles and procedures involved in implemented Integrated Coastal Management, Maritime Spatial Planning and the Ecosystem Approach are broadly similar and each have the ultimate aim of delivering sustainable development (see Le Tissier 2020). All advocate the need for more holistic, integrated and adaptive management moving away from sectoral approaches, increased stakeholder participation, better use of scientific data and knowledge and integrated monitoring to adapt management actions. The challenges arising are that all approaches can be applied at different scales, however, the ecosystem approach is more suited to the ecosystem scale whereas ICM and MSP will most likely to tailored to the scale of the management problems they seek to address which may not necessarily correspond to the ecosystem level. Whilst the principles are similar the approaches differ and accordingly there is a need for greater policy coherence that takes into account a wider range of economic, environmental and social aspects. The policies adopted need to secure the delivery of ecosystem and harmonise management and conservation objectives, both in the immediate and longer term. ICM is not a binding legal requirement on all EU Member States although the 2002 ICZM Recommendation remains valid. MSP is only in the initial stages of implementation in the majority of coastal Member States hence how the Ecosystem Approach and ecosystem services will be reflected in final plans remains to be seen. It is probable that the marine strategies developed by Member States to meet the requirements of the MSFD are more likely to be ecosystem-based than either MSP or ICM given their focus but this goes to reaffirm the need for policy coherence if the ecosystem approach is to be implemented across the entire aquatic area.

3 Discussion

The preceding sections demonstrate that the Ecosystem Approach and ecosystem services have been incorporated into EU law and policy to varying extents. At the international level this has also continued to progress but the variety of definitions and methodologies for implementation of the concepts at the global level has very real implications for other governance scales, including regional, national and local implementation which are more utilised for aquatic management under EU law. It has been long recognised (e.g. Scheiber 1997) that there is a critical need for clarity on what EBM actually requires and how progress on implementation can be measured. The commitment to implementing EBM at EU level requires more joint implementation of legal instruments to take account of the necessary move away from management according to administrative boundaries. This is central to the WFD and echoed in the MSFD yet despite having a Common Implementation Strategy there are significant disconnects between both instruments. For example, marine litter is one of the descriptors used for determining GES under MSFD but there is no equivalent in the WFD, even though the majority of marine litter derives from the land as scientific evidence confirms. From a governance perspective there is a need for institutional structures that facilitate cross-sectoral management and decision-making. Competences for aquatic management are split according to the land and sea, jurisdictional boundaries and also sectorally which goes against the principles of ecosystem-based management. The same is true for monitoring, enforcement and compliance actions.

Many existing EU policies in their current form contradict the commitment to implementing EBM and preserving ecosystem services, such as measures and activities under CAP and CFP. The Fitness Check (EC 2016a, p. 91) of the nature conservation legislation found that EU financial support applies to agriculture and forestry, the main land uses in Natura 2000 as well as to prevent damage caused by protected species (e.g. under rural development for large carnivores) or to compensate for such damages (e.g. under fisheries policy for fish-eating birds). Figure 1 illustrates the levels of agricultural, fisheries and conservation across the EU. The main beneficiaries of financial support in the EU remain the economic sectors of fishing and agriculture. The overall EU co-funding for Natura 2000 during the 2007–2013 period represented only 9–19% of the estimated financing needs and national co-funding was unable to cover the remaining gap. This could lead one to question the commitment to halting the loss of biodiversity and the degradation of ecosystem services in the EU by 2020. Alternatively, it could signal the need to better consider trade-offs in management processes. There is no 'recipe' for balancing conservation and development but it is essential that we understand who will benefit and who will lose out if ecosystem services change. This necessitates better understanding of ecosystem services. From a societal perspective, greater stakeholder participation is required so as to understand their interests and expectations. In theory, in this way more suitable interventions could be made to deliver sustainable development.

4 Conclusions

The European Commission has endorsed the United Nations' Sustainable Development Goals (SDGs). Effective implementation of the SDGs will require fully coordinated policies that take into account the multiple relationships that exist between the different dimensions of sustainability, something that cannot be achieved currently given the preponderance of sectoral legislation, policies and associated institutional structures. The Marine Strategy Framework Directive has the ecosystem approach at its core but implementation of its objectives is largely left up to Member States themselves, whilst other legislation and policies refer to EA but it is not the overarching objective or priority for implementation. There will always be trade-offs between conservation and development and no form of decision-making will make this disappear. Attaining societal agreement on long-term goals

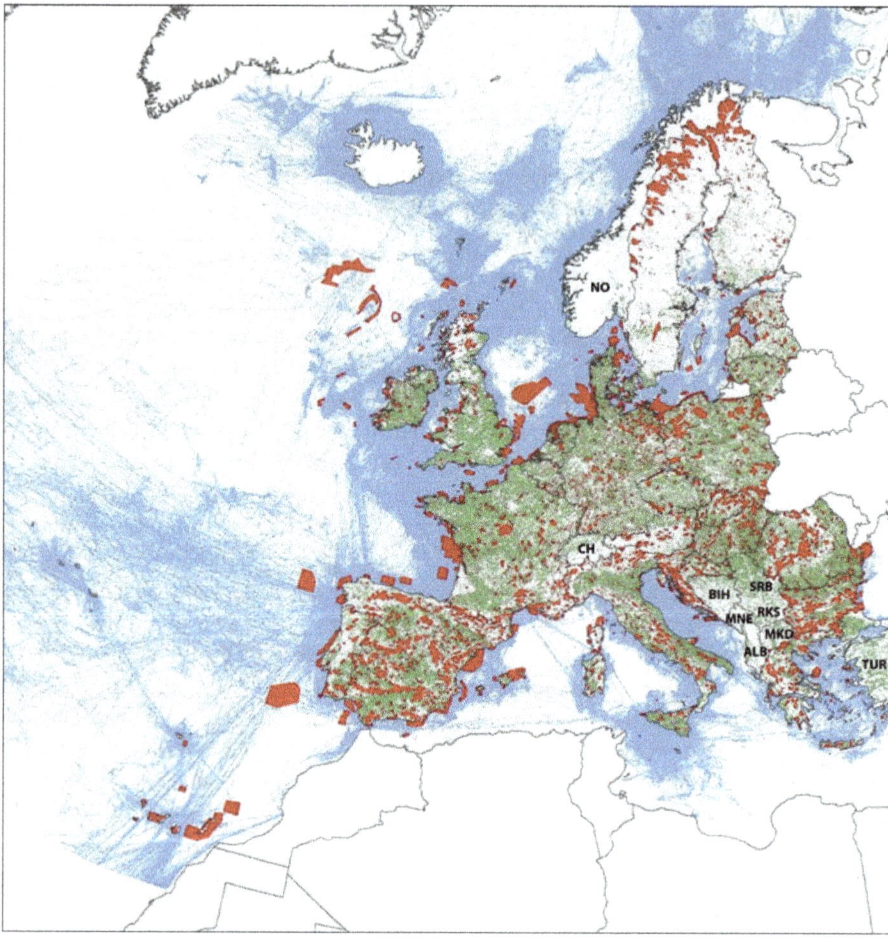

Fig. 1 Fishing vessel activity (blue), agriculture (green) and nature conservation (red) across the EU. Countries marked with ISO country codes are non-EU member states and do not participate in the EU Natura 2000 network, but have habitat data available under the Corine Land Cover inventory. Cartography by Tim O'Higgins

requires appropriate institutional mechanisms and these also need to acknowledge that societal goals change over time. Decision-making structures are still dominated by sectoral interests. This is clearly evident in the case of both the CFP and CAP. Reform of the CFP, for example, resulted in the creation of Advisory Councils as stakeholder bodies to address issues associated with participatory democracy (European Commission 2009). No measures affecting fishing activity can be adopted without reference to these Advisory Councils. Article 45(1) and Annex III (2) of Regulation 1380/2013 provides that 60% of the seats on the Advisory Councils are allocated to organisations representing the fisheries, processing and marketing sectors, with the remainder allocated to other interest groups. Wakefield (2019) states that the European Parliament had called for half the seats on the Advisory Councils to be reserved for interested parties outside the fishing industry but this was not accepted by the Member States in Council.

Like EBM, the conceptual basis for ecosystems services has been well expressed but implementation remains variable, perhaps explained by lack of agreement on definitions and uncertainties around the links between biodiversity and the services that flow from ecosystems. Each different ecosystem service also has a different legal status, some being public and others private. In effect, this means that rights will be held by individuals, groups, and the state, with further complexity arising between land and aquatic space and in transboundary contexts. Many of the SDGs are underpinned by the delivery of one or more ecosystem services, meaning policy-makers will need to manage for multiple ecosystem services, which cannot be achieved if a single policy 'lens' is taken. Neither EA or ecosystem services alone will be sufficient to deliver on all the SDGs, but will involve concerted efforts from the spheres of institutions, technology, science, politics and society generally. The effective implementation of EBM and ecosystem services is contingent on the necessary legal and regulatory frameworks being in place. Whilst there have been attempts to incorporate both concepts into the law and policy framework there is still a large degree of contradiction in terms of over-arching objectives and goals of many key instruments. Until these are resolved, EBM in Europe will continue to be an aspirational concept rather than a tangible and effective management approach.

Acknowledgement This contribution is based upon works supported by the Navigate project (Grant-Aid Agreement No. 842 PBA/IPG/17/01), carried out with the support of the Marine Institute and funded under the Marine Research Programme by the Irish Government, and by MaREI: the SFI Research Centre for Energy, Climate and Marine (12/RC/2302).

References

Bastmeijer, K. (2019). The ecosystem approach for the marine environment and the position of humans: Lessons from the EU natura 2000 regime. In D. Langlet & R. Rayfuse (Eds.), *The ecosystem approach in ocean planning and governance: Perspectives from Europe and beyond*. Leiden and Boston: Brill Nijhoff.

Berg, T., Fürhaupter, K., Teixeira, H., Uusitalo, L., & Zampoukas, N. (2015). The marine strategy framework directive and the ecosystem-based approach—pitfalls and solutions. *Marine Pollution Bulletin, 96*(1–2), 18–28. https://doi.org/10.1016/j.marpolbul.2015.04.050.

Bouwma, I., Schleyer, C., Primmer, E., Winkler, K. J., Berry, P., Young, J., Carmen, E., Špulerová, J., Bezák, P., Preda, E., & Vadineanu, A. (2018). Adoption of the ecosystem services concept in EU policies. *Ecosystem Services, 29*, 213–222. https://doi.org/10.1016/j.ecoser.2017.02.014.

Committee of the Regions. (2013). Opinion of the committee of the regions on proposed directive for maritime spatial planning and integrated coastal management (2013/C 356/18). *OJEU, C356*, 124–132. Retrieved from https://eur-lex.europa.eu/LexUriServ/LexUriServ.do?uri=OJ:C:2013:356:0124:0132:EN:PDF.

Convention on Biological Diversity Decision V6 of the Conference of the Parties to the CBD. (2000). 5th Meeting, Nairobi, Kenya, May 15–26, UNEP/COP/5/23. Retrieved from https://www.cbd.int/decision/cop/default.shtml?id=7148.

Convention on Biological Diversity Decision X/2 Annex of the Conference of the Parties to the CBD. (2010). 10th Meeting, Nagoya, Aichi Prefecture, Japan, October 18–29, UNEP/CBD/COP/10/27. Retrieved from https://www.cbd.int/decision/cop/default.shtml?id=12268.

EEA. (2010). EU 2010 biodiversity baseline EEA Report No. 12/2010 EEA, Denmark. Retrieved from https://www.eea.europa.eu/publications/eu-2010-biodiversity-baseline.

EEA. (2015a). State of nature in the EU: Results from reporting under the nature directives 2007–2012. EEA Technical Report No. 2/2015 EEA, Denmark. Retrieved from https://www.eea.europa.eu/publications/state-of-nature-in-the-eu.

EEA. (2015b). Spatial analysis of marine protected areas in Europe's seas. EEA Technical Report No. 17/2015 EEA, Denmark. Retrieved from https://www.eea.europa.eu/publications/spatial-analysis-of-marine-protected.

Enright, S.R., & Boetler, B. (2020). The ecosystem approach in international law. In T. O'Higgins, M. Lago, & T. H. DeWitt (Eds.), *Ecosystem-based management, ecosystem services and aquatic biodiversity: Theory, tools and applications* (pp. 333–352). Amsterdam: Springer.

European Commission. (2005). Proposal for a Directive of the European Parliament and of the Council establishing a framework for community action in the field of marine environmental policy (marine strategy directive). *COM,* 505 final. EC, Brussels, Belgium, pp. 2–3. Retrieved from https://eur-lex.europa.eu/legal-content/EN/TXT/?uri=CELEX:52005PC0505.

European Commission. (2007). Guidelines for the establishment of the Natura 2000 network in the marine environment. EC, Brussels, Belgium. Retrieved from https://ec.europa.eu/environment/nature/natura2000/marine/docs/marine_guidelines.pdf.

European Commission. (2009). Green paper on the reform of the common fisheries policy. *COM*, 163. Retrieved from https://eur-lex.europa.eu/LexUriServ/LexUriServ.do?uri=COM:2009:0163:FIN:EN:PDF.

European Commission. (2010). Commission decision of 1 September 2010 on criteria and methodological standards on good environmental status of marine waters (2010/477/EU). *Official Journal of the European Union, L232*(14), 14–24.

European Commission. (2011). Communication from the commission to the European parliament, the council, the economic and social committee and the committee of the regions. Our life insurance, our natural capital: An EU biodiversity strategy to 2020. *COM*, 0244 final. EC, Brussels, Belgium.

European Commission. (2012). Commission staff working document: The fitness check of EU freshwater policy. *SWD*, 393 final. EC, Brussels, Belgium, p. 10. Retrieved from https://ec.europa.eu/environment/water/blueprint/pdf/SWD-2012-393.pdf.

European Commission. (2013a). Overview of CAP reform 2014–2020. Agricultural policy perspectives brief No. 5, December. EC, Brussels, Belgium. Retrieved from https://ec.europa.eu/agriculture/sites/agriculture/files/policy-perspectives/policy-briefs/05_en.pdf.

European Commission. (2013b). Proposal for a Directive of the European Parliament and of the Council establishing a framework for maritime spatial planning and integrated coastal management. *COM*, 133 final. Retrieved from https://ec.europa.eu/environment/iczm/pdf/Proposal_en.pdf.

European Commission. (2015a). Report from the Commission to the European Parliament and the Council—the Mid-term Review of the EU biodiversity strategy to 2020. EC, Brussels, Belgium. Retrieved from https://ec.europa.eu/environment/nature/biodiversity/strategy/index_en.htm#mid.

European Commission. (2015b). Report from the Commission to the European Parliament and the Council on the progress in establishing marine protected areas (as required by Article 21 of the Marine Strategy Framework Directive 2008/56/EC). Retrieved from https://ec.europa.eu/environment/marine/eu-coast-and-marine-policy/implementation/pdf/marine_protected_areas.pdf.

European Commission. (2016a). Commission Staff Working Document Fitness Check of the EU Nature Legislation (Birds and Habitats Directives) Directive 2009/147/EC of the European Parliament and of the Council of 30 November 2009 on the conservation of wild birds and Council Directive 92/43/EEC of 21 May 1992 on the conservation of natural habitats and of wild fauna and flora. *SWD*, 472 final. EC, Brussels, Belgium. Retrieved from https://ec.europa.eu/environment/nature/legislation/fitness_check/docs/nature_fitness_check.pdf.

European Commission. (2016b). Special eurobarometer 440: Europeans, agriculture and the CAP. Survey requested by the European Commission, Directorate-General for Agriculture and Rural Development and co-ordinated by the Directorate-General for Communication. https://doi.org/10.2762/03171. Retrieved from http://ec.europa.eu/commfrontoffice/publicopinion/index.cfm/ResultDoc/download/DocumentKy/69756.

European Commission. (2019). Evaluation study of the impact of the CAP on climate change and greenhouse gas emissions: Final Report. https://doi.org/10.2762/54044. Retrieved from https://ec.europa.eu/agriculture/sites/agriculture/files/evaluation/market-and-income-reports/2019/cap-and-climate-evaluation-report_en.pdf.

European Parliament and European Council. (2013). Decision No. 1386/2013/EU of the European Parliament and of the Council of 20 November 2013 on a General Union Environment Action Programme to 2020 'Living well, within the limits of our planet' [Annex para.21 (p. 179)]. *Official Journal of the European Union L, 354*, 171–200. Retrieved from https://eur-lex.europa.eu/legal-content/EN/LSU/?uri=CELEX:32013D1386.

Helsinki and OSPAR Commissions. (2003). First Joint Ministerial Meeting of the Helsinki and OSPAR Commissions (JMM) Bremen Germany 25–26 June 2003, Statement on the Ecosystem Approach to the Management of Human Activities "Towards an Ecosystem Approach to the Management of Human Activities" Annex 5 Ref. §6.1. Retrieved from https://www.ospar.org/site/assets/files/1232/jmm_annex05_ecosystem_approach_statement.pdf.

ICES. (2005). Guidance on the application of the ecosystem approach to management of human activities in the European marine environment ICES cooperative research report no. 273. https://doi.org/10.17895/ices.pub.5477.

Le Tissier, M. (2020). Unravelling the relationship between ecosystem-based management, integrated coastal zone management and marine spatial planning. In T. O'Higgins, M. Lago, & T. H. DeWitt (Eds.), *Ecosystem-based management, ecosystem services and aquatic biodiversity: Theory, tools and applications* (pp. 403–416). Amsterdam: Springer.

Long, R. (2012). Legal aspects of ecosystem-based marine management in Europe. In A. Chircop, M. L. McConnell, & S. Coffen-Smout (Eds.), *Ocean yearbook* (Vol. 26, pp. 417–484). Boston: Brill Academic Publishers.

Pe'er, G., Zinngrebe, Y., Hauck, J., Schindler, S., Dittrich, A., Zingg, S., Tscharntke, T., Oppermann, R., Sutcliffe, L. M., Sirami, C., Schmidt, J., Hoyer, C., Schleyer, C., & Lakner, S. (2017). Adding some green to the greening: Improving the EU's ecological focus areas for biodiversity and farmers. *Conservation Letters, 10*, 517–530. https://doi.org/10.1111/conl.12333.

Pe'er, G., Zinngrebe, Y., Moreira, F., Sirami, C., Schindler, S., Müller, R., Bontzorlos, V., Clough, D., Bezák, P., Bonn, A., Hansjürgens, B., Lomba, A., Möckel, S., Passoni, G., Schleyer, C., Schmidt, J., & Lakner, S. (2019). A greener path for the EU common agricultural policy. *Science, 365*(6452), 449–451. https://doi.org/10.1126/science.aax3146.

Rouillard, J., Lago, M., Abhold, K., Röschel, L., Kafyeke, T., Mattheiß, V., & Klimmek, H. (2018). Protecting aquatic biodiversity in Europe: How much do EU environmental policies support ecosystem-based management? *Ambio, 47*, 15–24. https://doi.org/10.1007/s13280-017-0928-4.

Scheiber, H. N. (1997). From science to law to politics: An historical view of the ecosystem idea and its effects on resource management. *Ecology Law Quarterly, 24*(4), 631–651. https://doi.org/10.15779/Z385V74.

United Nations. (1972). Report of the United Nations Conference on the Human Environment Stockholm Sweden 5–June 16. UN Doc.A/Conf.49/14/Rev.1 Chapter 1 (Principles). Retrieved from https://www.un.org/ga/search/view_doc.asp?symbol=A/CONF.48/14/REV.1.

Wakefield, J. (2019). The ecosystem approach and the common fisheries policy. In D. Langlet & R. Rayfuse (Eds.), *The ecosystem approach in ocean planning and governance: Perspectives from Europe and beyond*. Leiden: Brill Nijhoff.

4

Ecosystem-Based Management to Support Conservation and Restoration Efforts in the Danube Basin

Andrea Funk, Timothy G. O'Higgins, Florian Borgwardt, Daniel Trauner, and Thomas Hein

Abstract Biodiversity and environmental integrity of river systems in the Danube catchment is threatened by multiple human alterations such as channelization, fragmentation or the disconnection of floodplains. Multiple human activities, including the construction of hydropower plants, expansion of agricultural use, and large-scale river regulation measures related to navigation and flood protection, are resulting in an ongoing loss of habitat, biodiversity and ecosystem service provision. Conservation and restoration of the systems biodiversity and ecosystem service provisioning is a key task for management but is challenging because the diversity of human activities and policy targets, scarcity of data compared to the complexity of the systems, heterogeneity of environmental problems and strong differences in socio-economic conditions along the Danube River hampers coordinated planning at the scale of the whole river basin and along the whole river from source to mouth. We evaluated three different implementations of an Ecosystem-Based Management (EBM) approach, which aims to support management efforts. This was done following the principles for EBM related to the resilience of ecosystems, the consideration of ecological and socio-economic concerns, the inclusion of multi-disciplinary knowledge and data addressing the ecosystem scale independent of administrative or political boundaries. This approach has been developed in the H2020 project AQUACROSS.

A. Funk (✉) · D. Trauner · T. Hein
University of Natural Resources & Life Sciences, Vienna, Austria

WasserCluster Lunz, Lunz am See, Austria
e-mail: andrea.funk@boku.ac.at

T. G. O'Higgins
University College Cork, National University of Ireland, Cork, Ireland

F. Borgwardt
University of Natural Resources & Life Sciences, Vienna, Austria

Lessons Learned

- Coupled modelling frameworks are a useful tool for modelling biodiversity restoration measures
- Multiple policy targets can be harmonized with this approach

Needs to Advance EBM

- Continued international cooperation informed by costed measures

1 Introduction

The core principle of Ecosystem-Based Management (EBM) is to concurrently consider biodiversity and human society as integral parts of the ecosystem and manage the socio-ecological system as a whole (Domisch et al. 2019; Langhans et al. 2019). Delacámara et al. (2020) review the many 'flavours' of EBM to identify six characteristics or principles, which set EBM apart from other types of management:

1. It considers ecological integrity, biodiversity, resilience and ecosystem services
2. It is carried out at appropriate spatial scales
3. It develops and uses multi-disciplinary knowledge
4. It builds on social-ecological interactions, stakeholder participation and transparency
5. It supports policy coordination
6. It incorporates adaptive management.

While these EBM principles are not proscriptive, i.e. any particular EBM activity is not required to have all these characteristics, they may offer useful criteria by which EBM activities may be practically assessed.

The Danube River Basin (DRB) is the most international river basin in the world shared by more than 80 million people across 19 countries (Fig. 1). The Danube River connects with 27 large and over 300 small tributaries on its way from the Black Forest to the Black Sea, covering a catchment size of approx. 800,000 km^2.

As a result, a huge variety of human activities and related pressures affect this area and a number of major environmental issues threaten the ecosystems of the Danube. As Europe's second longest river, the Danube has long been a major transport corridor. Today, it connects Europe's largest port of Rotterdam with the Black Sea via the Rhine-Main-Danube canal. Physical modifications of the river morphology to accommodate transport and power production have altered flow regimes with serious consequences for ecosystems including the disconnection of the river from its natural flood plains. Agricultural activities along the Danube have resulted in pollution by nutrients and pesticides. The combined effects of these and other pressures have resulted in overall degradation of the freshwater ecosystems and severe declines in iconic species such as different sturgeon species. The

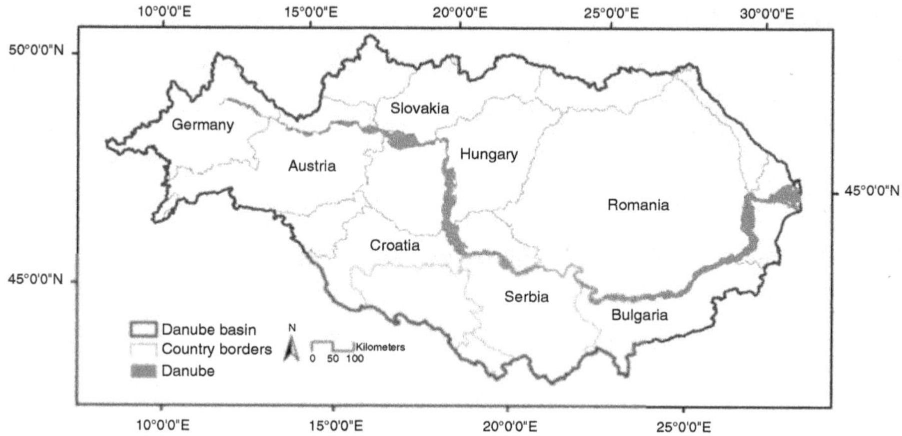

Fig. 1 The Danube River Basin and the corridor of the Danube river

International Commission for the Protection of the Danube River (ICPDR) provides a formal international mechanism for environmental management collaboration across the Danube Basin (detailed information on the many environmental issues can be found on their website (https://www.icpdr.org/main/).

Despite conservation efforts, ongoing and partially conflicting demands within and among the different neighboring countries, inconsistencies in legislation, high administrative and socioeconomic complexity as well as partially lack of on-site expert knowledge all hamper sustainable management (Hein et al. 2016, 2018; Habersack et al. 2016).

There are two major challenges for the management of the DRB. The multicultural setting makes transboundary issues extremely difficult and challenging. For example, the basin lies in the historical political border between capitalist and communist countries, which greatly influences the socio-economic situations, social behaviors, technical developments, as well as water uses and protections between the two former systems (Sommerwerk et al. 2010) and resulting in varying priorities towards, and capacities for, environmental protection (O'Higgins et al. 2014). In the DRB, this historical background is well reflected in the structural differences between the Upper Danube (capitalist countries) where hydro-morphological alteration is high but pollution is low, while in the Lower Danube (former communist countries) pollution is still a highly relevant issue but level of impact due to river engineering works is still relatively low (Sommerwerk et al. 2010). This phenomenon is also reflected in the ranking of stressors along the Danube River. Hein et al. (2018) found that for the Upper Danube hydro-morphological alterations due to hydropower generation, navigation, and flood protection has the highest importance followed by forestry, disturbance due to recreational activities, recreational fisheries and last by pollution, whereas the Lower Danube is mostly impacted by land use including forestry, agriculture and urbanization having an direct as well as an pollution effect on the system and last by hydro-morphological alterations of the river.

Table 1 Policies directives and initiatives with synergistic and antagonistic effects on conservation objectives in the DRB

Instrument type	Name	Targets and goals
Policy	EU Biodiversity Strategy	Full implementation of the Birds and Habitats Directives
	EU Strategy for the Danube Region	e.g. Sturgeon 2020 program for the protection and rehabilitation of sturgeon
Legally binding directives	Water Framework Directive (2000/60/EC; WFD)	Good Ecological Status—through implementation of the Danube River Basin Management Plan
	Flood Risk Directive	Danube Flood Risk Management plan
	Birds Directive (2009/147/EC)	Favorable conservation status (for selected species)
	Habitats Directive (92/43/EEC)	Favorable conservation status (for selected habitats)
	Renewable Energy Directive	Total of 20% of EU energy needs to be supplied by renewable sources (including hydro power).
Initiative	Trans-European Transport Network	Good navigability for important waterways, including the removal of obstacles

The second major challenge in DRB management is to establish synergies among multiple competing interests and policy targets including e.g. navigation, hydropower production, flood protection and nature conservation (Sommerwerk et al. 2010). Human stressors interact with the management goals of the Water Framework Directive (EC 2000) or Nature Directives (EC 1992) and the Biodiversity Strategy to 2020 (EC 2011), resulting in potential synergies and conflicts between the various management goals. The implementation of sectoral policies on hydropower (renewable energy), navigation, and flood protection may show significant synergies and antagonisms, and the interaction of their implementation significantly influences the actual type and extent of pressures on rivers. Table 1 lists some of the interrelated directives, policies and initiatives with specific relevance to the management of the Danube River and its associated ecosystems.

For example, the Flood Risk Directive (EC 2007) aims at reducing risk of flooding along water courses including natural water retention measures (e.g. dyke relocation to provide more space for rivers). Floodplains are therefore a key element of the EU Green Infrastructure Strategy (ICPDR 2016). Like-wise navigation projects might either have a synergistic effect on nature protection goals in already significantly altered river sections (if ecological restoration is supported within the project), or an antagonistic effect in intact river sections where every intervention may create a conflict with nature protection goals (DANUBEPARKS 2011). With a multitude of interacting environmental and other directives, management targets can have synergistic as well as antagonistic effects, which vary from place to place. Moreover, these interactions are complex and not sufficiently understood.

In this context, modern management concepts can neither exclusively focus on the mitigation of single pressures or stressors nor can they limit their measures to single ecosystem components, species groups or other single targets. In contrast, they have to consider complex interactions and feedback loops between the ecosystems and the society. Thus, for the future, explicit and well-defined ecosystem-based targets need to be formulated, and adequate measures need to be defined to achieve more resilient ecosystems, guarantee the provision of a broad range of ecosystem services, and increase the resilience against emerging stressors like climate change or invasive species (Hein et al. 2018). Given the need for holistic catchment scale management approaches (Hein et al. 2018; Seliger et al. 2016), EBM offers the potential to incorporate multiple objectives related to biodiversity, ecosystem services and socio-economic benefits into a single, harmonized management approach for the DRB. The Danube River, as one of the largest river-floodplain systems in Europe, is a highly complex, threatened and challenging socio-ecological system, and therefore an ideal system to test and apply an EBM approach. To this end, within the frame of the AQUACROSS research project a number of tools and techniques were combined and tested for application in the Danube catchment. In this paper we describe and discuss three different approaches and provide a qualitative assessment of how these methods relate to the EBM principles identified above.

2 The Studies

Other authors in this volume (Fulford et al. 2020; Lewis et al. 2020) have addressed the challenges of model design and selection and the potential for combining models to address particular situations. We evaluate three different quantitative and qualitative approaches that have been applied at the Danube catchment scale to describe and model the socio-ecological system. A linkage framework approach (Borgwardt et al. 2019; Teixeira et al. 2019; Robinson & Culhane 2020) was used to assess the relationships between different activities within the catchment and their relations to biodiversity and ecosystem services. The potential of EBM was also tested within two quantitative studies following an EBM planning framework based on a generic model-coupling approach proposed by Langhans et al. (2019). The workflow consists of three elements a spatial (model-based) representation of (1) biodiversity, (2) ecosystem services (ESS), and (3) a combined spatial prioritization of biodiversity and ESS supply and demand.

Finally, Domisch et al. (2019) combined the ARIES (Artificial Intelligence for Ecosystem Services) modelling framework (Villa et al. 2014) with the application of MARXAN (Ball et al. 2009) to identify a range of spatially explicit management zones and options. Funk et al. (2019) combined Bayesian Belief Network Modelling with the ARIES model to identify river reaches maintaining multiple ecological functions and support multiple services to prioritize individual areas for conservation incorporating a range multiple restoration criteria.

2.1 Linkage Frameworks

A Linkage Framework (LF) for the Danube Basin (Fig. 2) identified 53 specific human activities (or Drivers) occurring in the catchment (Borgwardt et al. 2019). Furthermore, 35 different pressures in five different categories (biological, chemical, physical, energy, and exogenous/unmanaged) were identified, as well as 33 ecosystem components (27 habitats and 6 biotic groups). These components were linked to 27 ecosystem services (ESS) and abiotic outputs. Over 23,000 impact chains relating drivers-pressures and ecosystem components were identified and categorized. To investigate the impact chains, their connectance was calculated and linkages were also weighted in terms of the extent, frequency, dispersal, severity and persistence of interactions to increase their explanatory power. Analysis of the impact risk of

1. Scientific; education; heritage; aesthetic; entertainment
2. Existence; bequest
3. In-situ recreational activities involving biota
4. Lifecycle maintenance, habitat and gene pool protection
5. Mediation of waste by biota
6. Symbolic; sacred and/or religious
7. Atmospheric composition and climate regulation
8. Mediation of waste by ecosystems
9. Water conditions
10. Soil formation and composition
11. Mass flows (erosion protection)
12. liquid flows (flood protection)
13. Maintenance of physical and chemical abiotic condition by natural chemical and physical processes (e.g climager regulation)
14. Mediation of air flows by biota (shelter)
15. Provision of biomass for nutrition
16. Mediation of waste by physcial and chemical processes
17. In-situ recreational activities involving landscapes
18. Provision of biomass
19. Pest and disease control
20. Provision of drinking water water
21. Provision water for other purposes
22. Mediation of flows by natural abiotic structures
23. Spiritual and emblematic value of landscapes
24. Biomass based energy sources
25. Existence and beqquest values for landsacpes and phyical settings
26. Abiotic reneable energy (eg. hydroelectricity)
27. Abstraction of rocks and minerals

Fig. 2 Flow diagram of the linkage framework depicting impact chains from habitat type to ecosystem services

pressures on ecosystem components revealed that physical change poses the highest threat to freshwater systems and to fish. Physical pressures are highly linked to environmental engineering and hydropower but also to the direct effects of land claim or land conversion activities (Borgwardt et al. 2019). Further along the impact chain, the ecosystem components within the Danube catchment were identified to have the capacity to supply 27 ESS (regulating and maintenance, provisioning, and cultural services and abiotic). Floodplains with their riparian forests and wetlands were the highest connected realms providing the greatest variety of ecosystem services.

2.2 Coupled Models: ARIES and MARXAN

Domisch et al. 2019 tested the EBM approach within the whole DRB by combining species distribution modelling for 85 fish species as a surrogate for biodiversity with four estimated ESS layers (carbon storage, flood regulation, recreation and water use) using the modeling platform ARIES. In a final step, multiple management zones were defined using the spatial prioritization tool Marxan with Zones to derive different spatially explicit management options for the whole region. In order to explore the transboundary challenges of the Danube catchment management the costs of establishing management zones were compared across nations using purchasing power parity (PPP) adjusted gross domestic product (GDP) per capita and the relative share of each country's area of the DRB. This approach therefore accounted for countries having limited financial resources (i.e. a proxy for social equity in the EBM approach) and less land area in the DRB as those might face additional challenges in financing EBM. Finally, they compared the spatial plan derived from an assumption where each country contributes equally to the EBM to one where the PPP-adjusted GDP and the percent area of each country in the basin were used as additional costs. The two analyses led to clear differences in the spatial configuration of management zones, in the GDP and percent area approach more conservation and critical management zones (with medium level of ecosystem service use) were allocated to the (wealthier) upper Danube region.

Domisch et al. 2019 used Marxan with Zones, to minimize the overall costs of a zoning plan, while ensuring that the predefined feature targets were met. Therefore four zones were characterized by different objectives and constraints (1) a "focal conservation zone", (2) a "critical management zone"—a buffer zone—, (3) a "catchment management zone" allowing for higher levels of ESS use potentially less compatible with protecting biodiversity (i.e., recreation), and (iv) a "production" zone with high use for ecosystem services (i.e., water use).

2.3 Coupled Models: Bayesian Belief Networks and ARIES

Funk et al. (2019) employed a coupled modelling approach at the scale of the Danube River to prioritize river-floodplain stretches of the navigable Danube for restoration and conservation, focusing on the river and its adjacent floodplains and riparian area (rather than the entire catchment). Bayesian Belief Networks (BBN: graphical models which represent the probabilistic relationships between different components of a system) were used to integrate different sources of information on Drivers and Pressures and their effects on environmental State (Elliott & O'Higgins, 2020). Open access GIS Datasets for Drivers and pressures included: land use data, potential riparian zone transport and navigation, and hydro-morphological pressures. This information was then used to inform weighting of the relationships within the BBNs.

Based on spatial information on conservation status based on the Habitats Directive reporting, BBNs were generated to spatially model likely species distribution in relation to the combinations of drivers and pressures for each of eleven indicator species representative of different habitat types (Table 2). The predictive power of these BBN models was tested statistically (using the R statistical computing package (see Funk et al for full details).

Table 2 Biodiversity indicators used by Funk et al. (2019)

Class	Species	Common name	Indicator
Fish	*Gymnocephalus baloni*	Danube ruffe	Fast moving waters
	Gymnocephalus schraetser	striped ruffe	Main stem larger river
	Rhodeus amarus	European bitterling	Stagnant water-connectivity
	Misgurnus fossilis	European weatherfish	Stagnant waters
	Zingel zingel	common zingel	Main stem large rivers, connected side arms
	Zingel streber	Danube streber	Main stem small to large rivers, connected sider arms
Amphibian	*Bombina sp.*	Fire-bellied toads	Fish free seasonal ponds
	Triturus dobrogicus	Danube crested newt	Temporary water bodies
Bird	*Haliaeetus albicilla*	White tailed eagle	Undisturbed wetlands
	Alcedo atthis	Common kingfisher	Active erosion and natural substrate
Mammal	*Lutra lutra*	Otter	Overall natural habitat conditions

Spatial mapping of ESS was conducted using the ARIES pollination, recreation and flood models submodels.

A spatial database combining the ARIES outputs with the outputs of the probabilistic species modelling was interrogated using clustering to identify multi-functional river and flood plain reaches supporting biodiversity and ESS supply. These multi-functional clusters were then mapped.

The model used a multi-objective optimization tool (e.g. Sacchelli et al. 2013), which enabled systematic optimization for different management objectives. One objective was to prioritize sections for conservation or restoration with a high remaining multi-functionality to reduce effort and costs, a second objective was to prefer sites with high reversibility (i.e. low level of human use) to increase probability of success, and finally to prefer semi-natural areas to reduce costs and loss of agricultural yield. Different weightings of the three objectives represent different possible management plans and therefore can be used as a basis for a more integrated and targeted planning. This process resulted in the development of a suite of potential target areas for restoration, conservation or mitigation efforts.

Consistent with other studies (Egoh et al. 2011; Maes et al. 2012), Funk et al. (2019) recorded a high overlap between areas important for biodiversity and areas important for ESS supply, pointing to a close interrelationship between biodiversity and ESS that is often greater in natural systems (Chan et al. 2011; Schneiders et al. 2012). Specifically, the multi-functionality approach tested by Funk et al. (2019) showed that in the study area, only natural and near-natural river-floodplain systems provided habitat for various aquatic species as well as multiple ESS.

In the study, sites with greater probability of restoration success, indicated by low level of driver intensity related to navigation, hydropower and flood protection constraints as well as sites with high level of remaining semi-natural area (compared to agricultural area) were prioritized. In this way the study addressed potential opportunity costs of restoration efforts across the entire Danube River. This approach afforded the ability to provide better cost-effectiveness in achieving large scale conservation and ESS targets at the catchment scale (Bladt et al. 2009; Egoh et al. 2014), and to potentially avoid conflicts with drivers.

3 EBM Principles

Overall the application of the LF to the Danube Basin, illustrated the complexity of interactions between human activities, ecosystem components and the ESS they provide, and is useful in identifying the most important ecosystem components with respect to ESS supply as well as the types of activities that most likely affect these components through pressures. With respect to the EBM principles, the LF can support the **first principle** in terms of communicating the links between ecological integrity, biodiversity (expressed at the habitat level) and ESS. The LF is not spatially explicit and can be transferred and adapted for use at in any similar system and applied to any spatial scale of interest thereby supporting the **second principle** (appropriate spatial scales) of EBM. The LFs are developed by 'experts' on a given location, who assess the activities and pressures, based on their knowledge. While LFs require an holistic view of a system, they do not necessarily integrate insights from a range of disciplines (**principle 3**) rather they characterize a suite of social-ecological interactions (**principle 4**). In its capacity to foster an understanding of the complexity of these links to promote understanding of policy synergies, they may also be used to facilitate and support policy coordination (**principle 5**). However, because the LF is a semi-quantitative and expert judgement based approach it is unlikely to carry sufficient confidence to justify any particular policy decision. Since the LF does not identify particular management options its current role in adaptive management (**principle 6**) is limited. Nevertheless, with its basis in the causal chain analysis of the DPSIR (see Elliott this volume) the linkages could potentially be extended to incorporate response options. For fully detailed accounts of development and analysis of the LF and comparison across regions, and aquatic ecosystem types, the reader is directed to Borgwardt et al. 2019, Teixeira et al. 2019, for a general description and discussion of the approach see Robinson and Culhane (2020).

The two integrated modelling studies (Domisch et al. 2019; Funk et al. 2019) exemplify how different holistic approaches can be used to identify management options which consider ecological integrity biodiversity resilience and ESS (**Principle 1**). Both implementations of the quantitative model coupling framework for EBM (Langhans et al. 2019), confirms how biodiversity and ESS estimates can be jointly simulated within the DRB given the availability of requisite data and models. It demonstrates that the method is very flexible and the criteria and models used are broadly applicable and the approach is transferable to other aquatic systems (Funk et al. 2019, Domisch et al. 2019).

Both approaches were spatially explicit and developed specifically to work at the appropriate spatial scales (**principle 2**). In the first study (Domisch et al. 2019) this included the entire catchment while the second study (Funk et al. 2019) had a more restricted focus specifically on rivers and the flood plain, nevertheless both studies worked across international borders which is a prerequisite for the work in the Danube.

Both model used a range of data sources, in particular Domisch et al. (2019) used truly multi-disciplinary, economic and environmental data (**principle 3**) to account for economic disparity, within the social part of the social-ecological system. This approach accounts for countries having limited financial resources (i.e. a proxy for social equity in the EBM approach) and land area in the Danube River Basin as those might face additional challenges in financing EBM in the basin.

In contrast, Funk et al. (2019) selected a method indirectly accounting for costs independent from country level's financial limitations, prioritizing sites with greater probability of restoration success at lower cost (i.e. indicated as lower loss of agricultural area). Therefore the multi-functionality approach accounts for the emerging view that ecological restoration requires restoring ecosystems for the sustainable and simultaneously provisioning of multiple goods and services such as water, flood protection, recreation, and biodiversity, among others to increase cost-effectiveness (Paschke et al. 2019).

One potential pitfall with both approaches is the stakeholder participation and transparency (**Principle 4**). Neither study directly used stakeholder input to inform the model building process, rather, the choices were made at the technical level by the modelling teams. To make the approach operational, participatory processes involving stakeholders across the catchment, member state and local levels would be a further important step. BBNs in particular are one promising technique which can be easily adapted to incorporate stakeholder input. It is possible to construct BBNs models based on stakeholder perceptions allowing co-design of modelling activities (see O'Higgins et al. 2020 for an example). In addition, the use of the AI approach included in the ARIES model may lack the transparency of more traditional deterministic environmental models, which may reduce the acceptability of model results. Elsewhere in this volume Fulford et al. 2020 discuss practical trade-offs inherent in model complexity.

Both the policy coordination potential (**principle 5**) and the adaptive management aspects (**principle 6**) are strong in both studies described above. Outputs from both models produced a suite of policy-relevant options enabling joint efforts to conserve the Danube.

Funk et al. 2019 accounted for this principle by using data and knowledge derived and used in the framework of different policies, directives and initiatives e.g. navigation and hydropower sector (e.g. TEN-T regulation), water management sector (Water Framework Directive), local data from protected areas (Birds and Habitats Directive) and spatial land use information. This includes a continuous hydro-morphological assessment for the navigable Danube River compliant with CEN standards (Schwarz 2014; ICPDR 2015), Land cover/Land use (developed to support e.g. EU Biodiversity Strategy to 2020) or sectoral data collected on the status of the waterway, critical locations for navigation and navigation class (Fairway 2016). Cause-effect relations within the network of interactions between driver, pressure and state variables along the Driver-Pressure-State chain were then analysed within a quantitative Bayesian Network approach. Therefore, the approach selected by Funk et al. 2019 provides the first large scale statistical proof of multiple relationships of biodiversity and human uses and pressures along the navigable stretch of the Danube River. Therefore, it has the potential to increase knowledge on the socio-ecological system across sectors and policies and is serving as a basis for a strategic and more integrated management approach.

The Domisch et al. (2019) study explicitly included consideration of regional inequalities and economic capacity and generated a more in-depth picture of the feasibility of particular conservation efforts, thus enabling the adaptation of plans to meet these real-world social constraints.

4 Conclusions

We developed and tested different qualitative and quantitative implementations of an EBM approach for a complex socio-ecological system, the DRB. The LF approach helped to understand the complex interaction within the social-ecological system and to describe the main human activities and pressures affecting the aquatic ecosystem components. The modelling approaches summarized in this paper have increased the consideration of ecological integrity and biodiversity, accounting for multiple species and different relevant ESS. These studies illustrate approaches considering cumulative impacts by multiple human activities including land use, navigation and hydropower and integrate this multidisciplinary data and knowledge. The prioritization approaches taken fosters integrated management planning across multiple policies by creating the opportunity to pursue different policy objectives simultaneously.

All three selected EBM application for the DRB were implemented at the ecosystem scale i.e. including the whole catchment or river independent of jurisdictional, administrative or political boundaries (Borgwardt et al. 2019, Domisch et al. 2019, Funk et al. 2019) and therefore have the potential to foster transboundary cooperation for a EBM of the DRB.

Both implementations of the quantitative model coupling framework for EBM (Langhans et al. 2019), showed how biodiversity and ESS estimates can be jointly simulated within the Danube River Basin given the availability of requisite data and models. This demonstrates that the method is flexible, the criteria and models used are broadly applicable, and the approach is transferable to other aquatic systems (Funk et al. 2019, Domisch et al. 2019). The EBM principles used for qualitative assessment of the modelling approaches may serve as a useful generic basis for the design of further EBM studies.

References

Ball, I. R., Possingham, H. P., & Watts, M. (2009). Marxan and relatives: Software for spatial conservation prioritisation. In *Spatial conservation prioritisation: Quantitative methods and computational tools* (pp. 185–195). New York: Oxford University Press.

Bladt, J., Strange, N., Abildtrup, J., Svenning, J. C., & Skov, F. (2009). Conservation efficiency of geopolitical coordination in the EU. *Journal for Nature Conservation, 17*, 72–86.

Borgwardt, F., Robinson, L., Trauner, D., Teixeira, H., Nogueira, A. J., Lillebø, A. I., et al. (2019). Exploring variability in environmental impact risk from human activities across aquatic ecosystems. *Science of the Total Environment, 652*, 1396–1408.

Chan, K. M. A., Hoshizaki, L., & Klinkenberg, B. (2011). Ecosystem services in conservation planning: Targeted benefits vs. co-benefits or costs? *PLoS One, 6*, e24378.

DANUBEPARKS. (2011). Strategy on conservation & navigation. Retrieved from http://www.danubeparks.org/files/781_DANUBEPARKS_ConservationNavigation.pdf

Delacámara, G., O'Higgins, T., Lago, M., & Langhans, S. (2020). Ecosystem-based management: moving from concept to practice. In T. O'Higgins, M. Lago, & T. H. DeWitt (Eds.), *Ecosystem-based management, ecosystem services and aquatic biodiversity: Theory, tools and applications* (pp. 39–60). Amsterdam: Springer.

Domisch, S., Kakouei, K., Martínez-López, J., Bagstad, K. J., Magrach, A., Balbi, S., et al. (2019). Social equity shapes zone-selection: Balancing aquatic biodiversity conservation and ecosystem services delivery in the transboundary Danube River Basin. *Science of the Total Environment, 656*, 797–807.

EC. (1992). Council Directive 92/43/EEC of 21 May 1992 on the conservation of natural habitats and of wild fauna and flora L 206/7.

EC. (2000). Establishing a framework for community action in the field of water policy. Directive 2000/60/EC of the European Parliament and of the Council. *Official Journal of European Communities, L327*, 1–72.

EC. (2007). EU Directive of the European Parliament and of the European Council on the estimation and management of flood risks (2007/60/EU).

EC. (2011). 'Our Life Insurance, Our Natural Capital: An EU Biodiversity Strategy to 2020' Communication from the Commission to the European Parliament, the Council, the Economic and Social Committee and the Committee of the Regions COM(2011)244, Brussels (17 pp).

Egoh, B. N., Reyers, B., Rouget, M., & Richardson, D. M. (2011). Identifying priority areas for ecosystem service management in South African grasslands. *Journal of Environmental Economics Management, 92*(6), 1642–1650.

Egoh, B. N., Paracchini, M. L., Zulian, G., Schägner, J. P., & Bidoglio, G. (2014). Exploring restoration options for habitats, species and ecosystem services in the European Union. *Journal of Applied Ecology, 51*(4), 899–908.

Elliott, M., & O'Higgins, T. G. (2020). From the DPSIR, the D(A)PSI(W)R(M) emerges... a butterfly-'protecting the natural stuff and delivering the human stuff'. In T. O'Higgins, M. Lago, & T. H. DeWitt (Eds.), *Ecosystem-based management, ecosystem services and aquatic biodiversity: Theory, tools and applications* (pp. 61–86). Amsterdam: Springer.

Fairway, D. (2016). Fairway rehabilitation and maintenance master plan for the Danube and its navigable tributaries. Tributaries. Retrieved from http://www.danube-navigation.eu/uploads/files/news/2016-05-31_FAIRway_National_action_plans_May_2016_final.pdf.

Fulford, R. S., Heymans, S. J. J., & Wu, W. (2020). Mathematical modelling for ecosystem-based management (EBM) and ecosystem goods and services (EGS) assessment. In T. O'Higgins, M. Lago, & T. H. DeWitt (Eds.), *Ecosystem-based management ecosystem services and aquatic biodiversity: Theory, tools and applications* (pp. 275–290). Amsterdam: Springer.

Funk, A., Martínez-López, J., Borgwardt, F., Trauner, D., Bagstad, K. J., Balbi, S., et al. (2019). Identification of conservation and restoration priority areas in the Danube River based on the multi-functionality of river-floodplain systems. *Science of the Total Environment, 654*, 763–777.

Habersack, H., Hein, T., Stanica, A., Liska, I., Mair, R., Jäger, E., et al. (2016). Challenges of river basin management: Current status of, and prospects for, the River Danube from a river engineering perspective. *Science of the Total Environment, 543*, 828–845.

Hein, T., Schwarz, U., Habersack, H., Nichersu, I., Preiner, S., Willby, N., & Weigelhofer, G. (2016). Current status and restoration options for floodplains along the Danube River. *Science of the Total Environment, 543*, 778–790.

Hein, T., Funk, A., Pletterbauer, F., Graf, W., Zsuffa, I., Haidvogl, G., et al. (2018). Management challenges related to long-term ecological impacts, complex stressor interactions, and different assessment approaches in the Danube River Basin. *River Research and Applications, 35*, 500–509.

ICPDR. (2015). Joint Danube survey 3. A comprehensive analysis of Danube water quality, Vienna, p. 369.

ICPDR (2016). The Danube River Basin District management plan. ICPDR—International Commission for the Protection of the Danube River, Vienna, p. 164.

Langhans, S. D., Domisch, S., Balbi, S., Delacámara, G., Hermoso, V., Kuemmerlen, M., et al. (2019). Combining eight research areas to foster the uptake of ecosystem-based management in fresh waters. *Aquatic Conservation: Marine and Freshwater Ecosystems, 29*, 1161–1173. https://doi.org/10.1002/aqc.3012.

Lewis, N. S., Marois, D. E., Littles, C. J., & Fulford, R. S. (2020). Projecting changes to coastal and estuarine ecosystem goods and services-models and tools. In T. O'Higgins, M. Lago, & T. H. DeWitt (Eds.), *Ecosystem-based management, ecosystem services and aquatic biodiversity: Theory, tools and applications* (pp. 235–254). Amsterdam: Springer.

Maes, J., Paracchini, M. L., Zulian, G., Dunbar, M. B., & Alkemade, R. (2012). Synergies and tradeoffs between ecosystem service supply, biodiversity, and habitat conservation status in Europe. *Biological Conservation, 155*, 1–12.

O'Higgins, T. G., Farmer, A. M., Daskalov, G., Knudsen, S., & Mee, L. (2014). Achieving good environmental status in the Black Sea: Scale mismatches in environmental management. *Ecology and Society 19*, 54.

O'Higgins, T. G., Culhane, F., O'Dwyer, B., Robinson, L., & Lago, M. (2020). Combining methods to establish potential management measures for invasive species *Elodea nutallii* in Lough Erne Northern Ireland. In T. O'Higgins, M. Lago, & T. H. DeWitt (Eds.), *Ecosystem-based management, ecosystem services and aquatic biodiversity: Theory, tools and applications* (pp. 445–460). Amsterdam: Springer.

Paschke, M. W., Perkins, L. B., & Veblen, K. E. (2019). Restoration for multiple use. *Restoration Ecology, 27*, 701–704. https://doi.org/10.1111/rec.12949.

Robinson, L., & Culhane, F. (2020). Linkage frameworks: An exploration tool for complex systems. In T. O'Higgins, M. Lago, & T. H. DeWitt (Eds.), *Ecosystem-based management, ecosystem services and aquatic biodiversity: Theory, tools and applications* (pp. 213–234). Amsterdam: Springer.

Sacchelli, S., De Meo, I., & Paletto, A. (2013). Bioenergy production and forest multifunctionality: A trade-off analysis using multiscale GIS model in a case study in Italy. *Applied Energy, 104*, 10–20.

Schneiders, A., Van Daele, T., Van Landuyt, W., & Van Reeth, W. (2012). Biodiversity and ecosystem services: Complementary approaches for ecosystem management? *Ecological Indicators, 21*, 123–133.

Schwarz, U. (2014). An extended method for continuous hydromorphological assessment applied in the joint Danube survey 3, 2013. *Acta Zoologica Bulgarica, 66*, 123–127.

Seliger, C., Scheikl, S., Schmutz, S., Schinegger, R., Fleck, S., Neubarth, J., et al. (2016). Hy: Con: A strategic tool for balancing hydropower development and conservation needs. *River Research and Applications, 32*(7), 1438–1449.

Sommerwerk, N., Bloesch, J., Paunović, M., Baumgartner, C., Venohr, M., Schneider-Jacoby, M., et al. (2010). Managing the world's most international river: The Danube River Basin. *Marine and Freshwater Research, 61*(7), 736–748.

Teixeira, H., Lillebø, A. I., Culhane, F., Robinson, L., Trauner, D., Borgwardt, F., et al. (2019). Linking biodiversity to ecosystem services supply: Patterns across aquatic ecosystems. *Science of the Total Environment, 657*, 517–534.

Villa, F., Bagstad, K. J., Voigt, B., Johnson, G. W., Portela, R., Honzák, M., & Batker, D. (2014). A methodology for adaptable and robust ecosystem services assessment. *PLoS One, 9*(3), e91001.

5

Using Stakeholder Engagement, Translational Science and Decision Support Tools for Ecosystem-Based Management in the Florida Everglades

Rebekah Gibble, Lori Miller, and Matthew C. Harwell

Abstract Managing water for competing human and environmental demands in the Greater Everglades is a multi-dimensional challenge that includes managing a complex ecological system while providing water supply and flood control for widespread high-density urban communities and nationally important agricultural lands. Ecosystem-Based Management (EBM) in the Florida Everglades is examined at multiple spatial and temporal scales. There is a corresponding increase in the number and diversity of stakeholders involved as the temporal and spatial scales of management across the landscape increases. Therefore, translational science, decision support tools, effective stakeholder engagement, and communication are paramount. This chapter provides a case study of EBM in an aquatic system facing ecological challenges, such as eutrophication and non-indigenous species management, which are framed by complex social, cultural, and political contexts. A framework for navigating multi-agency governance models and competing stakeholder visions using socio-ecological science (i.e., science of interlinked human and natural systems) to address practical and theoretical challenges for managing freshwater wetlands is discussed. By examining best practices in stakeholder engagement and linking translational science with multiple, science-driven decision support tools, important lessons learned can be carried forward in an effort for continually improved governance and collaboration for ecosystem management and restoration.

R. Gibble (✉)
U.S. Fish and Wildlife Service, Boynton Beach, FL, USA
e-mail: rebekah_gibble@fws.gov

L. Miller
U.S. Fish and Wildlife Service, Vero Beach, FL, USA
e-mail: lori_miller@fws.gov

M. C. Harwell
U.S. Environmental Protection Agency, Gulf Breeze, FL, USA
e-mail: harwell.matthew@epa.gov

Lessons Learned

- Real-time monitoring data, frequently updated and accessible modeling output and stakeholder communication are key to successful EBM in complex socio-ecological systems, such as the Everglades.
- EBM recommendations should consider agency-specific missions and goals across the key stakeholders involved.
- Coordinating short, intermediate, and long-term recommendations at both local and regional scales may improve EBM outcomes.
- Incorporating real-time monitoring data from across the landscape, along information from model output, enhances EBM.
- Decision support tools that integrate monitoring data with spatial habitat and wildlife models enhance assessment of conditions and development of recommendations that support EBM.

Needs to Advance EBM

- Additional development of the decision-making process framework and further integration of decision support tools for use in multiple spatial and temporal scales.
- Increased/enhanced incorporation of EBM approach into existing governance models/mandates, and methods of communication of recommendations to managing agencies.
- Perhaps most importantly, increased connection of operational decisions to measuring the resulting ecosystem responses to support adaptive management will further enhance the success of the EBM approach in the Everglades.

1 Introduction

Wetlands are productive ecological systems that provide habitat to many species that form complex and interdependent communities. Wetland systems collect water and sediment from across the landscape and regulate hydrologic cycles that provide ecosystem services such as water filtration and supply, flood control, coastal protection from storms, carbon sequestration, natural products (e.g., shellfish), and recreational opportunities. Humans often alter these systems by draining them to provide fertile farmland or to support development. The hydromorphological alterations resulting from these socio-ecological interactions often lead to impacts such as decreased wildlife populations and reduced ecological service. However, when adaptively managed, wetlands can be sustainable and provide a range of ecosystem services to humans while providing crucial habitat to wetlands.

This chapter outlines key socio-ecological interactions across the landscape and the Ecosystem-Based Management (EBM) approach used to promote restoration of the natural function of the Everglades system while also providing numerous ecosystem services, such as water supply and flood control. This approach integrates stakeholder engagement with translational science and decision support tools to inform multipurpose water management operations. Communication between various groups (e.g., stakeholders, managers, etc.) is critical to the EBM process and, in the Everglades, is based on a framework comprised of formal and informal requirements. Throughout this chapter, the various components and linkages that make up the EBM approach in the Everglades are presented as a case study to illustrate the practical application of EBM across a landscape.

2 Everglades Ecosystem

2.1 *Overview of an Ecosystem in Trouble (Through 2000)*

The historical Everglades were 9307 km^2 (2.3 million acres) of a vast wetland that began at a chain of lakes in the Kissimmee basin that flowed into Lake Okeechobee and stretched to the end of the Florida peninsula (Douglas 1947). Lake Okeechobee was historically 1891 km^2 (730 mi^2) in size and water depths fluctuated between 3–4 m (10–20 ft.) deep (McVoy et al. 2011). The lake acted as a natural reservoir storing water from the Kissimmee Chain of Lakes and during periods of high water from tropical cyclones. When full, the water spilled over into pond apple forests causing a wide swath of slow moving water in the form of sheetflow (Fig. 1) (McVoy et al. 2011). Water depths south of the lake in the sawgrass dominated Everglades ranged from 5–100 cm (2–40 inches) deep. The slow moving water with occasional periods of high pulse flows created a ridge and slough landscape that included tree islands as crucial habitat for wildlife and plant species. (Frederick and Ogden 2001). The slope of the land was so gradual at 5 cm per 1.5 km (2 inches per mile), that water only moved southward at 30 m (100 ft) per day (National Research Council 2010). This water eventually discharged into the mangroves of Florida Bay and the Ten Thousand Islands in southwestern Florida.

2.2 *Drying of the Marsh for Agriculture by Compartmentalization*

In the late 1800s through the 1930s, settlers sought to dry out the swamp south of Lake Okeechobee in order to use the rich muck and peat to grow crops. In 1947, major floods occurred over South Florida with over 2.5 m (100 in) of rain causing the United States Congress to authorize the Central and South Florida (C&SF) Project in 1948, which was intended to provide drainage and flood control for the croplands and the outlying communities (Fig. 2) (Light and Dineen 1994). In spite of the public outcry against compartmentalization, 1600 km (100 mi) of levees, 1160 km (720 mi) of canals, and 200 water control structures (Light and Dineen 1994) were designed.

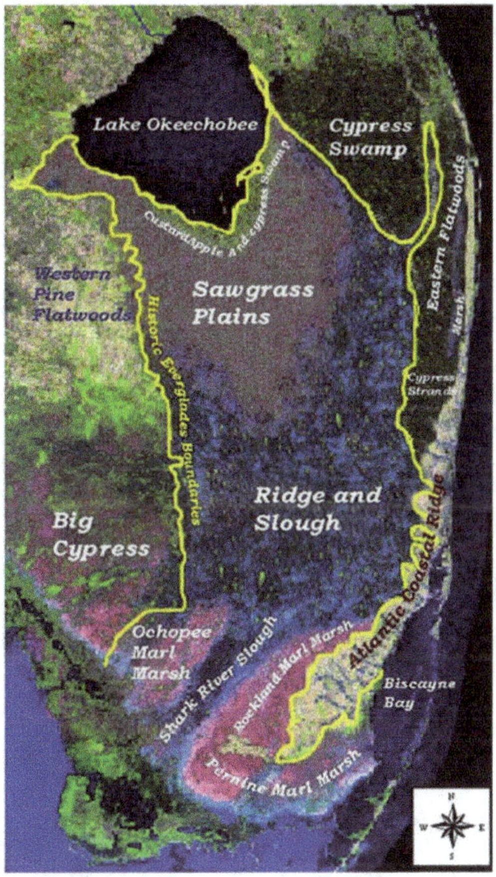

Fig. 1 The historical and present Everglades. (Reproduced from the National Academies of Sciences, Engineering, and Medicine 2018)

Although too late to stop damage to the Everglades, Marjory Stoneman Douglas published "*The Everglades: River of Grass*" in 1947 arguing for saving the Everglades.

The C&SF project, built during the 1950s and 1960s, completely disconnected the historical Everglades. The historical sheetflow lost its natural headwaters and became a complicated water management system designed to provide flood control and continual drainage of the system for agriculture (National Research Council 2010). The drained area south of Lake Okeechobee became known as the Everglades Agricultural Area (EAA) and is approximately 2850 km^2 (1100 mi^2) (Snyder and Davidson 1994). Multiple Water Conservation Areas (WCAs) covering approximately 3500 km^2 (1350 mi^2) were developed for managing water in the open areas of the Everglades.

The C&SF Project had multiple direct, but adverse, hydrological impacts on the Everglades (US Department of the Interior 1994):

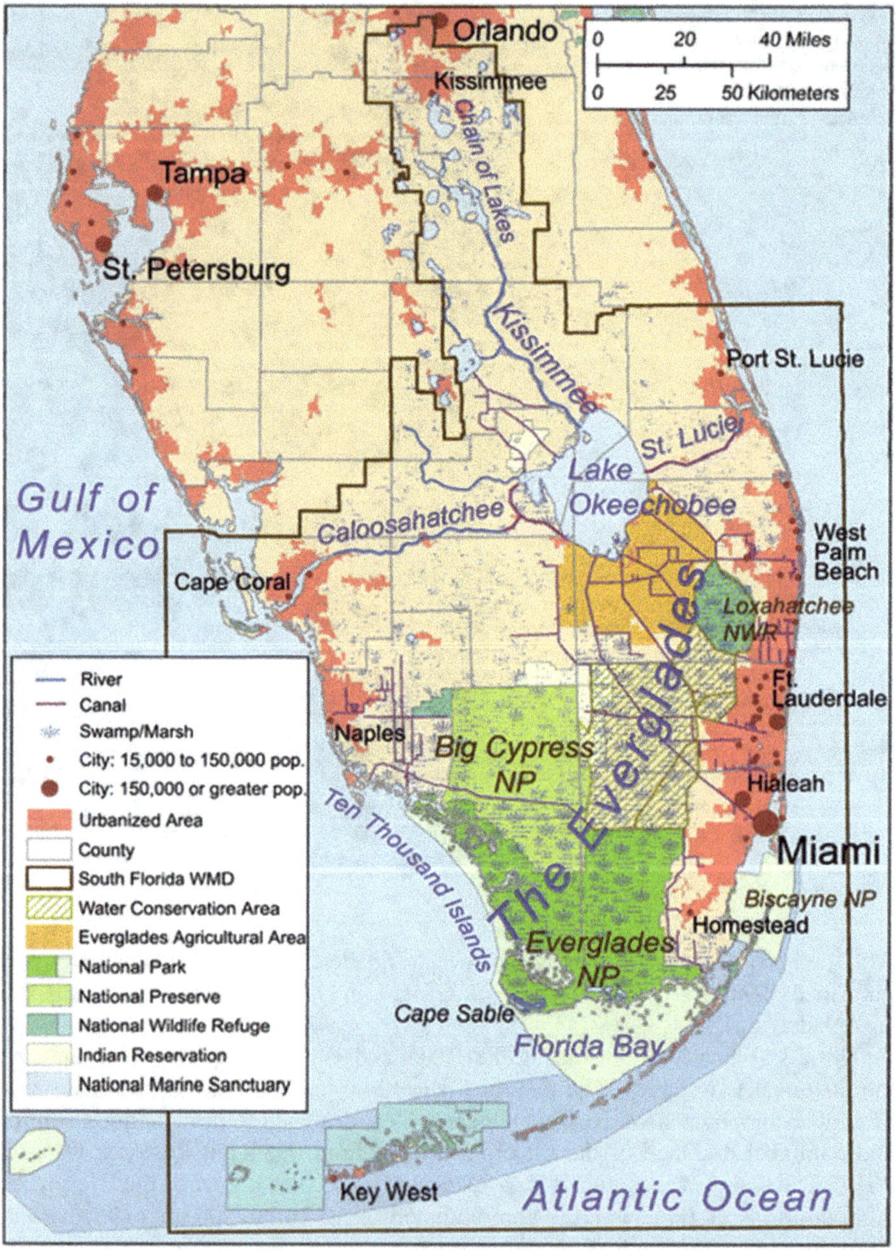

Fig. 2 C&SF project compartmentalization of the Everglades into the Everglades Agricultural Area, Water Conservation Areas, preserves, and parks. (Reproduced from Musser 2010)

- The Everglades were reduced in size by more than fifty percent (50%) to approximately 4000 km^2 (one million acres).
- Loss of an annual average of 2.7 billion m^3 (2.2 million acre feet) of fresh water that flowed into the coastal estuaries along Florida's east and west coast due to lack of conveyance and storage to send the water south into the Everglades. The lack of sheetflow and the resulting altered hydroperiod in the remaining Everglades changed native vegetation and habitat.
- Lack of variability in water delivery drastically altered the seasonal patterns of high and low flows to the remaining Everglades.
- The Everglades continued losing its unique ridge and slough landscape, along with tree islands, causing a change in wildlife population abundances and distribution.

In the 1970s, the larger U.S. environmental movement brought more attention to the Everglades. By the late 1980s, state and Federal agencies, citizen groups, and the Tribes began focusing on restoring the Everglades ecosystem and protecting its species and habitats. Currently, there are 68 threatened and endangered species in the Everglades (USFWS 1999). The most noted species change has been in the population, distribution, and habits of wading birds native to the Everglades (Ogden 1994). Even the most basic component of the Everglades—water—is at risk. Water depths, duration, and overall distribution across the landscape changed with drainage, creating a suite of ecosystem changes (Kushlan 1987; Ogden 2005). Water quality has been consistently deteriorating due to agricultural chemicals, urban runoff, and animal waste from ranchlands and dairies upstream in the EAA, Kissimmee Chain of Lakes, and urban areas such as Orlando. Successful restoration of the Everglades ecosystem requires the appropriate interaction between the quantity, quality, timing, and distribution of water (ENP 2015).

2.3 *An Ecosystem Managed for Multiple Purposes (2000–Current)*

At present, the remaining greater Everglades ecosystem is a human-shaped environment that is managed for multiple purposes, including: flood protection, water supply, water flows for the environment, and habitat supporting a variety of flora and fauna. Water flow is a foundational element of most Everglades ecosystem management, with two primary water management agencies, the U.S. Army Corps of Engineers (USACE) and the South Florida Water Management District (SFWMD), making and implementing operational water management decisions. Smaller government entities and utilities, such as the Lake Worth Drainage District and the Everglades Drainage District, influence hydrology at smaller scales. Additionally, a mosaic of federal, State and Tribal lands exist throughout the greater Everglades Ecosystem Management decisions in these land units are made by the U.S. National Park System, U.S. Fish & Wildlife Service, Florida Fish and Wildlife

Commission, the Miccosukee Tribe of Indians of Florida, or the Seminole Tribe of Florida. Overall, the greater Everglades landscape is a complex, socio-ecological system involving a range of governance models (Ankersen and Hamann 1996). Each governing body is operated under a different suite of legal and policy mandates, with different levels and types of stakeholder engagement. Juxtaposed against this complex range of governance models is a large body of scientific research underscoring the interconnectedness of the landscapes and ecosystems. From this research, an increased understanding of the system has led to the development of a diverse array of tools to evaluate, assess, and predict system responses to management operations and restoration projects.

3 Current Water Management

The USACE monitors and manages the multi-purpose operations of spillways, locks, pump stations, culverts, canals, reservoirs, and water conservation areas (USACE 2019) and is considered a federal partner to the State of Florida's water management districts. Among other activities, such as water quality monitoring and scientific research, the SFWMD manages water supply, flood control, and is the State partner in Everglades Restoration. The SFWMD operates approximately 2100 miles of canals and 2000 miles of levees/berms, 77 pump stations and more than 600 water control structures and 620 project culverts across central and southern Florida (SFWMD 2019). This extensive network of infrastructure (Fig. 3) encompasses three water conservation areas (WCAs), large wetland areas that are compartmentalized by berms and levees and receive and discharge water through water control structures. The WCAs make up the Greater Everglades and are located upstream of Everglades National Park (ENP). Historically, the Everglades were compartmentalized into WCAs to prevent catastrophic flooding as witnessed before the C&SF project was authorized in 1948. The WCAs store rainfall, Lake Okeechobee flood releases, and excess water runoff from the EAA, as well as recharge aquifers, reduce seepage into urban areas, and protect against salt water intrusion from rising sea levels. The WCAs also provide flow-through capacity of water cleaned (primarily by removal of phosphorus) by stormwater treatment areas (STAs). STAs are constructed wetlands that remove and store nutrients through plant growth and the accumulation of dead plant material. STAs are comprised of parcels of land utilizing different types of emergent (e.g., cattails, pickerel weed and bulrush) and submerged (e.g., hydrilla, southern naiad and chara) plants that take phosphorus directly from the water in STAs (SFWMD 2019). STAs are critical for providing environmental benefits to the Everglades landscape, species, and their habitats. (USGS 2013).

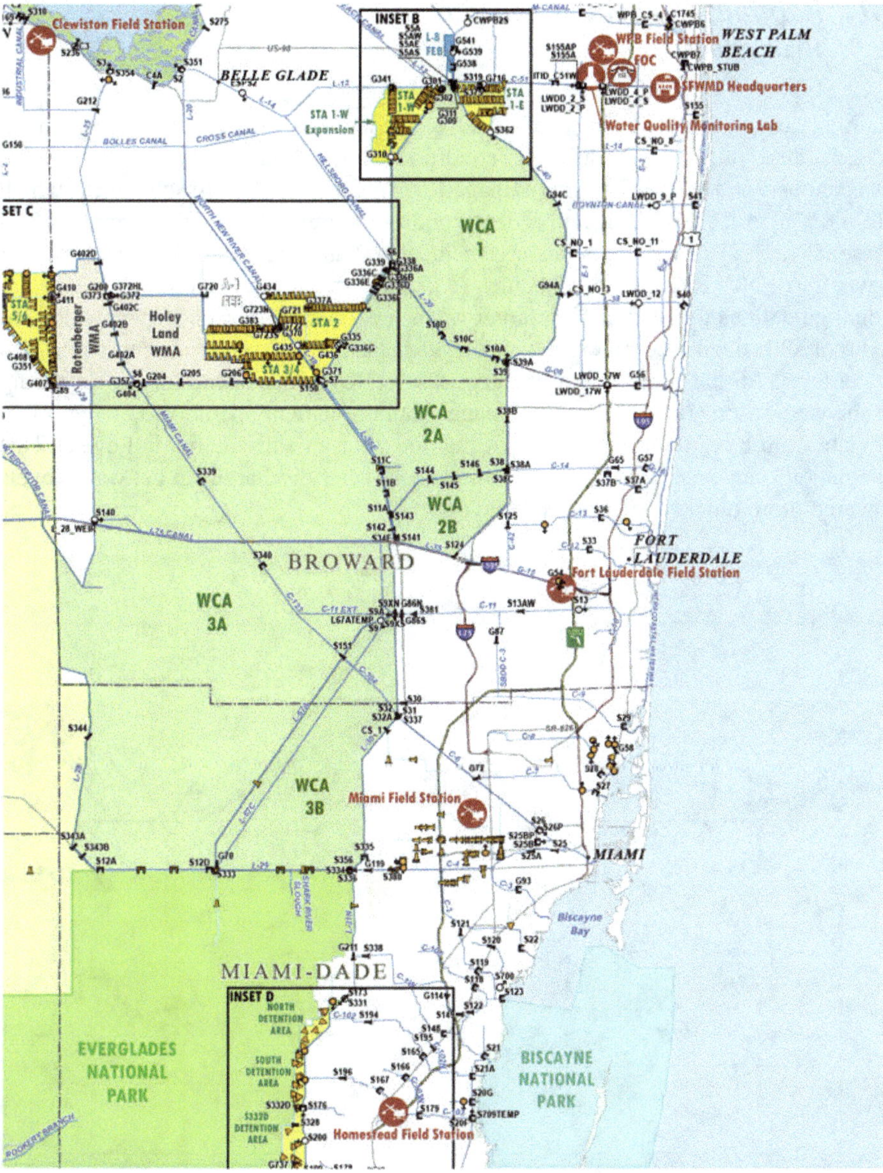

Fig. 3 Facility and infrastructure location index map indicating canals, pumps, weirs, spillways, and stormwater treatment areas in yellow. (Reproduced from SFWMD 2016)

3.1 *Rainfall, Regulation Schedules, and an Altered Ecosystem*

Rainfall drives the hydrology of the Everglades. However, water management actions also influence hydrologic conditions throughout the ecosystem. Lake Okeechobee and the WCAs are managed by Water Regulation Schedules, which are a set of rules based on antecedent conditions, rainfall formulas, monthly and seasonal rainfall based water management plans, and regulatory requirements for flows into ENP. Regulation schedules (e.g., Fig. 4) provide recommended operational guidelines for maintaining target water level ranges in each WCA, which are monitored through a complex network of gauges. They also provide recommended water levels for the beginning of the dry season (November 1) and for the beginning of the wet season (June 1). Water managers can implement operational changes that deviate from a regulation schedule during and after an extreme rainfall event. This "deviation" can exceed normally recommended water discharges to get water levels lowered more quickly to acceptable levels.

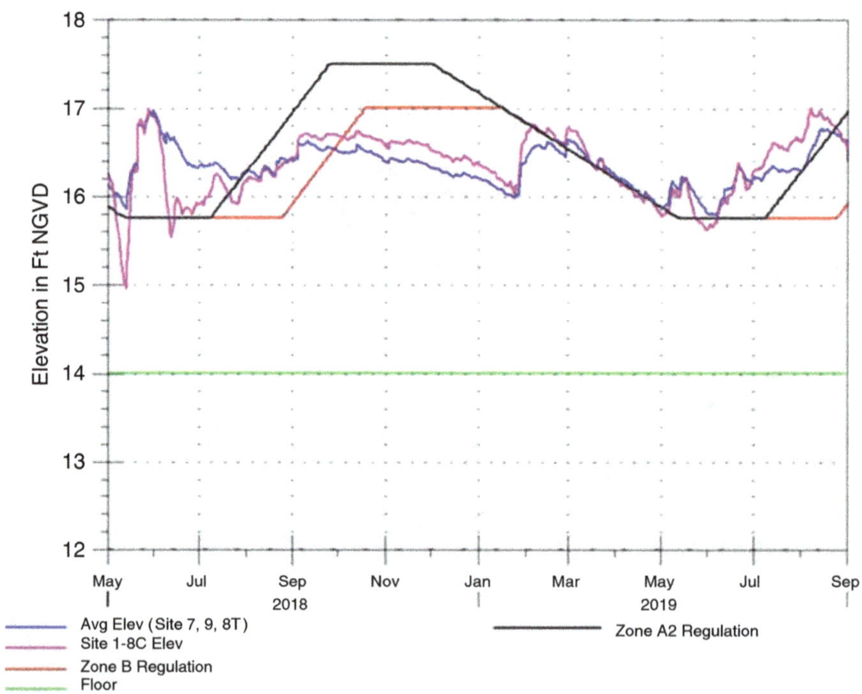

Fig. 4 Regulation schedule and water levels in 2018 for Water Conservation Area 1 within the Greater Everglades. Red and black lines indicate surface water elevation targets throughout the year. Purple and blue lines indicate actual surface water elevations resulting from water management operations and rainfall in Water Conservation Area 1 (USACE 2019)

At times, water management rules can be in conflict with the environmental needs of species and their habitats, although the intent of Everglades restoration projects is to mimic the natural hydrology of the Everglades. Hydrologic and ecological studies have shown that some rules could be altered to better mimic the Everglades' natural hydrology. This has been particularly evident with some of the State and federally listed threatened and endangered species (e.g., Snail Kites (*Rostrhamus sociabilis plumbeus*), Wood Storks (*Mycteria americana*), wading birds, and Cape Sable Seaside Sparrows (*Ammodramus maritimus mirabilis*) (USFWS 2016). Some of the habitat these species depend upon has become altered due to unnatural water depths, shorter- or longer-than usual hydroperiods, and altered rates and patterns of flow. The succession of short-hydroperiod marl prairie grasses to long-hydroperiod sawgrass in naturally occurring high ground in ENP is a prime example of areas becoming unnaturally wet due to water management operations. Management operations of water levels in the Kissimmee Chain of Lakes, Lake Okeechobee, and the Kissimmee River have also led to the deterioration of Snail Kite habitat in lake littoral zones and river floodplains (Cattau et al. 2008). Snail Kites breed and nest in the littoral zones and floodplains, which are also home to their primary prey, the apple snail. It has been documented that water management operations have at times reduced water levels too rapidly, causing damage to the habitat and forcing the apple snail to move into deeper water and leaving apple snail egg clusters without necessary water levels to survive (Bennetts and Kitchens 1997). For Wood Storks and wading bird colonies, 41 cm (16 in) of water depth is required to support their fish prey. Water management operations can, and have, either flooded or quickly dried out, core foraging areas for wading birds (USFWS 2014).

Because of the potential negative effects to listed threatened and endangered species, water management in central and southern Florida has occasionally been changed and amended by federal regulatory documents, such as a USFWS Biological Opinion (BO). A recent BO mandated that water managers coordinate with ecologists and biologists from state and federal government agencies, non-governmental organizations, and other interested parties prior to changing particular structural operations due to potential upstream or downstream ecological effects. These BOs are also responsible for changing the rate and timing of structural flows to better mimic natural rainfall and hydrology to protect certain habitats like the marl prairie in ENP (USFWS 2016). However, it was recognized in the 1999, and all subsequent BOs, that there would be times when unseasonal rainfall could and will overwhelm the water management system. During those times, it is critical to "share adversity" among stakeholders and thus, closely coordinate among agencies to provide the best operations for human health and safety, the Everglades landscape, and the biology and ecology of the system (USFWS 2016).

3.2 Monitoring and Current Status of the Ecosystem

Changes in hydrology (hydromorphological alterations), water quality, and water management are the principal stressors that affect the ecosystem (e.g., Walker 1999). Monitoring hydrology includes maintenance of hydrologic monitoring gauges, collection of data, and processing and dissemination of hydrologic data. The monitoring station network (Fig. 5) within the ecosystem is comprised of almost 300 gauges (USGS 2009) that measure water stages and water quality and are operated by the Big Cypress National Preserve (BCNP), ENP, SFWMD and United States Geological Survey (USGS) (2009).

These gauges are used to model surface water elevations and depths across the landscape, often as related to wildlife, (e.g., Fig. 6), and can be used for planning purposes. New gauges may occasionally be needed in strategic areas due to topographic variability. However, the need for new gauges is weighed against habitat impacts resulting from installation (USFWS 2016).

3.3 Planning for Future Water Management Operations

As a federal partner, USACE water managers monitor gauges, water depths, and regulation schedules of water conservation areas. These activities include analyzing past and predicted rainfall events and considering the requirements of a multitude of stakeholders, including, but not limited to, navigation, ecological needs, agriculture, and recreation. The SFWMD uses risk analysis to evaluate the "present condition" of the system, on which to base its water operations recommendations. The purpose of the risk assessment is to evaluate water resources and the risks associated with operational decisions (Hirsch 1978). This evaluation is accomplished by estimating the probability distribution function of select variables, conditional on the current, or otherwise specified, state of the system (SFWMD 2019). The SFWMD also provides historical rainfall on a monthly and seasonal basis. Other agencies provide products and input to operational planning including the Climate Prediction Center (CPC) with rainfall predictions out to a year, the National Weather Service (NWS) with quantitative precipitation forecasts out to seven days, USGS and ENP with EDEN (Everglades Depth Estimation Network) ecosystem-wide water depths, and USFWS with the Species Climate Outlook that focuses on the expected climate out to 12 months, but includes general weather expectations for species across the State, projected out to 2100. Each source of data, including model projections, provides value added information to the larger discussion that was not systematically incorporated into the historical discussion on how to best manage water operations.

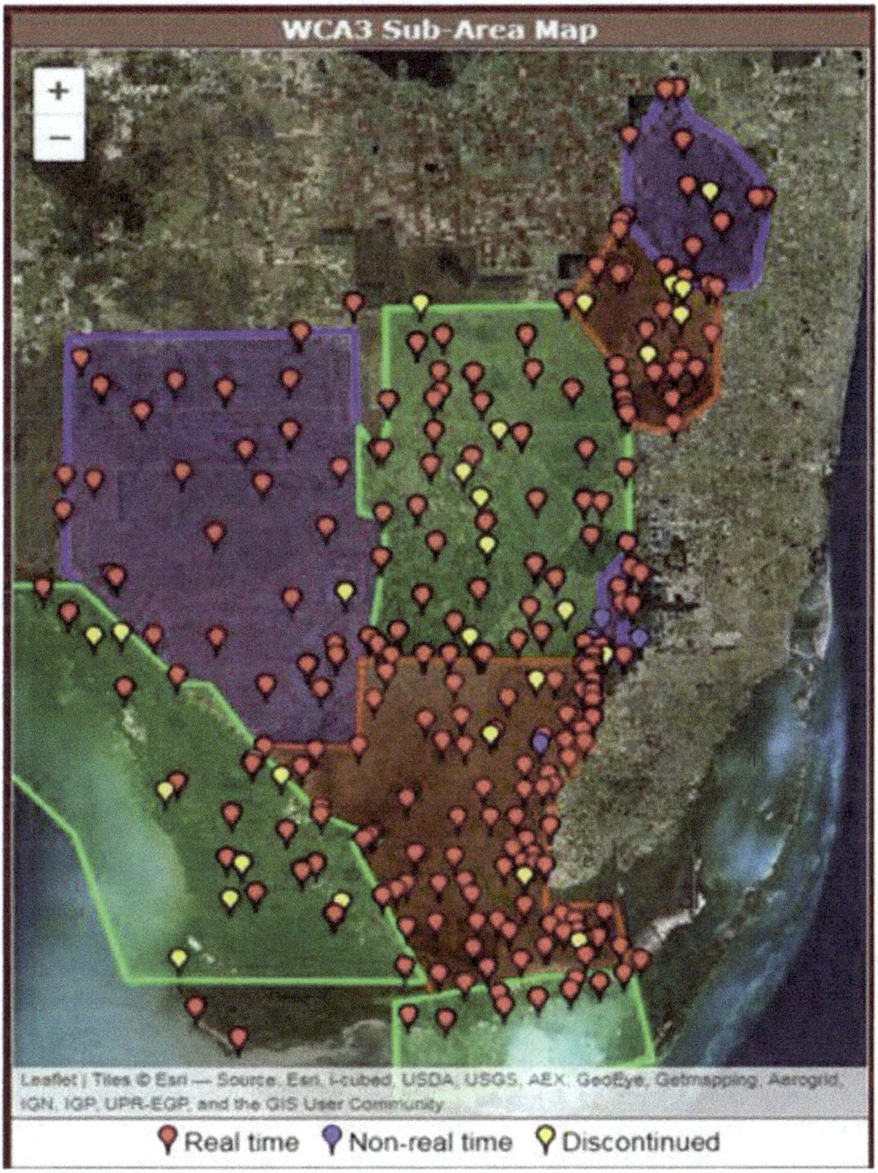

Fig. 5 Gauge locations for monitoring water levels and water quality in the Everglades. (Reproduced from USGS 2009)

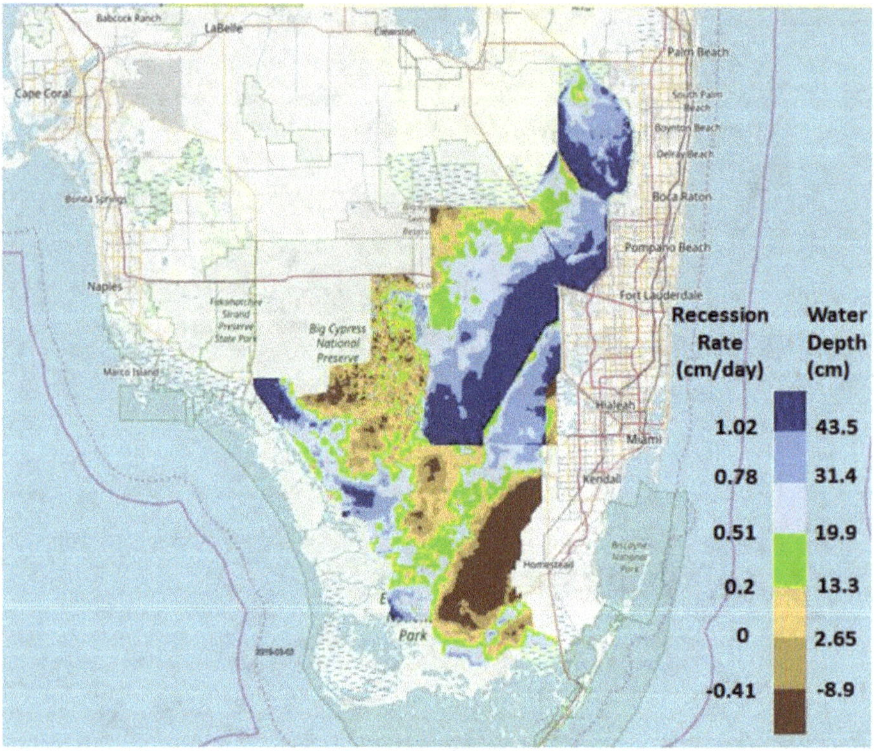

Fig. 6 Example of a water depth map, taken from the Wading Bird Depth Viewer. (Reproduced from USGS 2009; https://sofia.usgs.gov/eden/wadem/)

4 Translational Science

Translational science focuses on the importance of communicating scientific information to "connect end-users of environmental science to the field research carried out by scientists" (Schlesinger 2010). The use of a strategic communication approach (Harwell et al. 2020) thus can be useful in efforts to achieve effective EBM based on the EBM principle that, "decisions reflect societal choice" (Long et al. 2015). While "strategic communication" approaches were not explicitly identified in many EBM examples, a large EBM case study analysis by Mattheiß et al. (2018) concluded that "the better the communication strategy the likelier the demand for scientific knowledge from the social system." The translation of science within the context of ecosystem management is key to a broad understanding of both social and ecological systems and their interlinkages, which promote the development of innovative tools and management approaches to sustain biodiversity and the long-term delivery of ecosystem services (Piet et al. 2017).

As a complex socio-ecological system, the Greater Everglades involved numerous stakeholders from backgrounds including government agencies, universities,

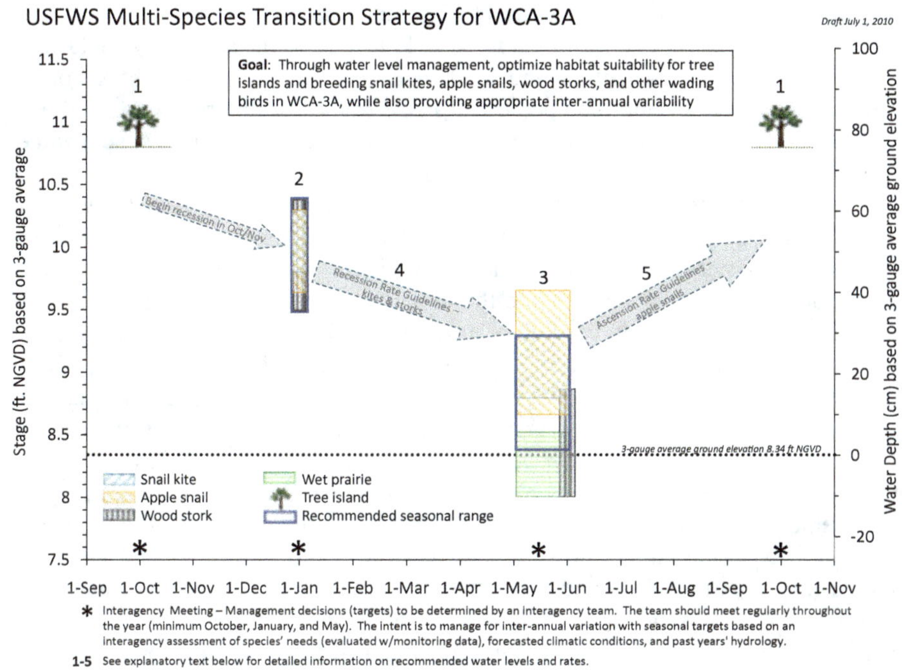

Fig. 7 USFWS Multi-Species Transition Strategy for Water Conservation Area 3A. Strategy includes recommending ranges and targets for species and habitats likely to be impacted by Everglades Restoration projects. *Denotes timing of intended interagency coordination meeting. (Reproduced from USFWS 2010)

non-profit conservation organizations, and tribes that rely on receiving ecosystem services. A variety of managing local, state, and federal agencies provide those services through ecosystem management of the adjoining Everglades. These same agencies, plus universities and other organizations, conduct research, collect monitoring data, and fill key roles in water management decisions and operations. Due to this complexity, communication among stakeholders that can be impacted by water management decisions, managers that make operational decisions and scientists collecting research and monitoring data is central to effectively applying EBM to an ecological system, particularly at the landscape scale. For the purposes of EBM of the Everglades, the need to strategically communicate the elements of translational science (e.g., Harwell et al., 2020) have been acknowledged for decades for both ecosystem management (Kushlan 1979) and ecosystem restoration purposes (Harwell 1997). Here, a translational science framework for EBM in the Everglades (Fig. 7), involves the communication of information among scientists, water managers, and stakeholders. Much of this EBM framework is foundationally defined in regulatory and/or planning documents related to Everglades Restoration (e.g., Biological Assessments under the Endangered Species Act (ESA), Environmental Impact Statements under the National Environmental Protection Act) and implemented at a high level through related programs such as the Restoration

Coordination and Verification (RECOVER) and the Congressionally mandated South Florida Ecosystem Restoration Task Force, as well as through field-level coordination between scientists and land managers. As such, these stakeholders helped define and develop the operational EBM framework presented here. We do recognize that other EBM frameworks have been developed for larger spatial and/or governance scales, such as the AQUACROSS framework (Piet et al. 2017).

At a field level, science generated by agency scientists (both associated with the operational management of the system and other agencies' supporting science) is translated into status and condition information for both ecosystem components (e.g., wading birds, tree islands) and the underlying hydrology and environmental conditions. This information is fed to stakeholders, including water managers and operational decision makers through a suite of communication forums (e.g., weekly and quarterly coordination meetings) and media (e.g., model output, informational graphics, narrative and numerical assessment statements). Water managers and operational decision makers translate this information into the context of agency mandates, goals, and operational constraints to make decisions that change the water management of the system. The resulting ecological outcomes, part of the monitoring effort for determining success, is folded back into the status and condition information as part of a larger adaptive management cycle. While water management decisions are still made by sector-focused operational managers, the Everglades EMB framework creates both the mechanism and the opportunity for other Everglades socio-ecological system goals and information to be served up for consideration.

This framework allows for real-time integration of operations data from discrete structures, knowledge of hydrodynamics throughout the system, modeled surface water elevations, and information on ecological envelopes (e.g., boundary conditions) for various species to develop water management recommendations for the best ecological outcome on multiple temporal and spatial scales. Recommendations are made with other uses, constraints, and regulations in mind but focus on ecological outcomes by aiming to identify where the system could use more/less water or; faster/slower water level ascension/recession rates, as well as highlight ecologically sensitive areas and/or species and habitats for a given operational decision.

5 Managing Eco-hydrology in the Everglades

Species typically used as indicators of Everglades ecosystem health and restoration success represent a range of habitats, behavioral characteristics, niches, and conservation status (Doren et al. 2009). Different species are used as indicators at different scales depending upon their response time to changes in environmental conditions, data availability, and the utility of relevant tools (Doren et al. 2009). Common indicator species include alligators, wading birds, Snail Kites, and apple snails. Typical habitats, such as tree islands, sawgrass ridges, and sloughs provide the basic structure of the Everglades and are sensitive to changes in water management.

All wildlife and habitats in the Everglades are adapted to annual patterns of rainfall and regional flooding. Typical Everglades habitats are characterized by a range of conditions largely driven by water levels and hydroperiods, which vary widely throughout the system because of a north-south elevation gradient, variations in landscape micro-topography that drive water levels and hydroperiods, and seasonal rainfall patterns that result in a typical dry season (November–April) and wet season (May–October).

Species that have evolved with the Everglades are able to survive in the highly dynamic system characterized by low nutrients and extremely variable intra- and inter-annual water levels using a variety of strategies, such as synchronization of breeding seasons with periods of suitable water levels and prey availability. In general, wading birds are colonial nesters and under typical conditions largely use the same flooded areas for nesting from year to year. Some species, such as Wood Storks, travel long distances to forage if conditions in their typical nesting spots are not ideal. However, other species, such as the non-wading Snail Kite, are more nomadic and select annual nesting sites based upon where conditions in the system are most conducive to successful nesting (i.e., appropriate nesting materials available and water levels that are 20–80 cm (50–200 in.) deep (Bennetts et al. 1988). These water depths protect nests from land-based predators and provide suitable habitat for their primary prey, apple snails, which prefer depths less than 50 cm (125 in.) (Darby et al. 2002). Species such as the endangered Cape Sable Seaside Sparrow nest near the ground yet require dry habitat to breed, and are therefore dependent upon areas of higher elevation with shorter hydroperiods. This diversity in habitat requirements are largely provided throughout the Everglades ecosystem by changes in ground surface elevation and extensive micro-topography, which results in a mosaic of habitats encompassing a range of water depths and hydroperiods across the landscape.

The Everglades ecosystem is heavily impacted by socio-ecological pressures, such as efforts to provide flood control and water supply to surrounding urban and agricultural development. Water managers strive to operate existing water control structures across the landscape in a coordinated fashion with the overall goal of providing conditions similar to those historically driven by rainfall and natural sheetflow, while providing crucial ecological services such as flood control and urban/agricultural water supply. Creating natural conditions in this highly impacted and managed system requires the complex integration of operational constraints, regulations, and policies, with the varied habitat requirements and population status of the suite of indicator species. For example, tree islands and ridge and slough habitat types, as well as the underlying peat substrate, are sensitive to hydrologic patterns such as water depth and hydroperiod. Science-based thresholds for water depths and hydroperiods are used to inform recommendations so that management is protective of these habitat features.

Because of the inextricable link between wildlife and water in the Everglades, consideration of past, current and future water conditions, habitat conditions, and wildlife population status at short, mid- and long-temporal scales are crucial for making water management recommendations that are protective of sensitive wildlife and habitats. The earliest formalized effort to summarize and integrate suitable

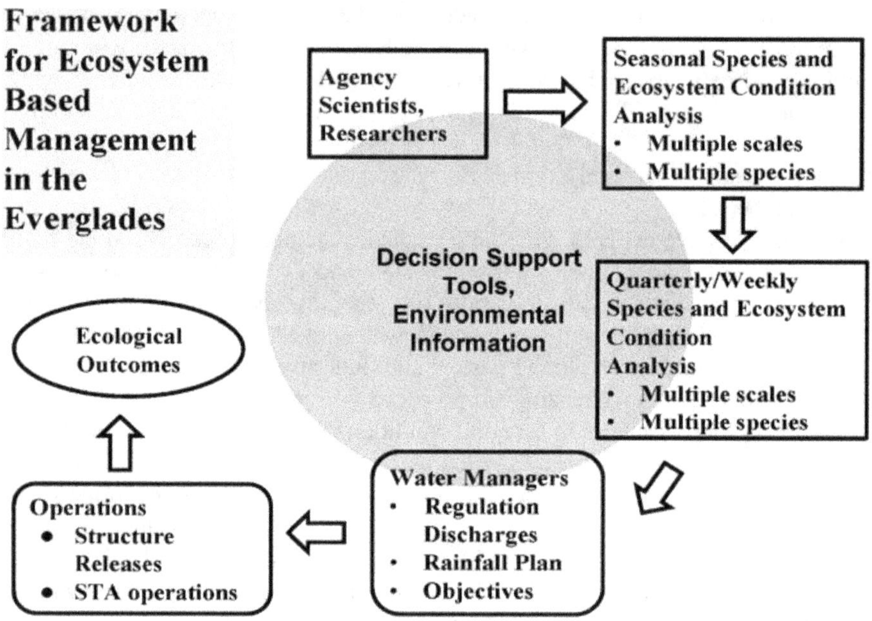

Fig. 8 Framework for Ecosystem-Based Management in the Greater Everglades ecosystem. Decision support tools and environmental information (*shaded circle* in background) provide an overarching anchor for EBM

conditions for a range of species was the Multi-Species Transition Strategy (MSTS) for WCA-3A in the Everglades Restoration Transition Plan (ERTP) developed by the Ecological Services branch of the USFWS as part of ESA Section 7 consultation (Fig. 8) (USFWS 2010).

The MSTS plan includes descriptions of the typical hydrologic ranges and timing of suitable conditions for multiple species considered indicators of Everglades Restoration. The plan, available to inform operations, management, and restoration decisions, directly compares and illustrates tolerance ranges for different sensitive habitat types and species, allowing identification of overlapping conditions and potential conflicts for management of these species. The information in this plan is combined with current, past, and projected future habitat conditions to inform daily/monthly/seasonal ecological recommendations for water management, with a particular focus on areas impacted by ongoing restoration construction projects by the USACE.

The ability to develop effective recommendations is dependent upon accurate, real-time data regarding habitat conditions and the ability to predict the likely response of wildlife populations to management operations. Much of the available research and/or monitoring efforts are driven by a given species' conservation status, population status, and/or inclusion in USFWS' species recovery plans, the Monitoring and Assessment Plan (RECOVER 2009), and indicator species identified by the

Department of Interior's South Florida Ecosystem Restoration Task Force, which is part of the Comprehensive Everglades Restoration Plan (CERP) (USACE and SFWMD 1999). Species designated with official conservation status, such as species protected by the ESA, or designated as a Florida Species of Special Concern, may have additional regulatory or recommended guidelines available to incorporate into a decision-making framework.

There are a variety of available tools and reports for determining and assessing antecedent, current, and potential future conditions such as USGS' EDEN, a landscape-scale surface water depth model, the National Oceanographic and Atmospheric Administration (NOAA) long- and short-term climate outlook predictions (Quantitative Precipitation Forecasts; QPFs), past rainfall trends, and the USFWS' Species Climate Outlook report, which characterizes forecasted conditions based on the requirements and tolerances of select species. A suite of scientists from universities, local, State, and federal agencies, Tribes, private consultants, and non-profit organizations provide wildlife population status and habitat requirement updates from ongoing research and monitoring efforts.

In addition to driving research and enabling partnerships, the CERP provides a critical framework for incorporating science, including new research and monitoring data, into management decisions using an adaptive management strategy (Loschiavo et al. 2013) creating opportunity for applying Ecosystem-Based Management (EBM) approaches to achieve restoration goals. A significant amount of resources have been invested in using results from scientific investigations to develop additional decision support tools to inform water operations and habitat management across the entire Everglades landscape (Table 1). These tools combine routinely collected environmental data with wildlife data and habitat condition information that can be used to assess past, current, and likely future conditions across the landscape to predict and evaluate potential impacts/benefits of water management and operations to wildlife and habitats.

Government and university partners developed the most easily accessible and frequently updated spatial modeling tools. These tools are used for making daily, weekly, and seasonal ecological recommendations. Many tools include maps that integrate known habitat preferences of a suite of wading birds, including the federally threatened Wood Stork (e.g., preferred water depths, rate and direction of change in water levels) with current and/or future conditions to indicate different levels of habitat suitability across the Everglades landscape (Wading Bird Depth Viewer, WADEM; Table 1). A similar tool based on recommended hydroperiod and water depths is available for the federally endangered Cape Sable Seaside Sparrow (Sparrow Viewer; Table 1), as well as other species such as Snail Kites and apple snails. While these tools and species updates are available to be individually considered by water managers when making operational decisions, most available tools do not provide a high-level integration of information or provide specific ecological recommendations.

Table 1 Names, descriptions, and sources of some of the spatial modeling tools used to develop multispecies water management recommendations in the Everglades

Species	Status	Time-scale	Tool description	Translational science information and source
Snail Kites[a]	Federally Endangered	Real-time; Long-term	EVERKite	Generates spatial maps of conditions for Snail Kites, either current or simulated under different hydrologic scenarios; specifically refers to targets defined in USFWS Biological Opinion. https://www.jem.gov/Modeling/EverKite
Apple Snails[a]	Least Concern; Primary prey of endangered Snail Kite	Long-term	EVERSnail	Generates estimated population size under different hydrologic scenarios. https://www.jem.gov/Modeling/AppleSnail
Wood Storks[a]	Federally Threatened	Real-time; Long-term	WADEM	Generates spatial maps of conditions for wading birds, including Wood Storks, either current or simulated under different hydrologic scenarios; specifically refers to targets defined in USFWS Biological Opinion. https://www.jem.gov/Modeling/WADEM
Wading Birds[b]	Varied	Real-time; Long-term	WADEM	Generates spatial maps of conditions for wading birds, including Wood Storks, either current or simulated under different hydrologic scenarios; specifically refers to targets defined in USFWS Biological Opinion. https://www.jem.gov/Modeling/WADEM
Alligators[b]	Threatened due to similarity of appearance (USFWS)	Long-term	Alligator Model	Generates spatial distribution maps of environmental condition related to Habitat, Breeding, Courtship & Mating, Nest Building, Nest Flooding, Overall Suitability, either current or simulated under different hydrologic scenarios. https://www.jem.gov/Modeling/Alligator
Cape Sable Seaside Sparrow[a]	Federally endangered	Real-time; Long-term	Sparrow Viewer	Generates maps of current or simulated conditions as related to Cape Sable Seaside Sparrow; either current or simulated under different hydrologic scenarios; specifically refers to targets defined in USFWS Biological Opinion. https://sofia.usgs.gov/eden/csss/index.php

For more on translational science aspects, the reader is directed to Sect. 4

[a]USFWS Multi-Species Recovery Plan (USFWS 1999)

[b]2009 Revised CERP Monitoring and Assessment Plan (RECOVER 2009)

6 Integrating Information Into Recommendations

Integrating relevant information into ecological recommendations at multiple temporal and spatial scales occurs through a series of seasonal, weekly, and/or daily meetings that occur at key times throughout the year. The scope of discussions and recommendations become more narrowly focused with increased meeting frequency. The utilization of available tools and monitoring information allow the characterization and assessment of species-specific habitat conditions as well as the potential impact of operational decisions on indicator species. Recommendations focus on addressing ecological needs of the Everglades more than specific operational decisions, although incorporating information about the feasibility of operations and regulatory guidelines strengthen recommendations. Some specific operational recommendations can be made, such as recommending preferred volumes and rates of inflows/outflows at specific structures, but only in areas that fall within existing regulatory frameworks and guidelines that can be feasibly considered (i.e., without formally updating regulatory guidelines).

Seasonal meetings generally occur at the beginning of the dry and wet seasons, as well as one meeting during the transition between wet and dry seasons (October, January, and May). These meetings focus on assessing conditions, characterizing desired ecological outcomes for the upcoming season, and defining water management guidelines for achieving those desired outcomes. Meetings are typically full-day, in-person workshops that include species updates (monitoring and research), overviews of current, past, and expected future climate conditions, a summary of ongoing and planned operations, and short- (7–10 days) and mid/long-term climate outlook discussions to develop recommendations for achieving desired ecologic (30–90 days) al outcomes. Meeting participants include key stakeholders such as agency, university, and non-governmental organization scientists. Factors such as short- and long-term stakeholder/managing agency goals and inter-annual variability of indicator species population dynamics are also considered when making recommendations. Seasonal recommendations are based on the overall assessment of conditions (past, present, and future) and typically include the identification of priority species and areas within the system, the characterization of desired ecological outcomes for the upcoming season, and the development of seasonal targets for rates of change in water depth and hydroperiod based on current conditions as indicated by monitoring data and modeled species habitat suitability. Recommendations can also include generalized recommendations such as to retain water where storage is available (per regulatory guidelines), to avoid particular operations that can have deleterious habitat impacts (e.g., degraded water quality), and/or suggestions that foster more natural (e.g., applying a ramping approach when adjusting structure inflows/outflows). Ecological thresholds and/or targets for other species and habitats, such as alligators and tree islands, also inform recommendations to benefit the greatest number, and/or highest priority (most sensitive or imperiled) of species across the landscape. Seasonal meetings tend to have the highest number of participants because a broader range of stakeholders, (many of which are identified

26.1.3), including non-governmental organizations, local municipalities, natural resource management agencies, and researchers from a range of organizations, are interested in providing species and habitat updates and input towards longer-term recommendations. Accounting for variability throughout the system, as well as agency-specific missions, directives, and/or priorities, recommendations are provided at multiple spatial and temporal scales as appropriate.

Once seasonal recommendations are developed, a smaller, core group of species experts and agency biologists evaluate, update, and communicate recommendations in real-time. Weekly meetings of a core group of primarily managing agency scientists develop habitat condition updates as well as any necessary updates to seasonal ecological recommendations. Weekly ecological recommendations typically include assessment of current and expected (short- and mid-term) conditions of habitat and wildlife populations, recent operations, and projected rainfall trends. Ultimately, the group identifies areas with sensitive habitats/wildlife populations (e.g., nesting wading bird colonies), and/or localized areas that could ecologically benefit from more/less water.

Weekly and/or daily coordination teleconferences typically take place during the dry season and coincide with wading bird, Snail Kite, and Cape Sable Seaside Sparrow nesting seasons. These coordination meetings partially focus on these indicator species to assess current conditions and short-term climate predictions to make targeted, short-term recommendations, often resulting in updating and refining seasonal recommendations to promote desired outcomes. The scope and focus of weekly and/or daily meetings tends to be narrower than in seasonal meetings. These meetings are often followed up with daily or semi-daily coordination meetings with agency scientists and managers during the occurrence of major events such as significant rainfall or changes in operations.

Group recommendations and updates are communicated to operations managers at the SFWMD and USACE through written reports developed by the group and presented during weekly operations meetings by agency scientists that participate in both the development of ecological recommendations and agency water management operations meetings. Routinely providing recommendations in other forums, such as regulatory Periodic Scientist Calls, which are monthly public coordination calls mandated by ESA consultation and hosted by USACE, as well as intra-agency management communication chains support coordinated recommendations across management agencies.

Available tools are currently integrated within the recommendation-making process to various degrees. Further integration of real-time population distribution and conditions would enhance the ability to understand and communicate short-term ecological needs. As additional decision support tools are developed and implemented, a combination of current conditions, past conditions, and past climate conditions over a long period can be used to determine the most likely wildlife response to water trends across the Everglades system for the upcoming season. These likely scenarios can be evaluated, and recommendations made, for individual species or a select group of species (e.g., wading birds). One highly anticipated tool currently under development is a species-forecasting tool (USGS) that considers habitat requirements and potential habitat suitability in the upcoming season for an entire collection of indicator species based on expected water level trends. Water trend scenarios can then be evaluated for their benefits to the greatest number or highest priority indicator species, which can help define priorities and potential outcomes. An early version of this tool is currently being tested and studied to determine the best way to integrate it into decision-making processes.

An infographic depicting information about current ecological conditions and species distributions, specifically designed for operations managers, is currently in development. This graphic is intended to provide a visual summary of recommendations and the ecological conditions that influenced them in order to more effectively communicate ecological needs and document the decision process to operations managers. An additional in-depth annual review of how recommendations influenced water operations and promoted desired ecological outcomes would allow further understanding of how ecological components are incorporated into the decision-making process and support refinement of how recommendations are made to maximize EBM effectiveness. Finally, additional benefits could be gained from the available tools and communication strategies if they were incorporated into existing and future regulatory guidance.

7 Conclusion

A suite of EBM activities support multi-purpose management of the Everglades ecosystem to provide ecosystem services as well as support plant and wildlife communities. With a focus on stakeholder engagement, communication, and the development/use of tools to integrate a wide range of conditions, operations, climate, and wildlife population data, the EBM framework presented here promotes consistent and effective management and restoration of the Everglades to meet a wide range of complex goals, needs, and ecological targets.

Stakeholder participation and communication is key to the effectiveness of this multiple element, decision-making EBM process in a complex socio-ecological system. Engagement by interested parties, researchers, and managing agencies enable integration of local and regional priorities to support healthy wildlife and habitats, as well as provide critical ecological services (e.g., flood control, water supply). Documenting the process and results for developing recommendations promotes communication with stakeholders and provides a record that can be used for adaptive learning.

This multi-agency/stakeholder approach to using integrative tools and real-time monitoring data for coordinating and developing comprehensive and effective water management recommendations is superior to previous approaches because this method is inclusive, transparent, comprehensive, and provides a landscape context to recommendations for individual management areas. The coordination of ecological recommendations among scientists from the various land management agencies across the landscape integrates research, monitoring, and stakeholder interests, and provides water managers with a more holistic and cohesive set of recommendations for supporting wildlife and habitats, even when there is not full consensus regarding ecological recommendations among the agencies due to area/agency-specific goals and objectives. A dedicated focus on stakeholder engagement facilitates the inclusion of local expertise (representing those stakeholders identified in Sect. 2.3) provided by agency biologists, who manage individual areas, with information provided by university and agency scientists that monitor regional conditions and wildlife. Stakeholder engagement allows for the incorporation of valuable insights from these and other stakeholders into the development of recommendations that integrate specific needs of individual areas with the needs of the Everglades ecosystem as a whole.

Acknowledgements We would like to thank Steve Henry, Miles Meyers, Darryl Marois, Manual Lago, and Ana I. Lillebø for valuable reviews of earlier versions of this manuscript. The views expressed in this chapter are those of the authors and do not necessarily reflect the views or policies of the U.S. Fish and Wildlife Service, U.S. Department of Interior, or the U.S. Environmental Protection Agency. Mention of trade names or commercial products does not constitute endorsement or recommendation for use.

References

Ankersen, T., & Hamann, R. (1996). Ecosystem management and the Everglades: A legal and institutional analysis. *The Journal of Land Use & Environmental Law, 11*, 473–536.

Bennetts, R. E., & Kitchens, W. M. (1997). Population dynamics and conservation of snail kites in Florida: The importance of spatial and temporal scales. *Colon Waterbird, 20*(2), 324–329.

Bennetts, R. E., Collopy, M. W., & Beissinger, S. R. (1988). *Nesting ecology of snail kites in water conservation area 3A* (pp. 1–174). Department of Animals and Range Sciences, University of Florida, 32.

Cattau, C. E., Kitchens, W. M., Reichert, B. E., Bowling, A., Hotaling, A., Zweig, C., Olbert, J., Pias, K., & Martin, J. (2008). *Demographic, movement, and habitat studies of the endangered snail kite in response to operational plans in Water Conservation Area 3*. Gainesville, FL: US Geological Survey Biological Resources, Division, Florida Cooperative Fish and Wildlife Research Unit.

Darby, P. C., Bennetts, R. E., Miller, S. I., & Percival, H. F. (2002). Movements of Florida apple snails in relation to water levels and drying events. *Wetlands, 22*(3), 489–498.

Doren, R. F., Trexler, J. C., Gottlieb, A. D., & Harwell, M. C. (2009). Ecological indicators for system-wide assessment of the greater Everglades ecosystem restoration program. *Ecological Indicators, 9*(6), S2–S16.

Douglas, M. S. (1947). *The Everglades: River of Grass*. New York: Rinehart.

Everglades National Park (ENP). (2015). Hydrologic monitoring program. Retrieved March 20, 2019, from https://www.nps.gov/ever/learn/nature/hydromon.htm.

Frederick, P. C., & Ogden, J. C. (2001). Pulsed breeding of long-legged wading birds and the importance of infrequent severe drought conditions in the Florida Everglades. *Wetlands, 21*(4), 484–491.

Harwell, M. A. (1997). Ecosystem management of South Florida: Developing a shared vision of ecological and societal sustainability. *Bioscience, 47*(8), 499–512.

Harwell, M. C., Molleda, J. L., Jackson, C. A., & Sharpe, L. (2020). Establishing a common framework for strategic communication in ecosystem-based management and the natural sciences. In T. O'Higgins, M. Lago, & T. H. DeWitt (Eds.), *Ecosystem-based management, ecosystem services and aquatic biodiversity: Theory, tools and applications* (pp. 165–188). Amsterdam: Springer.

Hirsch, R. M. (1978). Risk analysis for a water-supply system – Occoquan reservoir, Fairfax and prince William counties, Virginia. *Hydrological Sciences Bulletin, 23*(4), 476–505.

Kushlan, J. A. (1979). Design and management of continental wildlife reserves: Lessons from the Everglades. *Biological Conservation, 15*(4), 281–290.

Kushlan, J. A. (1987). External threats and internal management: The hydrologic regulation of the Everglades, Florida, USA. *Environmental Management, 11*(1), 109–119.

Light, S. S., & Dineen, J. W. (1994). Water control in the Everglades: A historical perspective. In S. M. Davis & J. C. Ogden (Eds.), *Everglades: The ecosystem and its restoration* (pp. 47–84). Boca Raton, FL: St. Lucie Press.

Long, R. D., Charles, A., & Stephenson, R. L. (2015). Key principles of marine ecosystem-based management. *Marine Policy, 57*, 53–60.

LoSchiavo, A., Best, R., Burns, R., Gray, S., Harwell, M., Hines, E., McLean, A., St. Clair, T., Traxler, S., & Vearil, J. (2013). Lessons learned from the first decade of adaptive management in comprehensive Everglades restoration. *Ecology and Society, 18*(4), 70.

Mattheiß, V., Strosser, P., Krautkraemer, A., Charbonnier, C., McDonald, H., Röschel, L., Hoffmann, H., Lago, M., Delacámara, G., Gómez, C. M., Piet, G., Schuwirth, N., Kuemmerlen, M., & Reichert, P. (2018). Evaluation of ecosystem-based management responses in case studies: AQUACROSS Deliverable 8.2. European Union's Horizon 2020 Framework Programme for Research and Innovation Grant Agreement No. 642317. Retrieved October 20, 2019, from www.aquacross.eu.

McVoy, C., Said, W. P., Obeysekera, J., VanArman, J. A., & Dreschel, T. W. (2011). *Landscapes and hydrology of the pre-drainage Everglades* (pp. 1–31). Gainesville, FL: University Press of Florida.

National Academies of Sciences, Engineering, and Medicine. (2018). *Progress Toward Restoring the Everglades: The Seventh Biennial Review—2018*. Washington, DC: The National Academies Press.

National Research Council. (2010). *Progress toward restoring the Everglades: The Third Biennial Review—2010*. Washington, DC: The National Academies Press.

Ogden, J. C. (1994). A comparison of wading bird nesting colony dynamics (1931–1946 and 1974–1989) as an indication of ecosystem conditions in the Southern Everglades. In S. M. Davis & J. C. Ogden (Eds.), *Everglades: The ecosystem and its restoration* (pp. 533–570). Boca Raton, FL: St. Lucie Press.

Ogden, J. C. (2005). Everglades ridge and slough conceptual ecological model. *Wetlands, 25*(4), 810–820.

Piet, G., Delacamara, G., Lago, M., Rouillard, J., Martin, R., & van Duinen, R. (2017). *Making ecosystem-based management operational*. Deliverable 8.1, European Union's Horizon 2020 Framework Programme for Research and Innovation grant agreement No. 642317.

RECOVER. (2009). Monitoring and Assessment Plan (MAP). Restoration Coordination and Verification, c/o U.S. Army Corps of Engineers, Jacksonville, Florida, USA, and South Florida Water Management District, West Palm Beach, Florida, USA. Retrieved March 20, 2019, from http://www.evergladesplan.org/pm/recover/recover_map.aspx.

Schlesinger, W. H. (2010). Translational ecology. *Science, 329*(5992), 609.

Snyder, G. H., & Davidson, J. M. (1994). Chapter 5. Everglades agriculture: Past, present, and future. In S. M. Davis & J. C. Ogden (Eds.), *Everglades: The ecosystem and its restoration*. Boca Raton, FL: St. Lucie Press.

South Florida Water Management District (SFWMD). (2016). *Facility and infrastructure location index map*. West Palm Beach, FL. Retrieved March 20, 2019, from https://www.sfwmd.gov/sites/default/files/documents/facility_map_overview.pdf.

South Florida Water Management District (SFWMD). (2019). *Operational planning*. Retrieved March 20, 2019, from https://www.sfwmd.gov/science-data/operational-planning.

U.S. Army Corps of Engineers (USACE). (2019). *Water management*. Retrieved March 20, 2019, from https://www.saj.usace.army.mil/Missions/Civil-Works/Water-Management/.

US Department of the Interior. (1994). *The Everglades, Coastal Louisiana, Galveston Bay, Puerto Rico, California's Central Valley, Western Riparian Areas, Southeastern and Western Alaska, The Delmarva Peninsula, North Carolina, Northeastern New Jersey, Michigan, and Nebraska,*

p. 123, vol. II of The Impact of Federal Programs on Wetlands. A Report to Congress by the Secretary of the Interior. Washington, DC: Department of the Interior.

U.S. Fish and Wildlife Service (USFWS). (1999). *South Florida multi-species recovery plan*. U.S. Fish & Wildlife Service, Atlanta, Georgia. Retrieved March 20, 2019, from https://www.fws.gov/verobeach/listedspeciesmsrp.html.

U.S. Fish and Wildlife Service (USFWS). (2010). *U.S. fish and wildlife service biological opinion for Everglades Restoration Transition Plan*, Phase 1. Vero Beach, FL.

U.S. Fish and Wildlife Service (USFWS). (2014). *Central Everglades Planning Project biological opinion*. Vero Beach, FL.

U.S. Fish and Wildlife Service (USFWS). (2016). *Biological opinion for the Everglades Restoration Transition Plan—2016*. Vero Beach, FL. Retrieved March 20, 2019, from https://www.fws.gov/verobeach/NewsReleasesPDFs/20160722ERTPJeopardyBO_FAQs.pdf.

U.S. Geological Survey (USGS). (2009). *Everglades Depth Estimation Network (EDEN) Applications: Tools to view, extract, plot, and manipulate EDEN data*. Retrieved March 20, 2019, from https://pubs.usgs.gov/fs/2009/3052/pdf/fs2009-3052_spread.pdf.

U.S. Geological Survey (USGS). (2013). *Water Conservation Areas (WCAs)*. Retrieved March 20, 2019, from https://archive.usgs.gov/archive/sites/sofia.usgs.gov/virtual_tour/controlling/wca.html.

USACE and SFWMD. (1999). *Central and Southern Florida Project Comprehensive Review Study: Final Integrated Feasibility Report and Programmatic Environmental Impact Statement*. U.S. Army Corps of Engineers, Jacksonville, Florida, USA, and South Florida Water Management District, West Palm Beach, Florida, USA. Retrieved March 20, 2019, from http://www.evergladesplan.org/docs/comp_plan_apr99/summary.pdf.

Walker, W. W., Jr. (1999). Long-term water quality trends in the Everglades. In K. R. Reddy, G. A. O'Connor, & C. L. Schelske (Eds.), *Phosphorus biogeochemistry in sub-tropical ecosystems: Florida as a case example*. Boca Raton, FL: CRC/Lewis Publishers.

6

From DPSIR the DAPSI(W) R(M) Emerges... a Butterfly – *'protecting the natural stuff and delivering the human stuff'*

Michael Elliott and Timothy G. O'Higgins

Abstract The complexity of interactions and feedbacks between human activities and ecosystems can make the analysis of such social-ecological systems intractable. In order to provide a common means to understand and analyse the links between social and ecological process within these systems, a range of analytical frameworks have been developed and adopted. Following decades of practical experience in implementation, the Driver Pressure State Impact Response (DPSIR) conceptual framework has been adapted and re-developed to become the D(A)PSI(W)R(M). This paper describes in detail the D(A)PSI(W)R(M) and its development from the original DPSIR conceptual frame. Despite its diverse application and demonstrated utility, a number of inherent shortcomings are identified. In particular the DPSIR model family tend to be best suited to individual environmental pressures and human activities and their resulting environmental problems, having a limited focus on the supply and demand of benefits from nature. We present a derived framework, the "Butterfly", a more holistic approach designed to expand the concept. The "Butterfly" model, moves away from the centralised accounting framework approach while more-fully incorporating the complexity of social and ecological systems, and the supply and demand of ecosystem services, which are central to human-environment interactions.

M. Elliott (✉)
Department of Biological & Marine Sciences, University of Hull, Hull, UK

International Estuarine & Coastal Specialists Ltd., Leven, UK
e-mail: Mike.Elliott@hull.ac.uk

T. G. O'Higgins
MaREI Center for Marine and Renewable Energy Ireland, Environmental Research Institute, University College Cork, Cork, Ireland
e-mail: tim.ohiggins@ucc.ie

1 Introduction

There is only one big idea in environmental management, especially that for aquatic ecosystem, that is '*how to maintain and protect the natural ecological structure and functioning and the resultant ecosystem services while delivering the societal goods and benefits*' (Elliott 2011). This is exemplified in Ecosystem-Based Management (EBM) which seeks to "integrate the connections between land air water and all living things including human beings and their institutions" (Mee et al. 2015) and to manage complex adaptive systems towards particular goals and targets. In essence, this is a risk assessment and risk management approach which requires monitoring to determine whether management actions have worked (Cormier et al. 2019). Hence in order to assess whether specific policy decisions are effective management efforts need to agree measurable targets on pre-defined indicators. Yet the complexity of interactions and feedbacks between human activities and ecosystems can make the analysis of such systems intractable and so we need an underlying accepted framework to link together the causes and consequences of change and their management. Systematic methods of accounting for socio-ecological interactions are required to assess the effectiveness of management in such complex adaptive systems. Here we follow environmental accounting as the process of systematically organising and presenting information relating to interactions between the economy and the environment in a standardised way in order to support policy making (UNSD 2012).

While there are many legal instruments which are required to manage the environment, and so fulfil the 'big idea' above (Boyes and Elliott 2014), in Europe there are two major framework environmental policies relating to aquatic environments: the Water Framework Directive (WFD) (European Commission 2000) and the Marine Strategy Framework Directive (MSFD) which both promote an ecosystem approach to freshwater, estuarine, coastal and marine management (see O'Hagan 2020, and Borja et al. 2010 and references therein for detailed accounts). These policies seek to apply common standards, equally, to the conservation of aquatic environments across the European Union Member States (European Commission 2008). This results in a particular need for standardised methodologies because each Member State has its own unique geographic and environmental conditions, as well as traditions of monitoring the environment based on indicators. For example, Teixeira et al. (2016) generated a catalogue of 611 indicators of biodiversity proposed or in use for monitoring European marine environments. Under these conditions, accounting frameworks enable the intercomparison of different Member State efforts and activities. This should enable assessment on a continental scale of whether environmental policies are reaching their objectives and whether EU member states are complying with these and other Directives. Experiences in the implementation of these directives have been instrumental in developing the state-of the art science addressing socio-ecological systems in Europe and has led to many valuable insights into Ecosystem-Based Management.

One environmental accounting framework, which has developed in parallel with the application of European aquatic environmental law is the Driver Pressure State

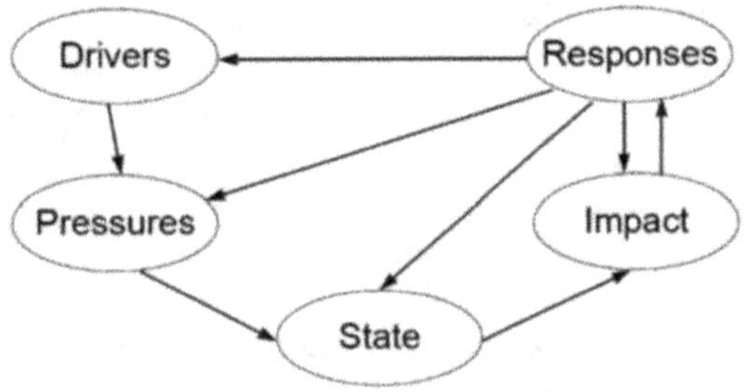

Fig. 1 Original depiction of the DPSIR framework. Source: European Environment Agency (1999)

Impact Response (DPSIR) framework (EEA 1999) which developed from the earlier PSI framework adopted by the OECD in the early 1990s. In brief, human activities or *Drivers* place *Pressures* on the environment resulting in changes to the environmental *State* and *Impacts* which result in a management *Response*. This causal chain in Fig. 1 uses the original diagram of the European Environment Agency (EEA 1999). DPSIR is a conceptual framework for analysing socio-ecological systems which encapsulates the interactions between anthropogenic and natural components of socio-ecological systems (Turner and Schaafsma 2015) where each element in the cause-consequence-management chain can be described by an indicator (see below). As a conceptual framework, the DPSIR is used to compartmentalise and thereby simplify and analyse socio-ecological systems.

With origins in the field of environmental risk assessment and accounting (Rapport and Friend 1979), the DPSIR was first formally published in 1999 (EEA 1999) prior to adopting the WFD (European Commission 2000) and has widely been used and adapted over the intervening 20 years (Patrício et al. 2016). The Web of Science indicates DPSIR has been the subject of 577 peer-reviewed papers, principally in the fields of environmental science ($n = 406$), water resources (88) and ecology (66). As an overarching conceptual frame, the DPSIR has been employed to analyse a broad variety of environmental problems in diverse environments and geographic settings from the study of water scarcity in Oman (Al Kalbani et al. 2015) to biodiversity loss (Maxim et al. 2009). Yet the DPSIR has been most commonly used in Europe (Patrício et al. 2016) and the most highly cited/influential DPSIR papers relate to the European aquatic environmental conservation directives, the WFD (Borja et al. 2006) and the MSFD (Atkins et al. 2011).

Early iterations of DPSIR (EEA 1999), did not rigorously define the various information categories, rather it described their interrelationships:

> ...social and economic developments [Driving forces] exert Pressure on the environment and, as a consequence, the State of the environment changes, such as the provision of adequate conditions for health, resources availability and biodiversity. Finally, this leads to Impacts on human health, ecosystems and materials that may elicit a societal Response that feeds back on the Driving forces, or on the state or impacts directly, through adaptation or curative action. (EEA 1999)

As a result, through practical application, the DPSIR has been continually refined and redefined in the context of European Marine environments. DPSIR has been applied to the Mediterranean and Baltic Seas (Lundberg 2005; Karageorgis et al.

2005; Skoulikidis 2009). Studies in the Black Sea (Langmead et al. 2009; Knudsen et al. 2010) employed a modified DPSIR (mDPSIR) approach, while, Cooper (2013) made a variety of changes to develop DPSWR (where W is Welfare) which has been used in a variety of geographic contexts from the Black Sea (O'Higgins et al. 2014 to the North East Atlantic (O'Higgins and Gilbert 2014) as well as to explore socio-ecological scale mismatches in marine sectors (O'Higgins et al. 2019) and to explore intertemporal trade-offs in activity and environmental quality (O'Higgins et al. 2014). This evolution of DPSIR is fully detailed by Patrício et al. 2016) which indicated anomalies in the way the concept has been used; this culminated in the development of DAPSI(W)R(M) which incorporated 20 years of the evolution of the concept and attempted to resolve the confusion in the use of the forerunners (as well as being the more pronounceable "*dap- see–worm*") (described in detail in Elliott et al. 2017, but used in their previous papers, e.g. for the Baltic Sea by Scharin et al. 2016 and the Arctic by Lovecraft and Meek 2019).

In this volume, the DPSIR and its variants are critical to many of the case studies. For example, these frameworks form the basis of the linkage framework techniques (Culhane et al. 2020); and were used as an organisational framework for case studies in the Azores (McDonald et al. 2020), the Danube (Funk et al. 2020) and in Lough Erne (O'Higgins et al. 2020). In this paper we first set out the major features of the DAPSI(W)R(M) framework (Elliott et al. 2017). We then consider the broader characteristics of ecosystem-based management (Langahns et al. 2019; Delacamara et al. 2020) and introduce a derivation of the framework which we call "the butterfly". The latter is designed to expand the concept, moving away from the centralised accounting framework approach while more-fully incorporating the complexity of social and ecological systems. We present the butterfly not as a replacement to the DPSIR-family (which has been so successfully applied to the existing marine policy) as a socio-ecological accounting framework, but as a potential tool to enable more fully integrated approaches to the development and application of Ecosystem-Based Management.

2 The DAPSI(W)R(M) Framework

2.1 Drivers (D)

Previous DPSIR frameworks (e.g. Cooper 2013) had an in-built confusion as the Drivers were often synonymous with activities (what is done in the environment) or sectors (groups of activities such as fisheries) (Patrício et al. 2016). There may also be different interpretations of this between natural and social scientists—for example the former may regard Drivers as activities but social scientists regard them as 'basic human needs'. In order to resolve the confusion, and because the later elements in the framework refer to and separate Activities and Pressures, Elliott et al. (2017) emphasised that the Drivers should refer to Basic Human Needs and thus the work of Maslow (1943) who proposed a five-tier hierarchical structure of such Drivers

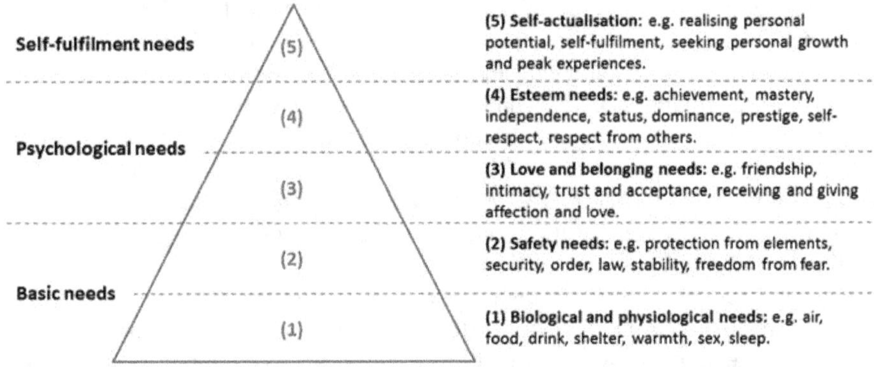

Fig. 2 Maslow's hierarchy of needs and human welfare (adapted from Maslow 1943)

(Fig. 2). At the lowest level these refer to an individual's biological and physiological needs which relate to survival (e.g. oxygen, food, drinking water) and safety and security (e.g. protection from external stressors). At its most simple, we can determine what are our basic human needs and then how do we get those from the natural and human environment.

The intermediate levels of basic human needs then cover psychological and emotional attributes which determine the societal structure and family relationships including love and belonging (e.g. friendship, intimacy, trust and acceptance) and esteem (e.g. prestige, achievement, self-respect). Hence, the lower four levels in Maslow's triangle cover the 'deficiency needs' i.e. what are required to motivate and satisfy people and for which desires and activities increase to satisfy these needs. The fifth and uppermost levels relate to meeting one's emotional fulfilment, such as the personal potential, emotional growth and satisfied experience—these may be regarded as 'growth needs'. Hence, there is the need for both an individual and society to satisfy the lower 'deficit needs' before achieving the higher 'growth needs' (Maslow 1943, 1970a, b). This includes three further intermediate categories of 'cognitive needs' (e.g. knowledge and understanding, curiosity, exploration, need for meaning and predictability—the scientific requirements), 'aesthetic needs' (e.g. appreciation and search for beauty, balance, form, etc. influencing our cultural appreciation) and 'transcendence needs' (e.g. altruistic behaviour helping others to achieve self-actualisation) (Maslow 1970a, b). Many of these needs are directly dependent on ecosystems, the 'basic needs' and safety needs require provisioning ecosystem services, while others for psychological and self-fulfilment needs depend on supply of ecosystem services and societal goods and benefits (Turner and Schaafsma 2015).

All of these basic human needs in turn dictate the way the marine space is used, the competition between individuals, tribes, societies and nations in the way in which the competition now occurs at larger scales from the regional to the global. Satisfying these needs and this competition (the basis of the economy) then leads to Activities inherent, for example, in international marine trade which is needed to deliver increasingly required/desired consumer products and services. All of this shows that the Drivers can ensure both economic benefits as well as high individual (physiological and psychological) and societal well-being and welfare. However, this unchecked growth in Drivers can lead to levels of Activities and Pressures which exceed the carrying and assimilative capacities of a particular system or even of the global system (Rockstrom et al. 2009) resulting in depletion of resources and damage to the natural system. This exceedance then requires measures which aim to minimise conflicts and partition that marine space in order to deliver the human and protect the natural aspects (Elliott et al. 2018).

The aggregate of these human wants and needs (or the urge to maximize utility) are the forces that drive economic development—the Blue Economy or Blue Growth. As set out above, they involve a range of social interactions between humans (e.g. respect from others, intimacy), as well as socio-ecological interactions between humans and the environment (e.g. food, shelter and warmth).

2.2 *Activities (A)*

As indicated above, there is the need to separate Drivers from Activities and Activities from Pressures. In essence, Activities are what we do in the seas to obtain those basic human basic needs, and the Pressures are the resulting mechanisms of change from those Activities (Elliott et al. 2017). Marine activities can be separated into 15 key marine sectors which then represent many individual, but often interlinked, activities which occur in most if not all seas (Table 1 gives 7 of these sectors). Elliott et al. (2017) lists all the Activities and Pressures and Burdon et al.

Table 1 Examples of sectors and activities in a marine environment (adapted from Elliott et al. 2017)

Sector	Examples of activities
Extraction of living resources	Benthic trawling (e.g. scallop dredging), netting, pelagic trawls, potting/creeling; bait digging, seaweed and harvesting, shellfish gathering
Transport and shipping	Mooring/beaching/launching, shipping, ferry operation, waste disposal.
Navigational dredging	Dredging, removal of substratum, disposal of dredge spoil.
Coastal infrastructure	Construction of artificial reefs and barrages, beach nourishment, laying of communication cables; constriction of transport infrastructure, dock and port facilities, land claim; construction of urban dwellings.
Land-based industry	Treatment and discharge of industrial; discharging particulate waste disposal, desalination effluent, sewage and thermal waste discharge
Agriculture	Animal waste disposal, crop fertilisation, farming, coastal forestry,
Tourism/recreation	Angling, boating/yachting, diving, public beach use, tourist resort and water sports operation, whale watching.
Research and education	Marine research; field education and training, research vessel cruises.

(2018) shows the complexity by giving a detailed breakdown of these relating to one main sector—oil and gas decommissioning. Hence the separation of sectors with nested activities may simplify the scheme. For example, the commercial fishing sector includes many types of fishing activity (trawling, potting, long-lines, etc.) and each produces Pressures which may or may not differ across Activities. This therefore creates operational marine management which is sufficiently specific for targeted problem-solving directed at specific activities (Cormier et al. 2019). Such a framework has to include historical marine activities (e.g. fisheries, oil and gas extraction), and newer offshore technologies (e.g. marine biotechnology, nodule mining) which together reflect the expanding global marine economy with its increasing human pressures (Stojanovic and Farmer 2013). Together this constitutes what may be termed the Blue Economy defined as 'smart, sustainable and inclusive economic and employment growth from oceans, seas and coasts' (e.g. marine energy extraction, aquaculture, maritime, coastal and cruise tourism, marine mineral resources, blue biotechnology) (European Commission 2012).

As indicated above, as the original DPSIR framework confused Drivers and Pressures, there was the need to separate them within the DAPSI(W)R (M) framework by adding Activities. As an example, the basic Driver is to obtain food and beam trawling is a fishery Activity which then leads to the Pressure of seabed abrasion caused by towed gear. It is then assumed that such abrasion would cause seabed habitat damage (e.g. a State change in the functional traits of the benthic community) which eventually leads to Impacts (on human Welfare) by reducing the fishable resource (see below). In turn, a Response (using management Measures), such as gear modifications or fishing-period limitation is needed to limit the Activities and hence minimise the Pressures.

The confusion between terms was noted in the otherwise important and seminal study by Halpern et al. (2008) giving the global analysis of 'human impact' on many marine ecosystems. This study listed 17 anthropogenic 'drivers' but, using the definitions here, these comprised seven Activities (including various forms of fishing, shipping and commercial activity) and 10 Pressures (including organic and inorganic pollutants, benthic structures, invasive species, sea temperature and ocean acidification) with none of the categories relating to Drivers as defined above. The global analysis of Halpern et al. (2008) appears to map Activities (and possibly Pressures) but terms them 'impacts'—it is emphasised here that this should not be assumed as mitigation measures may stop Activities creating impacts. This also reinforces the point that it may be easier to map Activities, which are often recorded on maps, photographs and databases, than Pressures. The latter requires the need to detect the spatial and temporal effects-footprints of each Activity which are much more difficult to determine (Elliott et al. 2018). Many of the assessment schemes used worldwide focus on Activities instead of Pressures (Borja et al. 2016), presumably because of the difficulty in determining these effects-footprints. Here the linkage Framework methodologies (Culhane et al. 2020) can be valuable in understanding and untangling activities and their resultant pressures.

2.3 *Pressures (P)*

Each Activity leads to one or more Pressures and each Pressure can result from one or more Activity thereby creating an interlinked matrix of Activities and Pressures (see Culhane et al. 2020 for approaches to analysing such linkages, and Burdon et al. 2018 for a precise example). A Pressure is defined as the mechanism of change, first to the natural system (State changes) and then to the social system (Impacts (on human Welfare)) (see below). Elliott (2011) then separated the pressures affecting a given sea area into Exogenic Unmanaged Pressures (ExUP) and Endogenic Managed Pressures (EnMP) (Fig. 3). In this way, exerting pressures on the ecosystem will reduce ecosystem services and ultimately affect societal goods and benefits.

The cause of Exogenic Unmanaged Pressures, as the name suggests, emanates outside the area, for example sea-level rise as the result of global climate change, and so management inside the area is only treating the consequences (such as building higher sea defences) (Elliott et al. 2016). Management actions to address these pressures therefore has to be at the large scale and even global level (such as the Paris COP meetings). In contrast, Endogenic Managed Pressures, by definition, occur within the management area and so both their causes and consequences need managing and can be managed. For example, increased infrastructure such as a new bridge or power plant in an area will cause pressures whereby the reduction of their consequences need to be incorporated into a management plan. Hence the

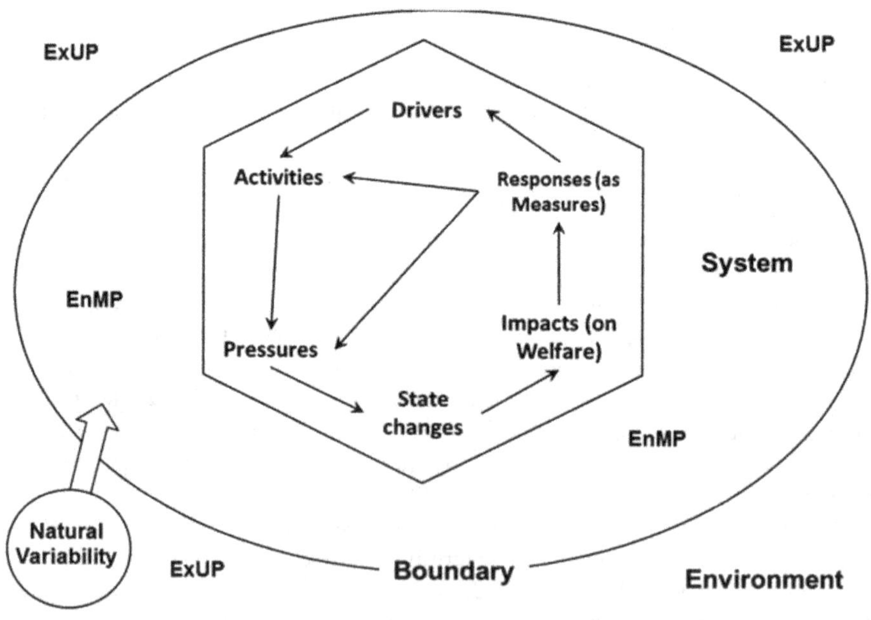

Fig. 3 The DAPSI(W)R(M) problem structuring framework (from Elliott et al. 2017). Key: ExUP = Exogenic Unmanaged Pressures; EnMP = Endogenic Managed Pressures (see text for explanation)

effects-footprints of all the endogenic pressures, both singly and cumulatively, need to be determined and managed in both space and time (Elliott et al. 2018).

2.4 *State or State Changes (S)*

In the original DPSIR framework, State, State change and Impact were often used interchangeably (Patrício et al. 2016) and often the natural scientists used State as the nature of the natural system and Impact as the change to the natural and social systems. In contrast, social scientists appear to have made the differentiation that State referred to the natural system and Impact referred to the social system. Because of this, the DAPSI(W)R(M) framework has used the term ‘State change’ (rather than ‘State’ as we are only interested in anthropogenic changes, i.e. a signal against a background of inherent variability (‘noise’)) to relate to the natural system (the ecology) due to single or multiple Pressures. This includes both the physico-chemical variables (i.e. sediment type, dissolved oxygen, organic matter, etc.) and biological health at all levels of organisation—the cellular system, individuals, populations, communities and ecosystems. Such changes can be referred to as structural characteristics (the features in each level at one time, for example the number of species in a community) or functioning variables (rate processes, such as productivity or ecosystem carbon flow) (Strong et al. 2015; de Jonge et al. 2003, 2012).

This then leads into the recent discussions regarding ecosystem services and societal goods and benefits which are derived from a healthy functioning natural system (Atkins et al. 2011; Turner and Schaafsma 2015). If the natural system has an appropriate structure and functioning then it is creating a healthy environment, for example if the waves, tides, substratum, etc. are appropriate then they will support the prey which sustains fishes (an ecosystem service). In turn, the natural State should provide the intermediate and final ecosystem services (as defined by Fisher et al. 2009, Turner and Schaafsma 2015, and Elliott et al. 2017) and human activities and pressures could then influence (in a positive or negative way) this natural state (i.e. State change) as well as the underlying marine ecosystem components and processes (Fig. 4, left hand side). The amount and fluxes of physical, chemical and biological materials may be regarded, in economic terms, as marine ecosystem stocks and flows which can be measured and which can have management targets and management measures to achieve those targets (Pinto et al. 2014; Atkins et al. 2015). In essence, the natural system can produce the ecosystem services but it then requires society to input ‘human complementary assets’ or ‘human capital’ (such as time, money, energy and skills) to extract societal goods and benefits (‘*the sea can produce fish but we need to learn how to catch and cook them*’!) (Elliott et al. 2017).

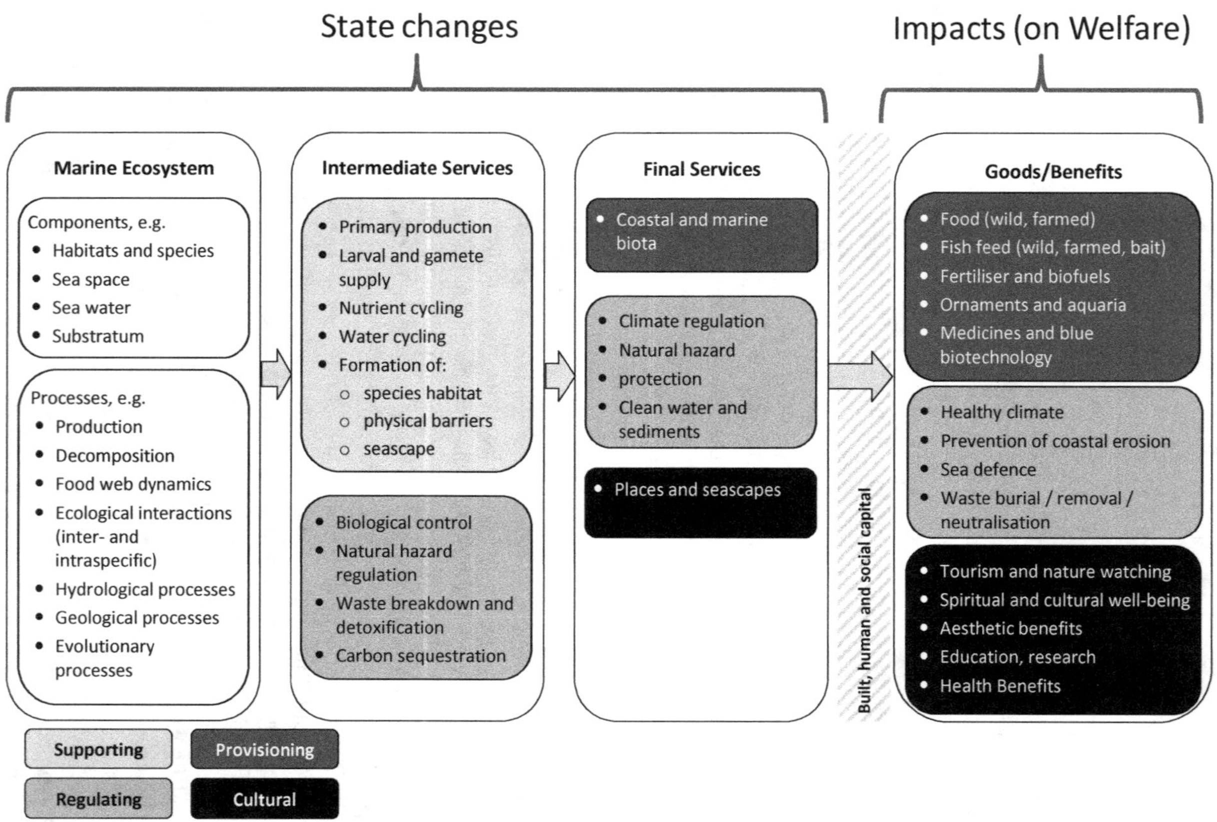

Fig. 4 State changes to the natural system reflected by changes in the marine ecosystem, intermediate and final ecosystem services (left hand side), and Impacts (on human Welfare) reflected by changes to the provision of Societal goods and benefits (right hand side) (modified and expanded from Turner et al. 2015)

2.5 *Impact (I) (on Human Welfare)*

Once the natural physico-chemical and ecological state has been changed by human activities and pressures then this could eventually influence the marine goods and benefits obtained by society. In DAPSI(W)R(M), this then is reflected as Impacts (on human Welfare). While Cooper (2013) changed the term 'Impact' to 'Welfare' this created a further confusion as it is the Impact *on* our Welfare that is of concern rather than Welfare *per se* (hence the use of parentheses). Hence, those Impacts (on Welfare) reflect the negative or even positive changes in the system providing societal goods and benefits (as defined by Turner et al. 2015, see below). Accordingly, it is necessary to derive and use quantitative indicators to detect and explain such changes in societal welfare (Fig. 5, right hand side). As indicated, societal goods and benefits are obtained by applying human complementary capital (social, human and human-made or built capital) to the natural environment (intermediate and final ecosystem services) (Atkins et al. 2011; Elliott et al. 2017). Therefore, any adverse Impacts on human Welfare are manifest as an inability of the marine system to provide those societal goods and benefits. For example, a healthy natural system is required to create natural places and seascapes which then influence our cultural appreciation of the sea. Society will then spend time and money to enjoy those benefits and, linking back to Maslow's work, this relies on us being sentient beings to appreciate the benefits—'*we need to expend energy to appreciate a blue whale even if we have never seen one*'!

The term Welfare in this element of DAPSI(W)R(M) also includes human well-being and happiness, two important Drivers in the upper part of Maslow's triangle.

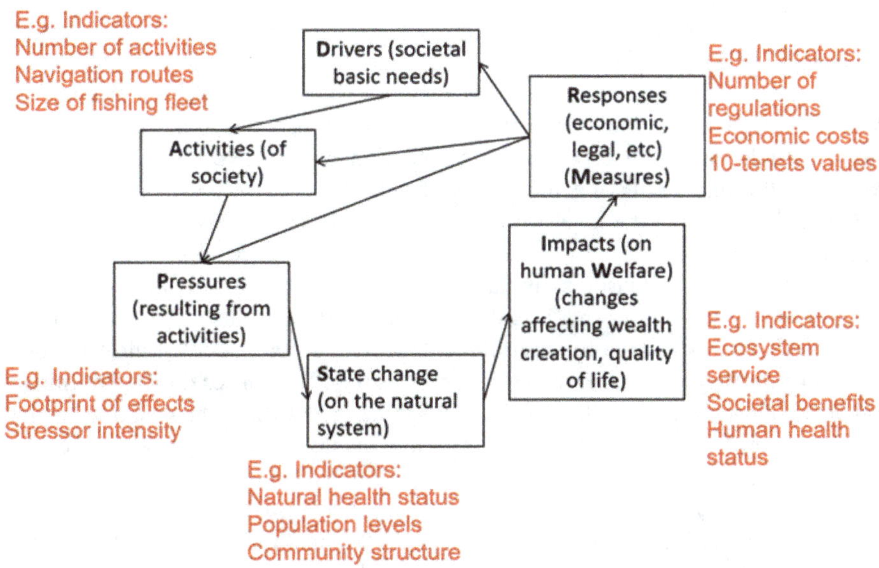

Fig. 5 Examples of indicators for each element of the DAPSI(W)R(M) framework

In addition, given the benefits that we extract from the sea and coastline, this part of the framework as the result of human Activities and Pressures reflects any adverse changes which affect Blue Growth and hence the Blue Economy as defined above. In addition, by adding complementary capital helps to identify the wider human and societal consequences, such as loss or gain in employment. In turn, such employment has a feedback loop to the Activities within the DAPSI(W)R(M) framework. As such, there is the need for operational indicators of the Impacts (on human Welfare) to show the impact of damaging Activities and Pressures assets valued by society (Turner and Schaafsma 2015). Some authors emphasise that changes in Welfare resulting from environmental state changes are the (environmental) costs which need to be traded against the benefits created by the Drivers in Cost-Benefit Analysis (Cooper 2013).

2.6 *Response (R) (as Management Measures)*

Marine management is dependent on a governance background in which governance relates to the marine policies, politics, administration and legislation (Boyes and Elliott 2014, 2015) as well as other management mechanisms such as economic instruments and technological developments. The management Responses which are required to overcome adverse effects on the natural and social systems (State Change and Impacts on human Welfare respectively) as the result of the Drivers, Activities and Pressures, then should include management Measures. The latter term is used as it appears widely in European Union Directives such as the MSFD and WFD (Borja et al. 2010).

From a regulatory perspective, Responses may be directed at any other element in the DAPSI(W)R(M) cycle but usually we measure the State change and Impacts (on human Welfare) but we use management. Measures to control the Drivers, Activities and Impacts (on human Welfare). A direct response to increasing Drivers might include curtailing economic or population growth. Responses may act on Activities through regulating the levels of activity (e.g. a ban on fishing) or on Pressures by reducing the levels of pressure resulting from a given level of activity (e.g. by modifying the type of fishing gear or employing mitigation and/or compensation). Restoration measures (replanting of saltmarsh, or stocking of fish) are ecoengineering Responses that acts directly on the State of the environment (Elliott et al. 2016). Compensation for environmental damage (for example following an oil spill) is of three types—to compensate the users (e.g. Responses directed at Impacts (on Welfare) of fishermen), the habitat (by habitat creation), or the resource (such as restocking and replanting)—the last two are management responses directed at rectifying State change.

There are many elements involved in defining successful and sustainable adaptive responses. Barnard and Elliott (2015) suggest 10-tenets for successful Measures—that our actions should be *ecologically sustainable, economically viable, technologically feasible, socially desirable/tolerable, administratively achievable, legally permissible, politically expedient, ethically defensible (morally correct), culturally inclusive* and *effectively communicated*. While directives and regulations may mandate specific measures and responses to a particular problem, the levels of Response to an environmental problem also depend on social and political perceptions of that problem. For example, increasing awareness of plastics in the marine environment has led to widespread public support, just as the protection of iconic species, the so-called charismatic megafauna, may resonate more easily with the public. However, detecting the problem is only the start of devising management measures (Borja and Elliott 2019).

All of the above relates to the natural and human-derived hazards in the environment and the way in which these become risks when they affect something which we value (Elliott et al. 2014), hence the DAPSI(W)R(M) framework becomes an integral part of a risk assessment and risk management framework (Cormier et al. 2019). However, it has to include adaptive management as in many cases the societal and management response to a particular event is unpredictable. For example the tsunami resulting from the exogenous unmanaged pressure of the 2011 earthquake off Japan and the resulting Fukishima nuclear disaster, resulted in sudden change in attitudes toward nuclear energy in Germany ultimately resulting in the sudden change of German nuclear energy policy (Strunz 2014).

As mentioned above, the elements in DPASI(W)R(M) framework are integral to the management of the seas. It is axiomatic that management cannot be achieved without measurement and that quantitative indicators are needed to determine the amount of each element and to determine whether the management has had the desired effect. Although Teixeira et al. (2016) shows the plethora of marine indicators in existence, Fig. 5 shows the types of indicators adopted for each of the elements.

Learning lessons from the evolution of DAPSI(W)R(M)—the evolving conceptual basis for EBM.

The description of the DAPSI(W)R(M) above illustrates how the simple DPSIR conceptual frame can be expanded, developed and applied to the MSFD. The evolution of DPSIR illustrates how information and concepts from different disciplines have informed the overall approach to analysis, as well as identifying and providing solutions to the problems of disciplinary silos in multidisciplinary research. The DPSIR and its successors (mDPSIR, DPSWR, and now DAPSI(W)R(M)) have for many European scientists and environmental managers been the basis of attempts to integrate our understanding of the ecological and social systems to develop an Ecosystem-Based Approach as mandated by the Directives. This has been a process of "learning by doing" (i.e. adaptive management *per se*) involving iterative improvement and refinement of the conceptual framework such that it now meets the regulatory needs of the Directives, and hence is embedded within the implementation process (European Commission 2017). Following three decades of interdisciplinary research, the framework has now been tailored to meet the requirements of the MSFD, integrating social and ecological information, linking cause-consequence-management, and providing an overarching frame for application and a standardisation of the Directive across European Member States.

The origins of DPSIR as a social environmental accounting framework and its application to a centralised European marine management policy have helped to elucidate, and overcome many of the conceptual difficulties with the practice of socio-ecological system research in the context of Europe-wide marine management. As a Framework Directive, the MSFD seeks to develop an Ecosystem Approach to management while also bringing together existing, more sectoral European marine environmental legislation including the Habitats & Species Directive (European Economic Community 1992) the Water Framework Directive (European Commission 2000) the legislation under the Common Fisheries policy amongst others. Hence while the implementation of the MSFD has developed rapidly since introduced in 2008 (e.g. Boyes et al. 2016), evidence for real improvements to the quality of European marine ecosystems is limited. There is increasing recognition that the sectoral policies for the environment and natural resource management within Europe are failing when it comes to the delivery of biodiversity objectives (Pe'er et al. 2019; O'Higgins 2017; Rouillard et al. 2018) and hence the requirement to develop more holistic approaches to European EBM (Lago et al. 2019). The question then arises, if one were to develop a conceptual framework for EBM *a priori*, what lessons can be learned from the DAPSI(W)R(M).

3 Standing on the Shoulders of Giants

Integration of social and ecological information relevant to stakeholders and managers is an essential component of any efforts which aim to remediate environmental impacts while reaching multiple policy goals. Such goals include those defined under EU Directives and strategies such as the WFD, the MSFD and the EU Biodiversity Strategy to 2020, and globally in the UNCED Sustainable Development Goals and various strategies are incorporated into different conceptual models to support EBM (Ogden 2005; Kelble et al. 2013). The first conceptual models, had a more environmental focus following the OECD pressure-state-impact (PSI). The Response dimension was added subsequently to incorporate policy responses, as in the PSR model (Gentile et al. 2001). In these first linear models, the entire social system was included in the 'Pressure' dimension and the ecological system under 'State'. These models gave little insight into the social processes that result in multiple pressures nor did they describe the entire management cycle. Furthermore, their conceptualisation of ecological systems was limited to the specific structural parameters described by the State term, rather than considering a comprehensive analysis of ecological processes and functions. The evolution of these model into DPSIR and its successors through to the DAPSI(W)R(M) resulted in a number of improvements.

Advances resulted from adding the basic human needs (anthropogenic 'Drivers') as the ultimate causes of ecosystem use and ecosystem change, incorporate a better understanding of the *raison d'etre* and functioning of the social system. Adding the

'Impact on (human Welfare)' dimension helped to provide deeper understanding of the consequences of human pressures on ecosystems (Sekowski et al. 2012). Finally, by incorporating the Response (using management Measures) these models have become important tools in the assessment of terrestrial, freshwater and marine ecosystems (e.g. Atkins et al. 2011; Tscherning et al. 2012; Kelble et al. 2013; Scharin et al. 2016).

This family of conceptual models has supported particular progress in the understanding and mainstreaming of impact pathways through which human activities affect the natural environment, both positively and negatively. However, many previous applications of the DPSIR have focussed on the single, pressures in a particular ecosystem. They are in danger of neglecting the simultaneously effects of multiple interacting pressures (Judd et al. 2015) and only rarely address the complexity associated with the assessment of multiple nested DPSIR causal chains running simultaneously (e.g. Atkins et al. 2011; Culhane et al. 2020). Burdon et al. (2018) shows the multiple links within these chains and the level of detail of the activities and pressures required by managers, in this case for oil and gas decommissioning. Hence there is the need for cumulative effects assessments which can accommodate the multiple activities, pressures, state changes and impacts on human welfare (Lonsdale et al. 2020).

All human activities occur within and are (directly or indirectly) entirely dependent on ecosystems (Boumans et al. 2002). The DPSIR models include information on the importance of nature for human welfare, i.e. integrating the linkage between ecosystem functions and ecosystem services and societal goods and benefits. Ecosystem services and societal goods and benefits are implicit in Maslow's (1943) hierarchy of needs. Within DAPSI(W)R(M), the link between ecosystem state change and human welfare is explicitly and negatively represented (generally by the costs, as economic externalities, in the cost-benefit analysis) (Fig. 3). While the importance of ecosystem services in assessing the human impacts of state changes has been recognised during the evolution of these models (see Fig. 4), it is not fully integrated within the model itself. Similarly, many of the indicators mentioned in Fig. 5 can be given monetary or other means of valuation.

As an extension of this, the 'Bow-tie' risk assessment and risk-management framework (Cormier et al. 2019) (Fig. 6) can then be merged with the DAPSI(W) R(M) framework. This shows that the central environmental concern (the red circle, a State change and/or Impact (on human Welfare)) has causes (the left-hand blue rectangles—the Drivers, Activities and Pressures) which in turn lead to consequences (the right-hand red boxes, also State changes and/or Impacts (on human Welfare)). The prevention, mitigation, compensation and adaptation controls then represent the Responses (using management Measures) (inserted between the causes and the consequences). In this sense, the model implicitly represents the centrality of the ecosystem in the production of human welfare and hence this aspect needs to be expanded to fully reflect EBM.

As a result, approaches based on the DPSIR need to incorporate feedback loops and cumulative forward and backward processes, hence favouring responses that are reactive and remedial rather than proactive and pre-emptive. Because of this, they

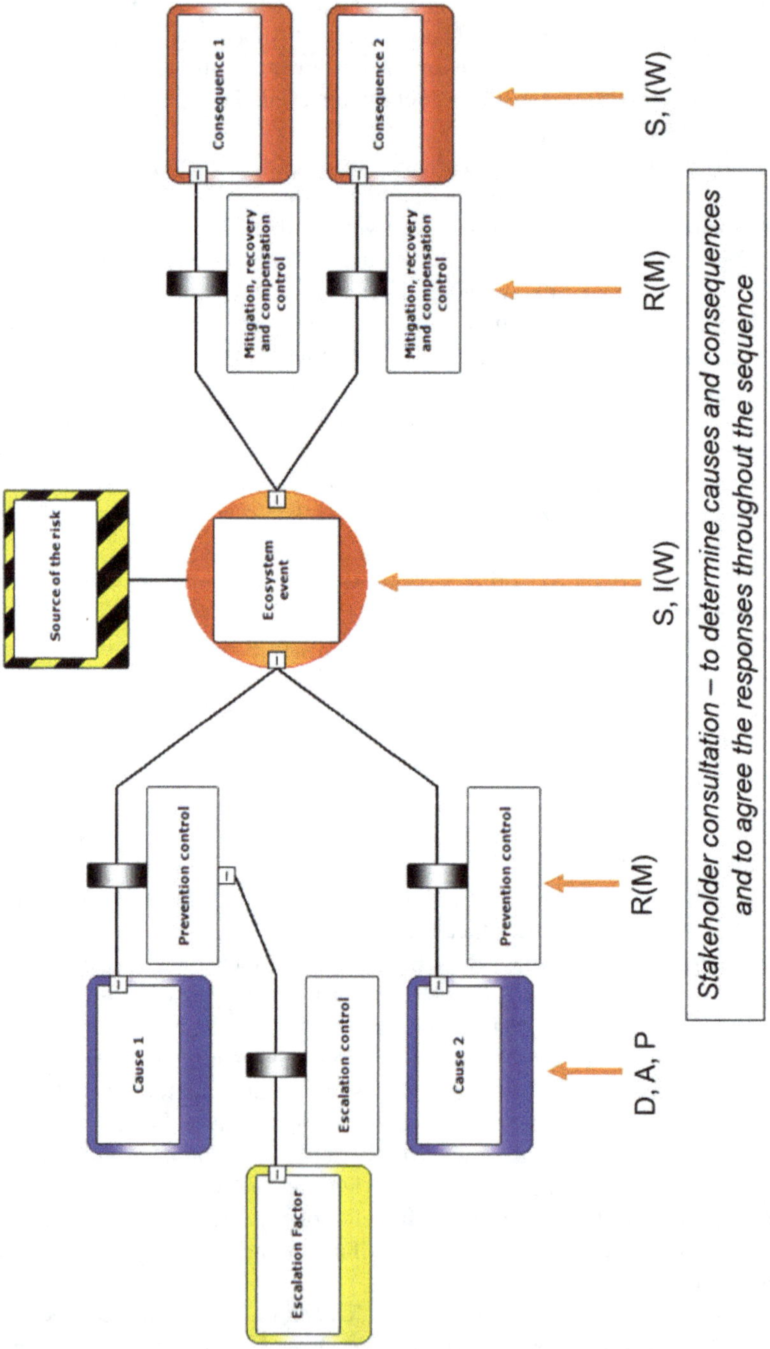

Fig. 6 A Bow-tie diagram (from Cormier et al. 2019, see text for explanation)

may be better suited to assess responses that reduce or modify pressures, regardless of how the socio-economic system and stakeholders adapt their decisions and behaviour and the drivers themselves of ecosystem change. Despite their adaptation to incorporate more fully social-ecological interactions, the causal chains of DAPSI (W)R(M) still reflect their origin in the field of environmental risk assessment. A geographic area and its management will thus need overlapping and interlinked DAPSI(W)R(M) cycles which also may need to be linked to similar cycles outside the immediate management area (see Elliott et al. 2017).

With the advent of the Ecosystem Approach in the 1990s, starting from the global Convention for Biological Diversity (see Enright and Boetler 2020 for a history of the term), the analysis of ecosystem services and their importance for human welfare and societal goods and benefits (Constanza et al. 1997; MEA 2003) shifted the perspective from "*what have we done to nature?*" towards "*what does nature do for us?*". This recognises the centrality of nature and ecosystems services in human well-being, a defining feature of EBM (Tallis et al. 2010). The ecosystem services approach led to a more comprehensive framework including economic perspectives and it called for more effective social action. This has led to new perspectives based upon the potential of ecosystems to provide society with the valuable goods and services they demand and to new conceptual frameworks to integrate these new concepts based on the previous DPSIR (Turner 2000; Cheong 2008; Weinstein 2009) leading to the DAPSI(W)R(M) described above. Ecosystem services and the resulting societal goods and benefits provided the missing analytical block to proceed from the biophysical to the human dimensions of science. Ecosystem services and the resulting societal goods and benefits are the main and most welfare-relevant outcomes from the interaction of social and ecological systems. Therefore, linking ecological and social systems to human welfare (and the goods and benefits it demands) through the notion of ecosystem services is essential to understand and assess the multiple trade-offs involved in individual and collective decisions in a clear and consistent manner and to the development of an EBM conceptual framework.

Ecosystem services and societal goods and benefits are the key emerging outcomes of the interaction between ecological and socio-economic systems (Biggs et al. 2012). They are 'produced' and delivered by ecosystems but are also continually shaped by their interaction with socio-economic systems and, in the case of societal goods and benefits, require an input of human capital in order to be realised. They may also favour detrimental, transformative or restorative processes (Biggs et al. 2015). Human actions and institutions shape ecosystem structures, processes and services in landscapes or seascapes by management and uses/users, which in turn shape human behaviour and institutional settings.

The integration of both traditions, impact pathway analysis and the ecosystem services/societal goods and benefits approach, has fostered the emergence of a growing number of alternative socio-ecological system analytical frameworks (Binder et al. 2013). Nevertheless, their success and their capacity for the smooth integration of knowledge may have been impaired by the mismatch resulting from mixing pieces created from slightly different conceptual directions and for different purposes. To some extent, both approaches share the drawbacks of common practice and may only offer a partial view of the complex links between the relevant and important parts of social and ecological systems.

In attempting to define a conceptual framework specifically for the analysis and design of EBM, there is a clear need to reflect the central nature of ecosystem services, the resulting delivery of societal goods and benefits and the costs and benefits of enhancing natural assets or ecosystems to improve resilience and adaptability. Hence there is the need conceptually to represent how ecosystems function in connection with the socio-economic system, delivering ecosystem services and contributing to social goods, benefits and welfare.

4 The Butterfly

Taking the valuable lessons learned through the application of the DPSIR/DAPSI(W)R(M) framework, we seek to develop a transferable framework with the aim of developing, a-priori, a more holistic methodological approach to implementing Ecosystem-Based Management. This is illustrated in the 'Butterfly' diagram (Fig. 7) (see Gómez et al. 2016 for the full explanation).

Two sets of relationships between humans and nature are implicit in the cyclic DAPSI(W)R(M). Drivers are dependent on the supply of ecosystem services and societal goods and benefits (see Maslow's hierarchy of Basic Human Needs above). In turn, these Drivers (through human Activities) cause Pressures and changes in environmental State altering the supply of ecosystems services. In turn, those State changes influence the Impacts (on human Welfare). These links can readily be conceptualised as supply (from nature) and demand (by humans) for ecosystem services and societal goods and benefits.

The supply side (Fig. 7, shown in blue) involves the links between ecosystems and human Welfare. In DAPSI(W)R(M), this is conceptualised as the (cost-benefit) feedback between Impacts on human Welfare and Drivers. The second set of relationships, 'the demand side' (Fig. 7, shown in yellow), refers to how social systems shape and change ecosystems (the links between Drivers and Activities to Pressures and then State changes). These are connected to each other through complex adaptive processes taking place in ecological and social systems.

The supply-side relationship goes from the ecological to the social system. It represents the potential of ecosystems to supply and effectively deliver ecosystem services to the social system, from which it gets goods and benefits. It includes the capacity of the social system to transform those services delivered into benefits for society through human built, financial and social capital (Fig. 3). This is all contingent on the ecosystem structure and on those processes/functioning taking place in the biophysical system from which ecosystem services lead to the most socially relevant outcomes.

The demand-side relationship goes from the social to the ecological system. It represents and explains the demand and the effective use of ecosystem services and

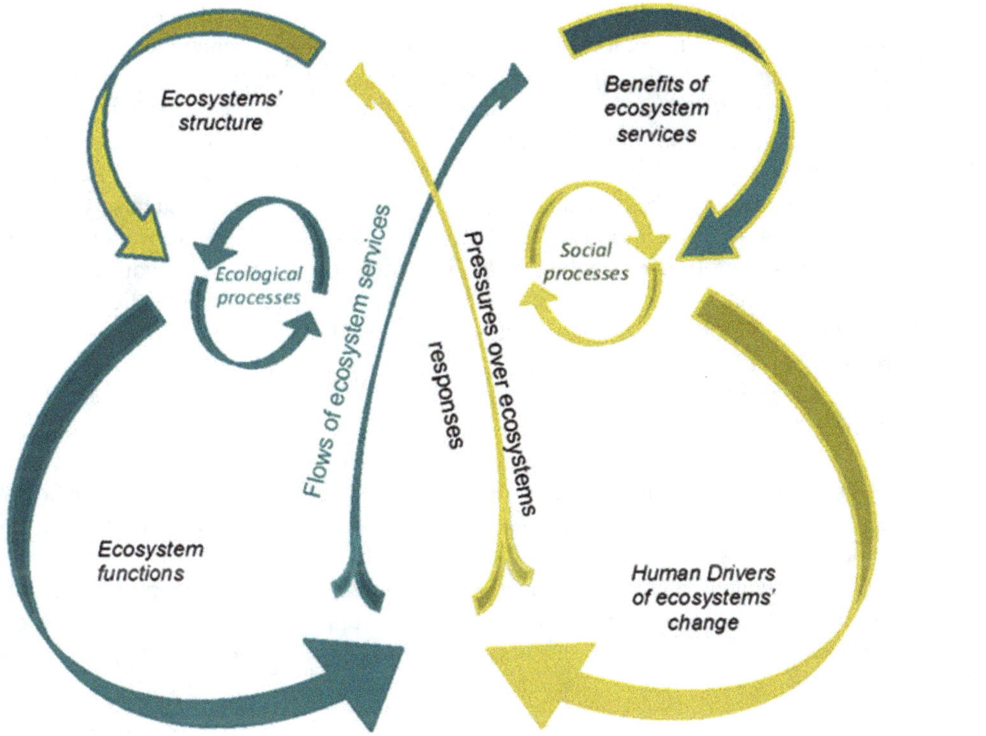

Fig. 7 The butterfly diagram, a new model for the assessment and design of Ecosystem-Based Management. Supply side is shown in blue, demand side is shown in yellow

the impacts on ecosystems. The demand for ecosystem services (and in turn goods and benefits) depends on income, tastes, technology, institutions, and other social and economic factors. Beyond pressures on ecosystems, this demand-side relationship also considers social and individual decisions towards protecting and restoring ecosystems in order to preserve their benefits depending on the governance institutions in place. Hence the need for the ten-tenets of sustainable ecosystem management which can incorporate all the facets of that management, both the natural and the societal. These more fully represent the variety of social processes by which society adapts to manage specific environmental and social conditions (Ostrom 1990) including new types of social innovation (Schor and Thomson 2014), going beyond Responses (and management Measures). This may be best suited to governance or economic instruments, best available technologies, cultural demands, etc. as summarised by the 10-tenets (Barnard and Elliott 2015).

5 Conclusions

Here we present the DAPSI(W)R(M) framework, the latest iteration of the well-known and widely-applied DPSIR framework, as an environmental risk management tool. We draw on advances over the past 3 decades in the analysis of socio-ecological systems, and now include elements such as ecosystem services and societal goods and benefits as well as more a holistic conception of Drivers. However, we emphasise that the causal chain mechanism (of cause-consequence-management) needs to be ideally suited to the cyclical application of adaptive management required under the framework of European environmental directives. The derived system has to fully incorporate the wealth of social and ecological process that result in the dynamics of supply and demand for ecosystem services and the resulting societally goods and benefits, the most socially relevant outputs from ecosystems. The Butterfly enhances the inherited conceptual framework of DAPSI (W)R(M). The Butterfly conceptual framework has been systematically tested through a suite of case studies in aquatic biodiversity management from freshwater rivers (Domisch et al. 2019) and lakes to estuarine and marine systems, some examples of which are presented in this volume (Piet et al. 2020, O'Higgins et al. 2020; Lillebø et al. 2020; Funk et al. 2020).

The ability to meet the predominant aim of satisfying Ecosystem-Based Management, covering both natural and social systems, requires the analysis of socio-ecological systems enabled by the cause-consequence-response chain, and the relationship to European (and other) environmental policies. The widely acknowledged biodiversity and climate crises requires holistic management approaches such as EBM which can then be translated into the supply and demand for ecosystem services and societal goods and benefits at the centre of the Butterfly conceptual framework.

References

Al Kalbani, M. S., Price, M. F., O'Higgins, T., Ahmed, M., & Abahussain, A. (2015). Integrated environmental assessment to explores water resources management in Al Jabla Al Akhdar, Sultanate of Oman. *Regional Environmental Change, 5*, 1345–1361.

Atkins, J. P., Burdon, D., Elliott, M., & Gregory, A. J. (2011). Management of the marine environment: Integrating ecosystem services and societal benefits with the DPSIR framework in a systems approach. *Marine Pollution Bulletin, 62*, 215–226.

Atkins, J. P., Burdon, D., & Elliott, M. (2015). Identification of a practicable set of ecosystem indicators for coastal and marine ecosystem services. In R. K. Turner & M. Schaafsma (Eds.), *Coastal zones ecosystem services: From science to values and decision making*. Cham: Springer.

Barnard, S., & Elliott, M. (2015). The 10-tenets of adaptive management and sustainability – Applying an holistic framework for understanding and managing the socio-ecological system. *Environmental Science & Policy, 51*, 181–191.

Biggs, R., Schlüter, M., Biggs, D., Bohensky, E. L., BurnSilver, S., Cundill, G., Dakos, V., et al. (2012). Toward principles for enhancing the resilience of ecosystem services. *Annual Review of Environment and Resources, 37*(1), 421–448.

Biggs, R., Schlüter, M., & Schoon, M. L. (2015). An introduction to resilience approach and principles to sustain ecosystem services. In *Principles for building resilience. Sustaining ecosystem services in social-ecological systems* (pp. 1–31). Cambrige University Press.

Binder, C. R., Hinkel, J., Bots, P. W. G., & Pahl-Wostl, C. (2013). Comparison of frameworks for analyzing social-ecological systems. *Ecology and Society, 18*(4), 26.

Borja, A., Galparsoro, I., Solaun, O., Muxika, I., Tello, E. M., Uriarte, A., & Valencia, V. (2006). The European Water Framework Directive and the DPSIR, a methodological approach to assess the risk of failing to achieve good ecological status. *Estuarine, Coastal and Shelf Science, 66*, 84–96.

Borja, Á., Elliott, M., Carstensen, J., Heiskanen, A.-S., & van de Bund, W. (2010). Marine management – Towards an integrated implementation of the European Marine strategy framework and the water framework directives. *Marine Pollution Bulletin, 60*, 2175–2186.

Borja, A., Elliott, M., Snelgraove, P. V. R., Austen, M. C. V., Berg, T., Cochrane, S., Cartensen, J., Danovaro, R., Greenstreet, S., Heiskannen, A.-S., Lynam, C. P., Mea, M., Newton, A., Particio, J., Uusitalo, L., Uyarra, M. C., & Wilson, C. (2016). Bridgin the gap between policy and science in assessing the health status of marine ecosystems. *Froniers in Marine Science*. https://doi.org/10.3389/fmars.2016.00175.

Borja, A., & Elliott, M. (2019). Editorial – So when will we have enough papers on microplastics and ocean litter? *Marine Pollution Bulletin, 146*, 312–316.

Boumans, R., Costanza, R., Farley, J., Wilson, M. A., Portela, R., Rotmans, J., Villa, F., & Grasso, M. (2002). Modeling the dynamics of the integrated earth system and the value of global ecosystem services using the GUMBO model. *Ecological Economics, 41*, 529–560.

Boyes, S. J., & Elliott, M. (2015). The excessive complexity of national marine governance systems – Has this decreased in England since the introduction of the Marine and Coastal Access Act 2009? *Marine Policy, 51*, 57–65. https://doi.org/10.1016/j.marpol.2014.07.019.

Boyes, S. J., & Elliott, M. (2014). Marine legislation – The ultimate 'horrendogram': International Law, European Directives & National Implementation. *Marine Pollution Bulletin, 86*(1–2), 39–47. https://doi.org/10.1016/j.marpolbul.2014.06.055.

Boyes, S. J., Elliott, M., Murillas-Maza, A., Papadopoulou, N., & Uyarra, M. C. (2016). Is existing legislation fit-for-purpose to achieve good environmental status in European seas? *Marine Pollution Bulletin, 111*, 18–32. https://doi.org/10.1016/j.marpolbul.2016.06.079.

Burdon, D., Barnard, S., Boyes, Elliott, M (2018). Oil and gas infrastructure decommissioning in marine protected areas: System complexity, analysis and challenges. Marine Pollution Bulletin, 135: 739–758.

Cheong, S.-M. (2008). A new direction in coastal management. *Marine Policy, 32*, 1090–1093.

Constanza, R., d'Arge, R., deGroot, R., Farber, S., Grasso, M., Hannon, B., Limburg, K., Naeem, S., O'Neill, R. V., Paruelo, J., Raskin, G. R., Sutton, P., & Van den Belt, M. (1997). The value of the world's ecosystem services and natural capital. *Nature, 387*, 253–260.

Cooper, P. (2013). Socio-ecological accounting: DPSWR, a modified DPSIR framework, and its application to marine ecosystems. *Ecological Economics, 94*, 106–115. https://doi.org/10.1016/j.ecolecon.2013.07.010.

Cormier, R., Elliott, M., & Rice, J. (2019). Putting on a bow-tie to sort out who does what and why in the complex arena of marine policy and management. *Science of the Total Environment, 648*, 293–305. https://doi.org/10.1016/j.scitotenv.2018.08.168.

Culhane, F. E., Robinson, L. A., & Lillebø, A. I. (2020). Approaches for estimating the supply of ecosystem services: concepts for ecosystem-based management in coastal and marine environments. In T. O'Higgins, M. Lago, & T. H. DeWitt (Eds.), *Ecosystem-based management, ecosystem services and aquatic biodiversity: Theory, tools and applications* (pp. 105–126). Amsterdam: Springer.

de Jonge, V. N., Kolkman, M. J., Ruijgrok, E. C. M., & de Vries, M. B. (2003). *The need for new paradigms in integrated socio-economic and ecological coastal policy making*. Proceedings of 10th International Wadden Sea Symposium, 247–270, Ministry of Agriculture, Nature Management and Fisheries, Department North, Groningen (272 pp.)

de Jonge, V. N., Pinto, R., & Turner, R. K. (2012). Integrating ecological, economic and social aspects to generate useful management information under the EU Directives' 'ecosystem approach'. *Ocean and Coastal Managagement, 68*(2012), 169–188.

Delacámara, G., O'Higgins, T., Lago, M., & Langhans, S. (2020). Ecosystem-based management: Moving from concept to practice. In T. O'Higgins, M. Lago, & T. H. DeWitt (Eds.), *Ecosystem-based management, ecosystem services and aquatic biodiversity: Theory, tools and applications* (pp. 39–60). Amsterdam: Springer.

Domisch, S., Kakouei, K., Martínez-López, J., Bagstad, K. J., Magrach, A., Balbi, S., Villa, F., et al. (2019). Social equity shapes zone-selection: Balancing aquatic biodiversity conservation and ecosystem services delivery in the transboundary Danube River Basin. *Science of the Total Environment, 656*, 797–807.

EEA. (1999). *Environmental indicators: Typology and overview*. Technical report No. 25, Copenhagen: European Environment Agency.

Elliott, M. (2011). Marine science and management means tackling exogenic unmanaged pressures and endogenic managed pressures – A numbered guide. *Marine Pollution Bulletin, 62*, 651–655.

Elliott, M., Burdon, D., Atkins, J. P., Borja, A., Cormier, R., de Jonge, V. N., & Turner, R. K. (2017). "And DPSIR begat DAPSI(W)R(M)!" – A unifying framework for marine environmental management. *Marine Pollution Bulletin, 118*(1–2), 27–40. https://doi.org/10.1016/j.marpolbul.2017.03.049.

Elliott, M., Cutts, N. D., & Trono, A. (2014). A typology of marine and estuarine hazards and risks as vectors of change: A review for vulnerable coasts and their management. *Ocean & Coastal Management, 93*, 88–99.

Elliott, M., Mander, L., Mazik, K., Simenstad, C., Valesini, F., Whitfield, A., & Wolanski, E. (2016). Ecoengineering with Ecohydrology: Successes and failures in estuarine restoration. *Estuarine, Coastal and Shelf Science, 176*, 12–35. https://doi.org/10.1016/j.ecss.2016.04.003.

Elliott, M., Boyes, S. J., Barnard, S., & Borja, Á. (2018). Using best expert judgement to harmonise marine environmental status assessment and maritime spatial planning. *Marine Pollution Bulletin, 133*, 367–377.

Enright, S. R., & Boetler, B. (2020). The ecosystem approach in international law. In T. O'Higgins, M. Lago and & T.H. DeWitt (Eds.), *Ecosystem-based management, ecosystem services and aquatic biodiversity: Theory, tools and applications* (pp. 333–352). Amsterdam: Springer.

European Commission. (2000). Directive 2000/60/EC of the European Parliament and of the Council of 23 October 2000 establishing a framework for Community actions in the field of water policy. Official Journal of the European Communities L327, 1.22.12.2000.

European Commission. (2008). Directive 2008/56/EC of the European Parliament and of the Council of 17 June 2008 establishing a framework for Community actions in the field of marine environmental policy (Marine Strategy Framework Directive). Official Journal of the European Communities L164/19 25.06.2008.

European Commission. (2012). Communication from the commission to the European parliament, the council, the European economic and social committee and the committee of the regions. Blue Growth opportunities for marine and maritime sustainable growth. Brussels, 13.9.2012 COM(2012) 494 final.

European Commission. (2017). Commission Directive 2017/845 amending Directive 2008/56/EC of the European Parliament and of the Council as regards the indicative lists of elements to be taken into account for the preparation of marine strategies. L 125/27.

European Economic Community. (1992). Council Directive 92/43/EEC of 21 May 1992 on the conservation of natural habitats and of wild fauna and flora. Official Journal of the EEC. 1992; L206/7 1.

Fisher, B., Turner, R. K., & Morling, P. (2009). Defining and classifying ecosystem services for decision making. *Ecological Economics, 68*, 643–653. https://doi.org/10.1016/j.ecolecon.2008.09.014.

Funk, A., O'Higgins, T. G., Borgwardt, F., Trauner, D., & Hein, T. (2020). Ecosystem-based management to support conservation and restoration efforts in the Danube Basin. In T. O'Higgins, M. Lago, & T. H. DeWitt (Eds.), *Ecosystem-based management, ecosystem services and aquatic biodiversity: Theory, tools and applications* (pp. 431–444). Amsterdam: Springer.

Gentile, J. H., Harwell, M. A., Cropper, W., Harwell, C. C., DeAngelis, D., Davies, S., Ogen, J. C., & Lirman, D. (2001). Ecological conceptual models: A framework and case study on ecosystem management for South Florida sustainability. *Science of the Total Environment, 274*, 251–253.

Gómez, C. M., Delacámara, G., Arévalo-Torres, J., Barbière, J., Barbosa, A. L., Boteler, B, Culhane, F., et al. (2016). The AQUACROSS innovative concept. Deliverable 3.1, European Union's Horizon 2020 Framework Programme for Research and Innovation grant agreement No. 642317.

Halpern, B. S., Walbridge, S., Selkoe, K. A., Kappel, C. V., Micheli, F., D'Agrosa, C., Bruno, J. F., Casey, K. S., Ebert, C., Fox, H. E., Fujita, R., Heinemann, D., Lenihan, H. S., Madin, E. M.,

Perry, M. T., Selig, E. R., Spalding, M., Steneck, R., & Watson, R. (2008). A global map of human impact on marine ecosystems. *Science, 319*, 948–952.

Judd, A. D., Backhaus, T., & Goodsir, F. (2015). An effective set of principles for practical implementation of marine cumulative effects assessment. *Environmental Science and Policy, 54*, 254–262.

Karageorgis, A. P., Skourtos, M. S., Kapsimalis, V., Kontogianni, A. D., Skoulikidis, N. T., Pagou, K., et al. (2005). An integrated approach to watershed management within the DPSIR framework: Axios River catchment and Thermaikos Gulf. *Regional Environmental Change, 5*, 138–160. https://doi.org/10.1007/s10113-004-0078-7.

Kelble, C. R., Loomis, D. K., Lovelace, S., Nuttle, W. K., Ortner, P. B., Fletcher, P., et al. (2013). The EBM-DPSER conceptual model: Integrating ecosystem services into the DPSIR framework. *PLoS One, 8*, e70766. https://doi.org/10.1371/journal.pone.0070766.

Knudsen, S., Zengin, M., & Koçak, M. H. (2010). Identifying drivers for fishing pressure. A multidisciplinary study of trawl and sea snail fisheries in Samsun, Black Sea coast of Turkey. *Ocean and Coastal Management, 53*, 252–269. https://doi.org/10.1016/j.ocecoaman.2010.04.008.

Lago, M., Boteler, B., Rouillard, J., Abhold, K., Jähnig, S. C., Iglesias-Campos, A., et al. (2019). Introducing the H2020 AQUACROSS project: Knowledge, assessment, and management for AQUAtic Biodiversity and Ecosystem Services aCROSS EU policies. *Science of the Total Environment, 652*, 320–329.

Langhans, S. D., Domisch, S., Balbi, S., Delacámara, G., Hermoso, V., Kuemmerlen, M., Martin, R., Martinez-Lopez, J., Vermeiren, P., Villa, F., & Jähnig, S. C. (2019). Combining eight research areas to foster the uptake of ecosystem-based management in fresh waters. *Aquatic Conservation: Marine and Freshwater Ecosystems, 297*, 1161–1173.

Langmead, O., McQuatters-Gollop, A., Mee, L. D., Friedrich, J., Gilbert, A. J., Gomoiu, M.-T., Jackson, E. L., Knudsen, S., Minicheva, G., & Todorova, V. (2009). Recovery or decline of the northwestern Black Sea: A societal choice revealed by socio-ecological modelling. *Ecological Modelling, 220*, 2927–2939. https://doi.org/10.1016/j.ecolmodel.2008.09.011.

Lillebø, A. I., Teixeira, H., Martínez-López, J., Genua-Olmedo, A., Marhubi, A., Delacámara, G., Mattheiß, V., Strosser, P., O'Higgins, T., & Nogueira, A. A. J. (2020). Mitigating negative unintended impacts on biodiversity in the Natura 2000 Vouga estuary (Ria de Aveiro, Portugal). In T. O'Higgins, M. Lago, & T. H. DeWitt (Eds.), *Ecosystem-based management, ecosystem services and aquatic biodiversity: Theory, tools and applications* (pp. 461–498). Amsterdam: Springer.

Lonsdale, J. A., Nicholson, R., Judd, A., Elliott, M., & Clarke, C. (2020). A novel approach for cumulative impacts assessment for marine spatial planning. *Environmental Science & Policy, 106*, 125–135.

Lovecraft, A. L., & Meek, C. L. (2019). Arctic Coastal systems: Evaluating the DAPSI(W)R (M) framework. In E. Wolanski et al. (Eds.), *Coasts and estuaries: The future*. Elsevier. https://doi.org/10.1016/B978-0-12-814003-1.00039-3.

Lundberg, C. (2005). Conceptualizing the Baltic Sea ecosystem: An interdisciplinary tool for environment decision making. *Ambio, 34*, 433–439. https://doi.org/10.1579/0044-7447-34.6.433.

Maslow, A. H. (1943). A theory of human motivation. *Psychological Review, 50*(4), 370–396.

Maslow, A. H. (1970a). *Motivation and personality* (2nd ed.). New York: Harper and Row.

Maslow, A. H. (1970b). *Religions, values, and peak experiences*. New York: Penguin.

Maxim, L., Spangenberg, J. H., & O'Connor, M. (2009). An analysis of risks for biodiversity under the DPSIR framework. *Ecological Economics, 69*, 12–23.

McDonald, H., Hoffman, H., Ressurreição, A., Röschel, L., Gerdes, H., Lago, M., Boetler, B., & McFarland, K. (2020). Ecosystem-based management for more effective and equitable marine protected areas: A case study on the Faial-Pico channel marine protected area, Azores. In T. O'Higgins, M. Lago, & T. H. DeWitt (Eds.), *Ecosystem-based management, ecosystem services and aquatic biodiversity: Theory, tools and applications* (pp. 499–516). Amsterdam: Springer.

Millennium Ecosystem Assessment (MEA). (2003). *Ecosystems and human well-being: A framework for assessment*. Washington, DC: Island Press.
Mee, L., Cooper, P., Kannen, A., Gilbert, A. J., & O'Higgins, T. (2015). Sustaining Europe's seas as coupled social-ecological systems. *Ecology and Society, 20*(1), 1. https://doi.org/10.5751/ES-07143-200101.
Ogden, J. C. (2005). Everglades ridge and slough conceptual ecological model. *Wetlands, 25*, 810–820.
O'Hagan, A. M. (2020). Ecosystem-based management (EBM) and ecosystem services in EU law, policy and governance. In T. O'Higgins, M. Lago, & T. H. DeWitt (Eds.), *Ecosystem-based management and ecosystem services: Theory, tools and practice* (pp. 353–372). Amsterdam: Springer.
O'Higgins, T. G. (2017). You can't eat biodiversity: Agency and irrational norms in European aquatic environmental law. *Challenges in sustainability, 5*, 43–51.
O'Higgins, T. G., & Gilbert, A. J. (2014). Embedding ecosystem services into the Marine Strategy Framework Directive: Illustrated by eutrophication in the North Sea. *Estuarine Coastal and Shelf Science, 140*, 146–152.
O'Higgins, T. G., Cooper, P., Roth, E., Newton, A., Farmer, A., Goulding, I., & Tett, P. (2014). Temporal constraints on ecosystem management: Definitions and examples from Europe's regional seas. *Ecology and Society, 19*(3).
O'Higgins, T. G., Alexander, K. A., & Graziano, M. (2019). Mismatches in spatial scale of supply and demand and their consequences for local welfare in Scottish aquaculture. *Anthropocene Coasts, 2*, 261–278.
O'Higgins, T. G., Culhane, F., O'Dwyer, B., Robinson, L., & Lago, M. (2020). Combining methods to establish potential management measures for invasive species *Elodea nutallii* in Lough Erne Northern Ireland. In T. O'Higgins, M. Lago, & T. H. DeWitt (Eds.), *Ecosystem-based management, ecosystem services and aquatic biodiversity: Theory, tools and applications* (pp. 445–460). Amsterdam: Springer.
Ostrom, E. (1990). *Governing the commons: The evolution of institutions for collective action*. Cambridge: Cambridge University Press.
Patrício, J., Elliott, M., Mazik, K., Papadopoulou, K.-N., & Smith, C. J. (2016). DPSIR – Two decades of trying to develop a unifying framework for marine environmental management? *Frontiers in Marine Science, 3*, 177. https://doi.org/10.3389/fmars.2016.00177.
Pe'er, G., Zinngrebe, Y., Moreira, F., Sirami, C., Schindler, S., Müller, R., Bontzorlos, V., Clough, D., Bezák, P., Bonn, A., Hansjürgens, B., Lomba, A., Möckel, S., Passoni, G., Schleyer, C., Schmidt, J., & Lakner, S. (2019). A greener path for the EU common agricultural policy. *Science, 365*(6452), 449–451. https://doi.org/10.1126/science.aax3146.
Piet, G., Delacamara, G., Kraan, M., Röckmann, G. C., & Lago, M. (2020). Advancing aquatic ecosystem-based management with full consideration of the social-ecological system. In T. O'Higgins, M. Lago, & T. H. DeWitt (Eds.), *Ecosystem-based management, ecosystem services and aquatic biodiversity: Theory, tools and applications* (pp. 17–38). Amsterdam: Springer.
Pinto, R., da Conceição Cunha, M., Roseata-Palma, C., & Marques, J. C. (2014). Mainstreaming sustainable decision making for ecosystem: Integrating ecological and social-economic targets with a decision support system. *Environmental Processes, 1*, 7–19.
Rapport, D., & Friend, A. (1979). *Towards a comprehensive framework for environmental statistics: A stress-response approach*. Statistics Canada Catalogue 11–510. Ottawa: Minister of Supply and Services Canada.
Rockstrom, J., Steffen, W., Noone, K., Persson, A., Chapin, F. S., Lambin, E. F., et al. (2009). A safe operating space for humanity. *Nature, 461*, 472–475.
Rouillard, J., Lago, M., Abhold, K., Roschel, L., Kafaeyke, T., Mattheiss, V., & Klimmek, H. (2018). Protecting aquaticbiodiversity in Europe: How much do EU environmental policies support ecosystem-based management? *Ambio, 47*, 15–24. https://doi.org/10.1007/s13280-017-0928-4.
Scharin, H., Ericsdotter, S., Elliott, M., Turner, R. K., Niiranen, S., Rockström, J., Blenckner, T., Hyytiäinen, K., Ahlvik, L., Heini Ahtiainen, H., Artell, J., Hasselström, L., & Söderqvist,

T. (2016). Processes for the sustainable stewardship of marine environments. *Ecological Economics., 128*, 55–67.

Schor, J. B., & Thomson, C. J. (2014). *Sustainable lifestyles and the quest for plentitude* (p. 280). Connetticut: Yale University Press.

Sekowski, I., Newton, A., & Dennison, W. C. (2012). Megacities in the coastal zone: Using a driver-pressure -state-impact-response framework to address complex environmental problems. *Estuarine Coastal and Shelf Science, 96*, 48–59.

Skoulikidis, N. T. (2009). The environmental state of rivers in the Balkans a review within the DPSIR framework. *Science of the Total Environment, 407*, 2501–2516.

Stojanovic, T. A., & Farmer, C. J. Q. (2013). The development of world oceans and coasts and concepts of sustainability. *Marine Policy, 42*, 157–165.

Strong, J. A., Andonegi, E., Bizsel, K. C., Danovaro, R., Elliott, M., Franco, A., Garces, E., Little, S., Mazik, K., Moncheva, S., Papadopoulou, N., Patrício, J., Queirós, A. M., Smith, C., Stefanova, K., & Solaun, O. (2015). Marine biodiversity and ecosystem function relationships: The potential for practical monitoring applications. *Estuarine Coastal Shelf Science, 161*, 46–64.

Strunz, S. (2014). The German energy transition as a regime shift. *Ecological Economics, 100*, 150–158.

Tallis, H., Levin, P. S., Ruckelhaus, M., Lester, S. E., McLeod, K. L., Fluharty, D. L., & Halpern, B. S. (2010). The many faces of ecosystem-based management: Making the process work today in real places. *Marine Policy, 34*(2), 340–348. https://doi.org/10.1016/j.marpol.2009.08.003.

Teixeira, H., Berg, T., Uusuatalo, L., Furhapter, K., Hesikanen, A.-S., Mazik, K., Lynam, C. P., Neville, S., Rodriguez, J. R., Papadopaulou, N., Moncheva, S., Churilova, T., & m Kryvenko, O., Krause-Jensen, D., ZZaiko, A., Verissimo, H., Pantazi, M., Cravalho, S., Particio, J., Uyarra, M.C., and Borja, A. (2016). A catalogue of marine biodiversity indicators. *Frontiers in Marine Science*. https://doi.org/10.3389/fmars.2016.00207.

Tscherning, K., Helming, K., Krippner, B., Sieber, S., & Gomez y Paloma, S. (2012). Does research applying the DPSIR framework support decision making? *Land Use Policy, 29*, 102–110. https://doi.org/10.1016/j.landusepol.2011.05.009.

Turner, R. K. (2000). Integrating natural and socio-economic science in coastal management. *Journal of Marine Systems, 25*, 447–460.

Turner, R. K., & Schaafsma, M. (Eds.). (2015). Coastal zones ecosystem services: From science to values and decision making. In *Springer ecological economic series*. London: Springer.

UNSD. (2012). System of environmental-economic accounting, Central Framework. New York: United Nations Statistical Division.

Weinstein, M. P. (2009). The road ahead: The sustainability transition and coastal research. *Estuaries and Coasts, 32*, 1044–1053.

7

Mathematical Modeling for Ecosystem-Based Management (EBM) and Ecosystem Goods and Services (EGS) Assessment

Richard S. Fulford, Sheila J. J. Heymans, and Wei Wu

"All models are wrong, but some models are useful"
George E.P. Box

Abstract There is a rich history in predicting ecological interactions in nature going back to the seminal work of Robert May. Historically, the use of models in ecological decision making has been centered on tools that require less data and are easier to communicate into policy. However, the explosion in the availability of ecological data, as well as the ready access to computer power, has opened the door to more detailed computation tools. This in turn has created a suite of questions about how and how much model-based projections can influence decision making in management of ecological resources. Here we address some of these issues particularly those created by model complexity, data quality and availability, and model acceptance by policy makers. Our goal is to use specific examples to critically examine these issues for both gaps and opportunities in how models can be used to inform decision making at the level of ecosystems and in the currency of ecosystem services to people.

R. S. Fulford (✉)
US Environmental Protection Agency, Gulf Ecosystem Measurement and Modeling Division, Gulf Breeze, FL, USA
e-mail: Fulford.Richard@epa.gov

S. J. J. Heymans
European Marine Board, Oostende, Belgium

W. Wu
University of Southern Mississippi, Hattiesburg, MS, USA

Lessons Learned

- Model complexity should fit the focal management question and the available data.
- The most useful ecosystem-based models are transparent with output that is easily translated into the language of policy. These endpoints are best accomplished with stakeholder engagement in model development.
- Multiple-model approaches, such as ensemble modeling, are better and more transferable than a single modeling tool.
- Model utility for Ecosystem-Based Management (EBM) decision making should be demonstrated through short-term forecasting efforts to improve the acceptance of models as a decision tool.

Needs to Advance

- Greater investment in ecological forecasting based on models as a testbed for model performance and improvement.
- Standardization of process for communicating model output in the language of policy including effective communication of model uncertainty as risk.
- Focus on improvement of existing models rather than new model development. This should be done in cooperation with policy makers early to improve model utility.

1 Introduction

Mathematical modeling has a rich history in ecology, but its foundations come from the physical sciences where well-characterized relationships such as reaction-diffusion are repeatable and well-validated (Getz 1998). The use of these relationships to address ecological questions has sprouted from multiple disciplines including population ecology (May 1974b), food web ecology (Whipple 1999), and nutrient dynamics (Clark and Gelfand 2006). What all these pathways have in common is the desire to use observable relationships to either predict or describe important but complex ecological outcomes. May (1974a) championed this concept in his seminal work using models to understand patterns in population dynamics. The work of May and others fostered debate on the utility of models vs. observation and fostered a modeling discipline in ecology that has grown steadily into a functional tool for management (Griffith and Fulton 2014; Nielsen et al. 2018; Piroddi et al. 2017).

Ultimately, we need models that are more than the sum of their parts and produce novel observations or theory useful for understanding nature. Nowhere is this more visible than in the effort to shift from management of single issues, such as nutrient load reduction, to a paradigm of ecosystem management (Heymans et al. 2018). Ecosystem dynamics are complicated and largely unobservable making models a pivotal tool for forecasting change (Heymans et al. 2016; Rose et al. 2010). The concept of ecosystem services further complicates the question, since social and economic dynamics not just ecological dynamics must also be considered (Austen et al. 2019). Models that incorporate all these dynamics are rare and largely unvalidated, yet the roadmap has been laid for such efforts and like the early days of ecological modeling the debate has shifted to incorporate models that predict ecosystem services (Nielsen et al. 2018; Zvoleff and An 2014). In this chapter we explore some of the issues related to ecosystem service modeling, first by defining it and then by exploring its possibilities and limits. This is not intended to be a thorough review of available ecosystem models, rather it is a discussion of where we are as a discipline and where we need to go to fully operationalize models for ecosystem decision making.

Ecosystem Goods and Services (EGS) are those elements of nature that directly benefit people. Models are well-suited to inform on an ecosystem good or service as the core value of models is to synthesize scientific information into a targeted endpoint. However, a focus on EGS as an endpoint for decision making requires socio/economic components, which are rarely included in ecosystem modeling (Bagstad et al. 2014; Fulford et al. 2015; Sanchirico and Mumby 2009). One example of interdisciplinary modeling for decision making is the integration of an economic model (Resource Investment Optimization System—RIOS) and an Ecosystem Services (ES) modeling tool (Integrated Valuation of Ecosystem Services and Tradeoffs—InVEST) to evaluate ES costs and benefits impacted by land use/land cover planning in the Thadee watershed in southern Thailand (Trisurat 2013, www.ipbes.net/resources/assessment reports). The more common approach is ecological modeling that forecasts EGS production without considering beneficiaries or assumes service delivery to be unrelated to changes in production (Craft et al. 2009; Redhead et al. 2018). Such limitations are not realistic and there is a need to broaden cross-disciplinary work in modeling to bridge this gap. The key stumbling block is to measure and communicate utility of models from different disciplines (Fulton 2010).

Model-based decision making is highly dependent on the accessibility and reliability of the chosen model. Accessibility encompasses ownership of the model framework, data availability as well as technical expertise, all which can affect a model's usability. Reliability considers the accuracy of model output, but also considers model acceptance based on how well the output is understood and the model's prior track record solving similar problems. Most of the issues to be discussed in this chapter find their origins in the need to demonstrate either model accessibility or reliability prior to application. There is typically a continuum of decision making from a purely policy-based to a data-based decision process. From the policy perspective, decisions are based largely on authoritative opinion. As the need for expert opinion grows, the decision begins to shift towards technical guidance and finally, when specific data are available, towards numerical decision rules. More complicated decisions are naturally shifted to the data end of the spectrum, however as problems become more complicated, so do the data. Ultimately the synthesis of that data into the language of policy can become a limiting factor. Here is where the value of models in decision making is most evident, as they are a powerful tool for synthesizing data. Yet, the process of data synthesis also yields a new challenge, that of communicating to non-technical decisions-makers. There is a balance to be struck between the amount of synthesis possible in any situation (e.g., how complicated is the problem?) and the ability to communicate the result well-enough to make it useful. In this chapter we take three examples of model applications to EGS in environmental decision making to illustrate how this balance was struck in each case.

The first example is taken from marine fisheries management and applies a population dynamics model to the challenge of maintaining a sustainable harvest in the context of both a variable fishery and a variable environment. Marine Fishery Stock Assessment (MFSA) has a long history of applying models to decision making. Stock assessment models such as the Stock Synthesis (SS) model (Methot 2009) are used to project stability of exploited populations in response to both fishery harvest and environmental variability (SEDAR 2016). Under MFSA models are usually combined with a formal process of stakeholder engagement to assure the right data and perspectives are utilized. The Gulf of Mexico Red Snapper (*Lutjanus campechanus*) stock (GMFMC 2018) provides an example. The sustainability of Gulf Red Snapper is assessed based on Maximum Sustainable Yield (MSY) i.e. the level at which harvest will not prevent the stock from replacing itself over the short term (~5 years). The SS model tracks population dynamics under different harvest scenarios and produces a prediction of MSY. Multiple model runs are used to account for several levels of uncertainty and model outcomes are combined to generate a best estimate of MSY, which is converted to a decision rule based on the uncertainty in the estimate and the level of acceptable risk of overfishing the stock. The models used in this assessment, as well as the input data for the model, are both well-validated and accepted for decision making (GMFMC 2018). In this example the ES is fishery harvest and the primary stakeholder (fishers) are engaged in the model-based decision process.

The second example comes from landscape management and involves a projection of how changes in land use practice in a watershed may impact the water quality of an estuary or other receiving water body (USDA-NRCS 2013). This example described the application of the Soil and Water Assessment Tool (SWAT) model (USDA NRCS; https://swat.tamu.edu/) to land conservation practices in Chesapeake Bay (CB) watershed USA. The SWAT model is a partially distributed hydrology model that plots water movement through a watershed to estimate the loading rate of nutrients and pollutants to the estuary. In the CB watershed, SWAT was used to examine changes in nutrient load between 2003–2006 and 2011 based on conservation practices in farm land mostly 'edge of field' retention of soil and nutrients. The model predicted a reduction delivered to streams in sediment (82%), nitrogen (44%), and phosphorus (75%). These results reinforce the value of land conservation on Bay water quality, but the information is not formally used to shape decisions. Rather the model output may be used to validate recommendations made to land owners for good conservation practice. In this example the ES is Bay water quality, which is an intermediate EGS leading to multiple benefits including recreational opportunities, shoreline land value, and aquatic biodiversity.

The third example comes from ecosystem management efforts in the Baltic Sea focusing on eutrophication over a large region using an ensemble modelling approach based on four different coastal eutrophication models (Skogen et al.

2014). The four models differed in coverage and approach but all projected coastal eutrophication and hypoxia responses to nutrient loading over time. The outcome was a map of eutrophication 'problem areas' in the Baltic Sea region that can be used to guide future conservation efforts. As in the case of the SWAT model, these results are not a part of a formal decision-making process but inform potential decisions aimed at reducing eutrophication. Ensemble models allow for different perspectives in modeling to be used and combined into a common recommendation (Skogen et al. 2014). In this example the ES is water qualtiy related to balanced primary production and a low incidence of hypoxia. These are intermediate EGS that contribute to fishery production, biodiversity, and human health.

Box 1 Central Questions in Making Model-Based Assessments Useful in Environmental Decision Making

1. Necessary complexity—How much detail is needed to address a management problem?
2. Operationalizing model output—Translating model output into the language of policy.
3. Transparent and transferable—Proper engagement with decision makers so they accept and use model-based information.

In this chapter we explore some of the dominant issues related to model-based assessments for environmental decision making based on ES assessments. We will use our examples from major areas of environmental decision making but we explore common ground by considering how models can be made more useful across important environmental issues. Three major issues exist in making models acceptable and reliable for model-based assessments of ES (Box 1). First models are meant to be simplified reflections of nature, but how much detail is needed? Modelers must address the question of how much model complexity is necessary for a given problem. Second, model output must be presented in a way that it is useful for decision making. This includes both translating model output into policy terms and properly communicating model uncertainty. The translation of model output and uncertainty into policy terms is called operationalizing the information. Graphic outputs and summarized metrics like ecological thresholds facilitate the operationalization process. Third, the model itself must be transparent and transferable to an issue and ecosystem. This means that not only is the model useful for a given problem, but it is also viewed as useful by policy makers and stakeholders, which may require early stakeholder engagement in model development. All these issues are necessary to make models useful for decision making, they are addressed here by examining some of the major issues in model choice and development.

2 Issues of Model Complexity

Quantitative models exist along a broad spectrum of complexity from simple empirical models (e.g. linear regression) to large mechanistic models that contain hundreds of parameters. By nature, simple empirical models are easier to conceptualize and fit as the parameters are simply components of a transformation from data to model output (X → Y). A good example is the exponential growth model (See Getz 1998), which describes exponential population growth and introduces the concept of carrying capacity in the case of limited resources. There is a tradition of parsimony in modeling that finds its roots in empirical models, since degrees of freedom are lost as fitting parameters are added (Getz 1998; May 1974a). In contrast, mechanistic models serve to recreate real pathways between data and model prediction in a way that allows for observable responses at the sub-model level but requires parameters that are themselves derived from empirical relationships. A simple example of a mechanistic model is the application of a functional response to predator-prey dynamics (Rosenweig and MacArthur 1963). The addition of data-derived parameters greatly adds to the complexity of the model, as well as the data requirements, but also provides a platform for examining the effects of real change that is not generally acceptable in empirical models where results cannot exceed the bounds of the data used to fit the model. If mechanistic models can be properly calibrated and validated for use, they become more useful for theoretical exploration of system change (Getz 1998). See Lewis et al. (2020) in this volume for additional examples of models across the complexity spectrum.

History of parsimony in modeling is derived largely from empirical modeling and its dependence on 'Goodness of Fit'. Empirical models are penalized based on inclusion of additional complexity if that complexity does not contribute significantly to the fit of the model (Burnham and Anderson 1998). However, that constraint is based on complexity that is not typically biologically interpretable. For instance, using a 3-parameter polynomial vs. a 2-parameter polynomial may improve fit, but adds very little to the interpretation of the results. This can be contrasted with mechanistic models commonly employed in ecological forecasting where new complexity may improve fit but also contributes to interpretation of the results (Scott et al. 2016). For instance, a well fit model predicting primary productivity in an estuary that does not consider plankton grazing is not *a priori* better as this is a significant loss term in real systems. In such a case, a modeler may accept a decline in the model fit to assure important dynamics are included. As data availability and computational power have increased, so has the attraction of increasing model complexity. Yet, neither of these improvements is a good reason to increase model complexity, rather that decision should be based on the trade-off between parsimony and realism (e.g., ARIES, Martínez-López et al. 2019). What rule should we use to justify model complexity in any particular situation?

The best approach is to match model complexity to the system and question in hand. Getz (1998) provided a review of model complexity relating it directly to the utility of models as a scientific tool. In that review, he highlighted the need to remain true to scientific principles of hypothesis testing to determine cause and the value of models for seeking a pattern that is believable and testable. Testability requires simplicity, but this goal is hard to apply in ecosystem science dealing with large complex systems. The question is how representative is the observed pattern in simplistic models? The second goal of predicting outcomes is much more aligned with ecosystem problems, but far less aligned with parsimony (Dietze et al. 2018). For example Bauer et al. (2019) assessed three models based on their ability to inform a comparison of fisheries management strategy, but their assessment was based not on information- theory comparisons of model accuracy (which reward simplicity) but on the range of disagreement among models in the predicted outcome. The review also considered model complexity in that the models differed both in approach and in the number of important issues each could consider (Bauer et al. 2019). The target was management guidance and the models were evaluated on how useful the results were for informing management. They concluded that complexity is needed where it applies to a known important dynamic in the ecosystem, such as competitive interactions between harvested and non-harvested prey.

The question of how much complexity is enough is a difficult one but is reminiscent of other threshold-based questions in computational science. The one common axiom is that you must pass over the threshold to identify it clearly (Hilborn et al. 1995). In modeling, the concept of necessary complexity has been suggested (Biebricher et al. 2012; Cahill and Mackay 2003) and we argue here that in ecological modeling it is necessary to overfit the problem at hand and then critically evaluate how much complexity is necessary to get a useable answer. Good examples of questioning simplifying assumptions comes from models that describe optimal behavior in animals (Alerstam 2011; Petchey et al. 2008). Response to change is a critical feature of forecasting ecological impacts of change in landscapes, water qualtiy, and food supply. Historically this type of modeling describes response to change based on the assumption that organisms will always respond to optimize their fitness in any novel situation (Doniol-Valcroze et al. 2011). Yet this approach is contrary to behavioral theory (McNamara and Houston 2009) and assumes an unrealistic level of knowledge regarding the spectrum of available resources (Fulford et al. 2011). Tools exist to predict non-optimal short-term behavior that both satisfy optimality theory and account for more realistic decision making in complex environments. Such tools pave the way for inclusion of additional complexity when necessary to the problem, such as for animal distributions in heterogenous landscapes (Rustigian et al. 2003) and mate choice predictions in fish reproductive models (Wooten 1984).

It is important to realize that the question of necessary complexity can be tied to the choice of model, such as with SWAT described above (USDA NRCS 2013). This mode-based analysis was intended to predict improvements in an ES (water quality) related to a management decision (land use conservation strategies) in the watershed. The model used was a semi-distributed hydrologic model with three effective vertical layers (surface flow, shallow ground water, and deep ground water). The conservation questions at hand all focused on surface flow, so loss to groundwater was a secondary element. In addition, the timescale was seasonal to annual and avoided examination of daily fluctuations in flow. This can be contrasted with fully-gridded models (e.g. Abdelnour et al. 2013), which are more detailed and optimized for measuring flow in multiple vertical layers at the scale of meters and days that might be needed to examine episodic storm events or annual forest harvesting strategies, rather than longer term landscape changes. The key is to approach any model problem by first critically evaluating the complexity that is needed to properly characterize the ecosystem and the issue at hand.

3 Communicating Model Uncertainty as Risk

A core objective of modeling applied to management is the meaningful assessment of risk. This is a key element of ES valuation and trade-off analysis among different competing services in the context of their impacts on social well-being (Spence et al. 2018). In the face of uncertainty and complex outcomes, models can and should communicate not just how the system will respond to change but also the probability and consequences of those responses in the context of policy. For example, in fishery management the goal is to reduce the probability of overfishing over a fixed time horizon (~5 years) (SEDAR 2016). Models are used to examine the probability of overfishing based on a suite of scenarios for future events (e.g., management limits, climate, recruitment variability), and these results are interpreted in the context of policy-driven risk thresholds (e.g., $P(\text{overfishing}) < 0.25$). Another example is the use of models in toxicological risk assessment to address the probability of adverse events (Forbes and Calow 2012). The challenge for traditional toxicological risk assessment is extrapolating from experimental data at the sub-organism level to population and ecosystem level effects. The use of models for toxicological risk assessment is controversial but shows promise as a method for extrapolation in cases where empirical data are unavailable (e.g., novel species) or unobtainable (e.g., ecosystems). The scientific community tends to vilify model uncertainty as a fault of the approach (Dietze et al. 2018), yet uncertainty is an intrinsic qualtiy of complex systems that necessitates a risk-based approach, and the quantifcation and communication of model uncertainty can contribute to the assessment of risk. For instance, in the fishery example from the Gulf of Mexico Red Snapper, the Stock Synthesis model is used to project the Over Fishing Limit (OFL) as the maximum allowable harvest rate for the stock. The goal here is to sustain the ES (fishery harvest) by managing fishing pressure in the context of other environmental influences (e.g., climate). Therefore, model output uncertainty is also reported and used to quantify the policy-based level of risk into a reduced maximum harvest value called Allowable Biological Catch (ABC). This value is the functional maximum rate used in management. See Lewis et al. (2020) in this volume for additional examples of quantifying uncertainty in specific models.

At the center of risk assessment based on model output is the proper quantification of model uncertainty. Model uncertainty can be parsed into uncertainty in data, parameter values, model functional choice, and underlying variability in the modeled system. Uncertainty within a model caused by input data and parameter uncertainty has received the most attention (LaDeau 2010; Nielsen et al. 2018). Several common tools exist for uncertainty assessment including Monte Carlo Simulation (Thorson et al. 2015; Zhang et al. 2012) and Bayesian Belief Networks (Schmitt and Brugere 2013). There is a strong need to standardize approaches for quantifying uncertainty so that non-modelers can better evaluate and compare results. Returning to the Gulf of Mexico Red Snapper example, the uncertainty analysis included examining sensitivity to parameter values, but also sensitivity to initial model conditions, timeseries dependencies from independent data, and model validation (SEDAR 52). These results were then converted to a cumulative estimate of uncertainty in the OFL value that was used to inform selection of ABC for the stock.

The key element for using models in decision making is the translation of model uncertainty into an estimation of risk. Risk has two components that must be considered in assessment of model output. First model output must be matched to a policy-based outcome (e.g., overfishing, HAB events). Frequently, this is the hardest step as model output is tied to the ecological dynamics not the policy-based objectives. For instance, in the case of Gulf Red Snapper the policy based risk component is standardized to a probability of 25% (SEDAR 2016). This value was derived from high level discussion of acceptable risk and all subsequent assessment outcomes must be reported in these terms by rule, which results in an operationalized model outcome. A comparable nutrient model might predict changes in nutrient concentrations through time and space but is less likely to synthesize those output into a time/space specific risk prediction. Policy makers must be engaged to make model output match the needs of management. In the nutrient case, that can be an estimate of the probability of exceeding Total Maximum Daily Load (TMDL) daily, which can be used to evaluate different strategies for reducing nutrient loading. Alternatively, a model might be adapted to estimate probability of harmful events given TMDL's are met, which is useful for assessment of different TMDL values. In either case, model uncertainty can then be evaluated as a component of estimating risk for the chosen outcome. This approach has been well-developed in fishery management where p(overfishing) has been defined so that it's incorporation into assessment models is straightforward. The same cannot be said for nutrient load management or habitat management, but this is largely a process limitation in that the tools exist but a mutually-acceptable process for inclusion does not.

4 Model Temporal and Spatial Scale

Model-based analyses are highly dependent on the choice of temporal and spatial scale and resolution. Models typically have an optimal resolution such as the SWAT model, which is a semi-distributed model designed to work best at medium resolution (e.g., Wellen et al. 2015) (seasons, km^2) vs. Ordinary Differential Equation Models designed to work at high temporal resolution but amalgamated over large spatial areas (e.g., Xu et al. 2011). The choice of a spatial and temporal resolution should be tied to the management question at hand and not the model. For instance, a fishery assessment of highly migratory coastal fishes does not require meter-scale resolution. In contrast, assessment of marine protected areas as a tool for increasing fishery recruitment may in fact require a high spatial resolution that can capture habitat features important to nursery production (e.g., Fulford et al. 2011). So, the choice to use a model also includes choices of scale that will impact the validity of the results. See Lewis et al. (2020) in this volume for additional examples of models optimized for particular spatial and temporal scales.

Simply considering both spatial and temporal variability in the same model is relatively novel in modeling as the issue of simultaneously quantifying both spatial and temporal uncertainty is challenging (Jager et al. 2005). For instance, in landscape management a wide variety of models with different spatial and temporal resolutions exist for examination of the impact of land use practices on nutrient loading into waterbodies. Historically, loading rates have been the focus of management, which do not require high spatial resolution (Hashemi et al. 2016). Yet, the impact of land use and the relative importance of surface flow vs. groundwater have greatly increased the need to examine nutrient loading at small spatial scales and in conjunction with short-term episodic events (Abdelnour et al. 2013). Models have had to keep up with this need. A potentially fruitful solution to scale issues is ensemble modeling or end-to-end modeling (Heymans et al. 2018; Lewis et al. 2020; Serpetti et al. 2017). Ensemble modeling was described earlier and end-to-end differs from ensemble modeling in that models are linked sequentially so that output from one model is input for another. Both approaches use multiple models and therefore open the door for examination of multiple temporal and spatial scales for a single system or problem. A good example of the application of multiple models to ES assessment is the ensemble model used to explore eutrophication problem areas in the Baltic Sea (Skogen et al. 2014). In this case, multiple spatial and temporal scales were involved in the collective analysis of eutrophication effects resulting from the interaction of nutrient loading and climatic forcing to generate common guidance for decision making. The result was the identification of 'problem areas' for eutrophication at several different scales, which makes the information more useful for decision making. As with other forms of complexity, spatial and temporal scale must be chosen deliberately to match the management problem of interest. This can be inhibited by fidelity to a particular model or limitations of available data for calibration and validation. Optimally, these three elements (model, problem, data) will be examined *a priori* to decide on what can be done so that operating at the wrong scale is not a criticism of the output. This is another opportunity for stakeholder engagement prior to any model-based analysis.

5 Connecting Science and Policy Objectives in Models

Models can be powerful tools for synthesizing scientific information to inform complex decision making if we can identify steps to operationalize model output (Box 2). Yet, for ecosystem-based management the use of models to justify decisions has not reached its full potential. If model-based assessments are needed to more fully inform decision making for ecosystem-based management, then modelers must meet the challenges described by making models more transparent and by communicating model output in the language of policy (e.g., ICES WGSAM 'Key run' model designation; https://www.ices.dk/community/groups/Pages/WGSAM.aspx). This will be greatly facilitated by early engagement with policy makers and other stakeholders in model choice and development, as well as formal processes for communicating uncertainty as a tool for risk management.

Box 2 Future Needs and Directions for Model-Based Decision Making

1. Standardize the process not the model—no model fits all problems and we should step away from promoting specific tools to promoting methods for tool selection. This includes a raised awareness of available tools and some standards regarding their evaluation.
2. Engage with policy makers early not late in the model development process.
3. Develop clear decision criteria that can be readily translated to and from model output.
4. Communication of model output should include uncertainty not as a limitation of the model but as a property of the decisional landscape useful for risk analysis.
5. Develop success stories based on ensemble modeling and provable short-term forecasting.

For models to be more widely used in decision making, scientists should also check the temptation to create new models. Rather, the focus should be on evaluating current models and providing practical suggestions on model selection for management decisions. Model libraries such as GULF TREE (http://www.gulftree.org/) and the EPA EcoService Models Library (https://esml.epa.gov/) provide search engines to help managers find the right tool for a specific problem. If new models are needed, decision makers should be involved in the development process and their opinions should be integrated into the modeling process (O'Higgins et al. 2020). Multiple model types can also be combined to improve usefulness of both predictive models, as well as utility models, such as neutral models to provide a benchmark for model performance and Bayesian Belief Networks, which can integrate qualitative information to make probabilistic predictions. The use of ensemble modeling holds particular promise for handling uncertainty, as it allows for multiple perspectives and the interactive exploitation of strengths of multiple models (Bauer et al. 2019).

There is also a need to develop proven track records for model-based assessments, which can only occur as a part of application to a practical problem. This can best be achieved through short-term forecasting exercises (Dietze et al. 2018) in real world situations. This approach demonstrates model abilities, but also provides a testbed for feedback and model improvement. Two key examples of this approach are in weather prediction (Bengtsson et al. 2019; Wu et al. 2019) and fishery stock assessment (GMFMC 2018). Future directions in model development for predicting ES should also include identification of such opportunities to engage in short-term forecasting in partnership with decision makers. Ultimately, the use of models in ecological decision making requires acceptance from policy makers and the public and that is where the most effort if needed.

Disclaimer This chapter has been subjected to Agency review and has been approved for publication. The views expressed in this paper are those of the author(s) and do not necessarily reflect the views or policies of the U.S. Environmental Protection Agency.

References

Abdelnour, A., McKane, R. B., Stieglitz, M., Pan, F., & Cheng, Y. (2013). Effects of harvest on carbon and nitrogen dynamics in a Pacific northwest forest catchments. *Water Resources Research, 49*, 1292–1313.

Alerstam, T. (2011). Optimal bird migration revisited. *Journal für Ornithologie, 152*, 5–23.

Austen, M. C., Andersen, P., Armstrong, C., Doring, R., Hynes, S., Levrel, H., Oinonen, S., & Ressurreicao, A. (2019). Valuing marine ecosystems - taking into account the value of ecosystem benefits in the blue economy. In J. Coopman, J. J. Heymans, P. Kellett, P. A. Munoz, V. French, & B. Alexander (Eds.), *Future science brief 5 of the European marine board* (p. 32). Belgium: Ostend.

Bagstad, K. J., Villa, F., Batker, D., Harrison-Cox, J., Voigt, B., & Johnson, G. W. (2014). From theoretical to actual ecosystem services: Mapping beneficiaries and spatial flows in ecosystem service assessments. *Ecology and Society, 19*(2), 64.

Bauer, B., Horbowy, J., Rahikainen, M., Kulatska, N., Muller-Karulis, B., Tomczak, M. T., & Bartolino, V. (2019). Model uncertainty and simulated multi-species fisheries management advice in the Baltic Sea. *PLoS One, 14*, e0211320.

Bengtsson, L., Bao, J. W., Pegion, P., Penland, C., Michelson, S., & Whitaker, J. (2019). A model framework for stochastic representation of uncertainties associated with physical processes in NOAA's next generation global prediction system (NGGPS). *Monthly Weather Review, 147*, 893–911.

Biebricher, A., Havnes, O., & Bast, R. (2012). On the necessary complexity of modeling of the polar mesosphere summer echo overshoot effect. *Journal of Plasma Physics, 78*, 225–239.

Burnham, K. P., & Anderson, D. R. (1998). *Model selection and inference: A practical information theoretic approach*. New York, NY: Springer.

Cahill, T. M., & Mackay, D. (2003). Complexity in multimedia mass balance models: When are simple models adequate and when are more complex models necessary? *Environmental Toxicology and Chemistry, 22*(6), 1404–1412.

Clark, J. S., & Gelfand, A. E. (2006). A future for models and data in environmental science. *Trends in Ecology & Evolution, 21*, 375–380.

Craft, C., Clough, J., Ehman, J., Joye, S., Park, R., Pennings, S., Guo, H. Y., & Machmuller, M. (2009). Forecasting the effects of accelerated sea-level rise on tidal marsh ecosystem services. *Frontiers in Ecology and the Environment, 7*, 73–78.

Dietze, M. C., Fox, A., Beck-Johnson, L. M., Betancourt, J. L., Hooten, M. B., Jarnevich, C. S., Keitt, T. H., Kenney, M. A., Laney, C. M., Larsen, L. G., Loescher, H. W., Lunch, C. K., Pijanowski, B. C., Randerson, J. T., Read, E. K., Tredennick, A. T., Vargas, R., Weathers, K. C., & White, E. P. (2018). Iterative near-term ecological forecasting: Needs, opportunities, and challenges. *Proceedings of the National Academy of Sciences of the United States of America, 115*, 1424–1432.

Doniol-Valcroze, T., Lesage, V., Giard, J., & Michaud, R. (2011). Optimal foraging theory predicts diving and feeding strategies of the largest marine predator. *Behavioral Ecology, 22*(4), 880–888.

Forbes, V. E., & Calow, P. (2012). Promises and problems for the new paradigm for risk assessment and an alternative approach involving predictive systems models. *Environmental Toxicology and Chemistry, 31*, 2663–2671.

Fulford, R. S., Peterson, M. S., & Grammer, P. O. (2011). An ecological model of the habitat mosaic in estuarine nursery areas: Part I-Interaction of dispersal theory and habitat variability in describing juvenile fish distributions. *Ecological Modelling, 222*, 3203–3215.

Fulford, R. S., Yoskowitz, D., Russell, M., Dantin, D. D., & Rogers, J. (2015). Habitat and recreational fishing opportunity in Tampa Bay: Linking ecological and ecosystem services to human beneficiaries. *Ecosystem Services, 17*, 64–74.

Fulton, E. A. (2010). Approaches to end-to-end ecosystem models. *Journal of Marine Systems, 81*, 171–183.

Getz, W. M. (1998). An introspection on the art of modeling in population ecology. *BioScience, 48*, 540–552.

GMFMC. (2018). *SEDAR 52 stock assessment report: Gulf of Mexico red snapper*. North Charleston, SC: Southeast Data Assessment and Review (SEDAR).

Griffith, G. P., & Fulton, E. A. (2014). New approaches to simulating the complex interaction effects of multiple human impacts on the marine environment. *ICES Journal of Marine Science, 71*, 764–774.

Hashemi, F., Olesen, J. E., Dalgaard, T., & Borgesen, C. D. (2016). Review of scenario analyses to reduce agricultural nitrogen and phosphorus loading to the aquatic environment. *Science of the Total Environment, 573*, 608–626.

Heymans, J. J., Coll, M., Link, J. S., Mackinson, S., Steenbeek, J., Walters, C., & Christensen, V. (2016). Best practice in Ecopath with Ecosim food-web models for ecosystem-based management. *Ecological Modelling, 331*, 173–184.

Heymans, J. J., Skogen, M., Schrum, C., & Solidoro, C. (2018). Enhancing Europe's capabiltiy in marine ecosystem modelling for societal benefit. In K. E. Larkin, J. Coopman, P. A. Munoz, P. Kellett, C. Simon, C. Rundt, C. Viegas, & J. J. Heymans (Eds.), *Future science brief 4 of the European marine board* (p. 32). Ostend, Belgium: European Marine Board.

Hilborn, R., Walters, C. J., & Ludwig, D. (1995). Sustainable exploitation of renewable resources. *Annual Review of Ecology and Systematics, 26*, 45–67.

Jager, H. I., King, A. W., Schumaker, N. H., Ashwood, T. L., & Jackson, B. L. (2005). Spatial uncertainty analysis of population models. *Ecological Modelling, 185*, 13–27.

LaDeau, S. (2010). Advances in modeling highlight a tension between analytical accuracy and accessibility. *Ecology, 91*, 3488–3492.

Lewis, N. S., Marois, D. E., Littles, C. J., & Fulford, R. S. (2020). Projecting changes to coastal and estuarine ecosystem goods and services - models and tools. In T. O'Higgins, M. Lago, & T. H. DeWitt (Eds.), *Ecosystem-based management, ecosystem services and aquatic biodiversity: Theory, tools and applications* (pp. 235–254). Amsterdam: Springer.

Martínez-López, J., Bagstad, K. J., Balbi, S., Magrach, A., Voigt, B., Athanasiadis, I., Pascual, M., Willcock, S., & Villa, F. (2019). Towards globally customizable ecosystem service models. *Science of The Total Environment, 650*, 2325–2336.

May, R. M. (1974a). Biological populations with nonoverlapping generations—Stable points, stable cycles, and chaos. *Science, 186*, 645–647.

May, R. M. (1974b). Patterns of species abundance—Mathematical aspects of dynamics of populations. *SIAM Review, 16*, 585–585.

McNamara, J. M., & Houston, A. I. (2009). Integrating function and mechanism. *Trends in Ecology & Evolution, 24*, 670–675.

Methot, R. D. (2009). Stock assessment: Operational models in support of fisheries management. In Beamish, R. J & Methot, R. D. (Eds.), *The future of fisheries science in North America. 137 fish and fisheries science series*. Berlin: Springer.

Nielsen, J. R., Thunberg, E., Holland, D. S., Schmidt, J. O., Fulton, E. A., Bastardie, F., Punt, A. E., Allen, I., Bartelings, H., Bertignac, M., Bethke, E., Bossier, S., Buckworth, R., Carpenter, G., Christensen, A., Christensen, V., Da-Rocha, J. M., Deng, R., Dichmont, C., Doering, R., Esteban, A., Fernandes, J. A., Frost, H., Garcia, D., Gasche, L., Gascuel, D., Gourguet, S., Groeneveld, R. A., Guillen, J., Guyader, O., Hamon, K. G., Hoff, A., Horbowy, J., Hutton, T., Lehuta, S., Little, L. R., Lleonart, J., Macher, C., Mackinson, S., Mahevas, S., Marchal, P., Mato-Amboage, R., Mapstone, B., Maynou, F., Merzereaud, M., Palacz, A., Pascoe, S., Paulrud, A., Plaganyi, E., Prellezo, R., van Putten, E. I., Quaas, M., Ravn-Jonsen, L., Sanchez, S., Simons, S., Thebaud, O., Tomczak, M. T., Ulrich, C., van Dijk, D., Vermard, Y., Voss, R., & Waldo, S. (2018). Integrated ecological-economic fisheries models-evaluation, review and challenges for implementation. *Fish and Fisheries, 19*, 1–29.

O'Higgins, T. G., Culhane, F., O'Dwyer, B., Robinson, L., & Lago, M. (2020). Combining methods to establish potential management measures for invasive species *Elodea nutallii* in Lough Erne Northern Ireland. In T. O'Higgins, M. Lago, & T. H. DeWitt (Eds.), *Ecosystem-based management, ecosystem services and aquatic biodiversity: Theory, tools and applications* (pp. 445–460). Amsterdam: Springer.

Petchey, O. L., Beckerman, A. P., Riede, J. O., & Warren, P. H. (2008). Size, foraging, and food web structure. *Proceedings of the National Academy of Sciences of the United States of America, 105*, 4191–4196.

Piroddi, C., Coll, M., Liquete, C., Macias, D., Greer, K., Buszowski, J., Steenbeek, J., Danovaro, R., & Christensen, V. (2017). Historical changes of the Mediterranean Sea ecosystem: Modelling the role and impact of primary productivity and fisheries changes over time. *Scientific Reports, 7*, 44491.

Redhead, J. W., May, L., Oliver, T. H., Hamel, I., Hamel, P., Sharp, R., & Bullock, J. M. (2018). National scale evaluation of the InVEST nutrient retention model in the United Kingdom. *Science of the Total Environment, 610*, 666–677.

Rose, K. A., Allen, J. I., Artioli, Y., Barange, M., Blackford, J., Carlotti, F., Cropp, R., Daewel, U., Edwards, K., Flynn, K., Hill, S. L., HilleRisLambers, R., Huse, G., Mackinson, S., Megrey, B., Moll, A., Rivkin, R., Salihoglu, B., Schrum, C., Shannon, L., Shin, Y. J., Smith, S. L., Smith, C., Solidoro, C., John, M. S., & Zhou, M. (2010). End-to-end models for the analysis of marine ecosystems: Challenges, issues, and next steps. *Marine and Coastal Fisheries, 2*, 115–130.

Rosenweig, M. L., & MacArthur, R. H. (1963). Graphical representation and stability conditions of predator-prey interactions. *The American Naturalist, 97*, 209–223.

Rustigian, H. L., Santelmann, M. V., & Schumaker, N. H. (2003). Assessing the potential impacts of alternative landscape designs on amphibian population dynamics. *Landscape Ecology, 18*, 65–81.

Sanchirico, J. N., & Mumby, P. J. (2009). Mapping ecosystem functions to the valuation of ecosystem services: Implications of species-habitat associations for coastal land-use decisions. *Theoretical Ecology, 2*, 67–77.

Schmitt, L. H. M., & Brugere, C. (2013). Capturing ecosystem services, stakeholders' preferences and trade-offs in coastal aquaculture decisions: A bayesian belief network application. *PLoS One, 8*, e75956.

Scott, E., Serpetti, N., Steenbeek, J., & Heymans, J. J. (2016). A stepwise fitting procedure for automated fitting of Ecopath with EcoSim models. *Software X, 5*, 25–30.

SEDAR. (2016, September). *Southeast data assessment and review—Data best practices: Living document* (p. 115). North Charleston, SC: SEDAR. Retrieved from http://sedarweb.org/sedar-data-best-practices.

Serpetti, N., Baudron, A. R., Burrows, M. T., Payne, B. L., Helaouet, P., Fernandes, P. G., & Heymans, J. J. (2017). *Impact of ocean warming on sustainable fisheries management informs the ecosystem approach to fisheries* (Scientific reports, 7 of the European marine board). Belgium: Ostend.

Skogen, M. D., Eilola, K., Hansen, J. L. S., Meier, H. E. M., Molchanov, M. S., & Ryabchenko, V. A. (2014). Eutrophication status of the North Sea, Skagerrak, Kattegat and the Baltic Sea in present and future climates: A model study. *Journal of Marine Systems, 132*, 174–184.

Spence, M. A., Blanchard, J. L., Rossberg, A. G., Heath, M. R., Heymans, J. J., Mackinson, S., Serpetti, N., Speirs, D. C., Thorpe, R. B., & Blackwell, P. G. (2018). A general framework for combining ecosystem models. *Fish and Fisheries, 19*, 1031–1042. https://doi.org/10.1111/faf.12310.

Thorson, J. T., Hicks, A. C., & Methot, R. D. (2015). Random effect estimation of time-varying factors in stock synthesis. *ICES Journal of Marine Science, 72*, 178–185.

Trisurat, Y. 2013. *Ecological assessment: Assessing condition and trend of ecosystem service of Thadee watershed*. Nakhon Si Thammarat Province, Bangkoknd: ECO-BEST Project, Faculty of Forestry, Kasetsart University.

USDA-NRCS. (2013). *Impacts of conservation adoption on cultivated acres of cropland in the Chesapeake Bay region 2003–06 to 2011* (p. 113). United States Department of Agriculture, Natural Resources Conservation Service. https://www.nrcs.usda.gov/wps/portal/nrcs/detail/national/technical/nra/ceap/na/?cid=stelprdb1240074.

Wellen, C., Kamran-Disfani, A. R., & Arhonditsis, G. B. (2015). Evaluation of the current state of distributed watershed nutrient water quality modeling. *Environmental Science & Technology, 49*, 3278–3290.

Whipple, S. J. (1999). Analysis of ecosystem structure and function: Extended path and flow analysis of a steady-state oyster reef model. *Ecological Modelling, 114*, 251–274.

Wooten, R. J. (1984). *A functional biology of sticklebacks*. New York, NY: Springer.

Wu, T. J., Min, J. Z., & Wu, S. (2019). A comparison of the rainfall forecasting skills of the WRF ensemble forecasting system using SPCPT and other cumulus parameterization error representation schemes. *Atmospheric Research, 218*, 160–175.

Xu, S., Chen, Z., Li, S., & He, P. (2011). Modeling trophic structure and energy flows in a coastal artificial ecosystem using mass-balance Ecopath model. *Estuaries and Coasts, 34*, 351–363.

Zhang, L., Yu, G., Gu, F., He, H., Zhang, L., & Han, S. (2012). Uncertainty analysis of modeled carbon fluxes for a broad-leaved Korean pine mixed forest using a process-based ecosystem model. *Journal of Forest Research, 17*, 268–282.

Zvoleff, A., & An, L. (2014). Analyzing human-landscape interactions: Tools that integrate. *Environmental Management, 53*, 94–111.

8

Predicting Future Vegetated Landscapes under Climate Change: Application of the Environmental Stratification Methodology to Protected Areas in the Lower Mekong Basin

John M. Johnston, Robert J. Zomer, and Ming-cheng Wang

Abstract There are 176 protected areas within the Lower Mekong Basin (LMB), comprising 183,000 km^2, almost 30% of the LMB. Climate change poses challenges to their management, because changes in the timing and amount of rainfall and maximum temperature from current (baseline) conditions alter vegetation growth and composition. The prediction of future climate, i.e., the pattern of temperature and rainfall expected 30–60 years from present, is accomplished using Earth System Models (ESMs). However, future ecosystem structure, including dominant vegetation, is less well studied. A successful approach is Environmental Stratification (EnS), involving statistical analysis of climate variables to identify relatively homogeneous spatial climate patterns (zones and strata) that are robust predictors of vegetation associations. Our objective was to predict changes in the spatial distribution of bioclimatic conditions across the LMB by the year 2030 and 2060, based on downscaled (1 km^2 resolution) ESM projections. Five major bioclimatic zones and twenty-five bioclimatic strata were identified using EnS, ranging from extremely hot and xeric at the lower elevations, to warm temperate and mesic at higher elevations. The largest expanse of area is extremely hot and moist (50% of total area), followed by extremely hot and xeric (24%), and extremely hot and mesic (18%), with mean annual temperature for the various zones ranging from 18.1 to 27.2°C. More than 9% to 29% of all protected areas are projected to shift to a different bioclimatic zone by 2030, and from 7% to over 77% by 2060.

J. M. Johnston (✉)
US EPA Office Research and Development, Center for Environmental Measurement and Modeling, Athens, GA, USA
e-mail: Johnston.JohnM@epa.gov

R. J. Zomer · M.-c. Wang
World Agroforestry Center, Kunming, China

Lessons Learned

- Up to 29% of protected areas will shift to a new bioclimatic zone by 2030, necessitating adaptive management to adjust the current boundaries and their level of protected status.
- Ecosystem-based adaptation (EbA), a type of ecosystem-based management (EBM), provides a holistic approach to habitat and species protection under climate change.
- Coordinated management efforts across national boundaries will be important in the future of protected areas in the Lower Mekong Basin (LMB).
- Both environmental stratification (EnS) and the analysis of projected change in spatial distribution of bioclimatic conditions provide reliable information for climate change adaptation planning.

Needs to Advance EBM

- Conservation research and monitoring efforts in the LMB have focused more on freshwater ecosystems and species, and a greater emphasis is needed in the future on terrestrial ecosystems and vegetation assemblages as provided by EnS.
- EBM of terrestrial ecosystems, including protected areas, must account for human uses and co-existence, possibly through emphasis on non-timber forest products for food and livelihoods, permitting sustainable harvest of renewable resources by local populations.
- The interaction of aquatic and terrestrial systems in the LMB is a needed research area to achieve greater sustainability of the food-energy-water nexus.
- EbA strategies are needed in agricultural areas of the LMB, especially the Khorat plateau, where temperatures are expected to increase and rainfall decrease, leading to crop stress and failures, increasing food insecurity and further stressing protected areas.

1 Introduction

1.1 Lower Mekong River Basin Context: Ecosystem, People and Challenges

The Mekong River Basin beginning in Lao People's Democratic Republic (PDR) is referred to as the Lower Mekong Basin (LMB), and it supports over 60 million people (MRC 2011c). The LMB includes portions of the riparian countries of Lao PDR, Vietnam, Cambodia and Thailand. It is estimated that over 60% of the LMB population is connected to the Mekong River for its food and livelihood (Fig. 1). Lao PDR and Cambodia plan to graduate from least developed country status by 2020, with Vietnam achieving middle-income status by 2030. Aquaculture is forecast to double to 4 million metric tons in 20 years, with dry season irrigation expansion planned to increase by 50% to 1.8 million ha, including mainstream water transfers planned by Thailand to alleviate drought conditions in the Northeast (MRC 2011a).

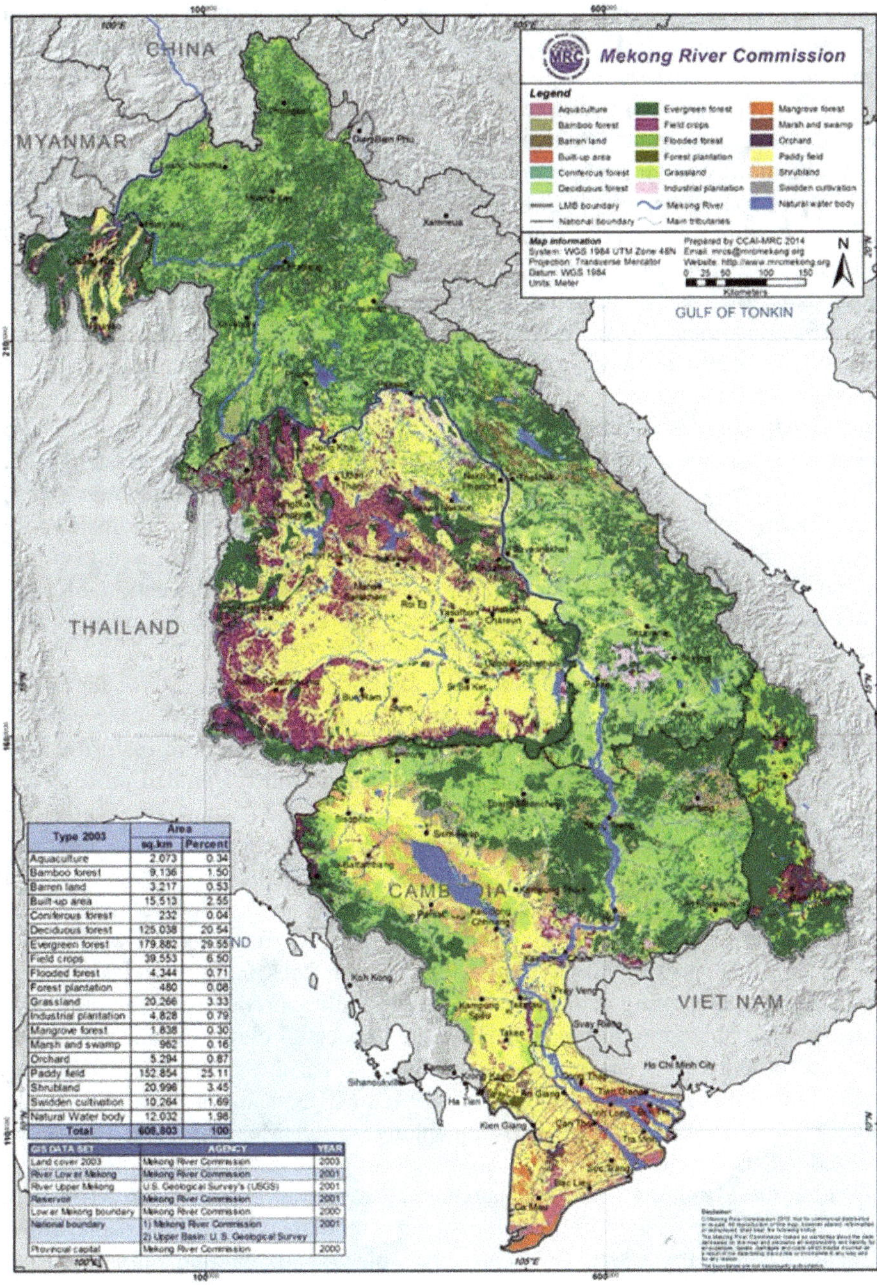

Fig. 1 Lower Mekong Basin study area. Land cover types are provided basinwide for all watersheds contributing to the Mekong River

The food, water, and energy needs of the population are therefore expected to increase pressure on LMB ecosystems, with climate change an additional long-term driver to be considered. The Mekong River Commission (MRC) was created in 1995 by the riparian governments of the LMB to promote sustainable development for the greater good of all countries (MRC 2011c).

Although the overall amount of water is not considered limiting in the LMB, the Southwest Monsoon results in strong seasonality of rainfall, with very little precipitation occurring in the dry season from November through May. Dry season food cultivation is dependent on irrigation from surface waters of the river and adjacent ecosystems. Because of their interdependence, water, energy, and food must be managed concomitantly, with careful consideration of tradeoffs that directly and indirectly affect the nexus (Linke 2014; Rasul 2014). Downstream ecosystems and national and local community dependence on flows from the northern basin are transboundary issues that require integrated planning and coordination. As a type of Ecosystem-Based Management (EBM), attention is being given to ecosystem-based adaptation (EbA) strategies for managing climate change impacts (Munang et al. 2013; Sierra-Correa and Cantera Kintz 2015; Vignola et al. 2013; Wamsler et al. 2014). An EbA strategy balances infrastructure development with investments in natural capital for more resilient and sustainable solutions. Protection of LMB watersheds contributing flow and material input to the Mekong is a primary EbA strategy in consideration (ISPONRE 2013; MONRE 2013).

These promising examples of national-level initiative within the LMB for EbA to climate change impacts were developed for Vietnam and Lao PDR. Each provides a national level operational framework for a participatory process of mainstreaming ecosystem services concepts and practices, outlining a fairly detailed process with example tools and methods to facilitate replication of the framework. A guide for decision makers is also recommended as a companion document with more detailed descriptions and examples of the concepts (Ranganathan et al. 2008). An important and necessary complement to the national level is a basin-wide strategy for transboundary adaptation planning in the LMB. Ecosystem services of water, nutrient, and sediment provisioning provided by upstream forest and wetland ecosystems are critical to maintaining delta fisheries and agricultural productivity and protection from saltwater intrusion, because many of the impacts of climate change cannot be managed effectively at the national level. Only when the suite of transboundary adaptation strategies are fully addressed will the MRC vision of "an economically prosperous, socially just, and environmentally sound Mekong River Basin" be realized (MRC 2010).

Within the MRC, the Climate Change and Adaptation Initiative (CCAI) has the mission of climate change impact assessment and adaptation planning and implementation within the LMB. CCAI has the objective of guiding climate change adaptation planning and implementation through improved strategies and plans at various levels and in priority locations throughout the LMB, which includes a mix of scoping, synthesis, outreach, and capacity building (MRC 2011b). A critical role of the MRC CCAI is to provide a consistent, basin-wide approach for evaluating transboundary climate change adaptation. The CCAI is dedicated to developing basin-wide datasets and assessment methods that can be used at multiple spatial scales (e.g., nationally and locally) whether data rich or data limited. This is an essential role of CCAI: no other group is committed to providing the basin-wide context for national studies and the ability to compare results consistently across the LMB. Outputs include basin-wide datasets, verified ecosystem modeling methods and knowledgebase for reuse by member countries, and the identification of data gaps and priorities for reducing uncertainty. The CCAI is responsible for providing data and modeling methods that are well documented (e.g., user's guides, peer-reviewed literature), freely available, and could be used at multiple spatial scales, from basin-wide assessments to province-level if needed within member countries. Forecasting the impacts of climate change on ecosystems and biodiversity was a focus of the CCAI (Trisurat et al. 2018), with multiple ecosystem service forecasting models in consideration for potential use in the LMB.

1.2 Landcover Change and Ecosystem Models for Scenario Analysis

Landcover data provide an efficient means of working with large geographic regions, and a number of models have been developed to utilize these data to characterize ecosystem services (Bagstad et al. 2013a, 2013b; Feng et al. 2011; Jackson et al. 2013; Vigerstol and Aukema 2011). The majority of these ecosystem services models utilize landcover data as input to quantify services and their value using biophysical models with a statistical or empirical formulation, a necessary strategy when working in data limited areas (Bagstad et al. 2013a; Daily et al. 2009). Attributes that differentiate the approaches, making some models more suitable than others, include availability, maturity, flexibility, and ease of use. The InVEST model developed by The Natural Capital Project is freely available, well documented, includes economic valuation methods, and has been applied around the world at multiple spatial scales to evaluate the impacts of land use and climate change. Examples include the impact of land use change on water yield, carbon storage, nutrient retention, sediment retention and biodiversity provisioning (Leh et al. 2013), the impact of hurricane and typhoon disturbance on water yield, water purification, soil conservation, carbon storage and biodiversity (Chiang et al. 2014), timber production and carbon sequestration losses due to hurricane damage (Delphin et al. 2013), sediment retention and water yield under land use and climate change (Su and Fu 2013), the impact of climate change on water yield for a low flow regime in a river basin (Marquès et al. 2013), and to assess the combined effects of land use and climate change scenarios on water yield, nutrient and sediment retention in 2050 (Hoyer and Chang 2014).

The InVEST model has also been used to evaluate payment for ecosystem services schemes and the value of hydropower energy production (Fu et al. 2014) and the economic value of wetland ecoservices water purification and carbon sequestration under three land use change scenarios (Harmáčková and Vačkář 2015). Water-related ecosystem services have received the most attention compared to other services, and Vigerstol and Aukema contrasted the InVEST and ARIES models for freshwater services provisioning with the watershed hydrologic models SWAT and VIC (Vigerstol and Aukema 2011). Sánchez-Canales et al. performed a sensitivity analysis of the InVEST water yield model and found it most sensitive to potential evapotranspiration and rainfall (Sánchez-Canales et al. 2012). Bai et al. used InVEST to evaluate the correlation and spatial overlap of sediment and nutrient retention, water yield and biodiversity services for possible management and restoration of regional hotspots (Bai et al. 2011). The InVEST model had been applied successfully within the LMB to address climate change impacts (ISPONRE 2013; MONRE 2013; Rosenthal et al. 2013). However, none of these models or approaches provides a forecast of future vegetation composition. Often, the most recent landcover data are used, held constant over the analysis period, or assumed to have a future composition as in scenario analysis. In this chapter, we explore Environmental Stratification (EnS) to develop bioclimatic strata and major bioclimatic zones for analyzing potential changes in LMB protected areas under different climate scenarios. EnS was selected by the CCAI to forecast future bioclimatic zones and the related vegetation structure in 2030 and 2060 and used to support the protected area analysis basinwide.

1.3 Environmental Stratification

Environmental Stratification (EnS) is a statistical technique for analyzing climate variables to identify relatively homogeneous spatial climate patterns (i.e., strata) that are in turn strongly associated with plant species and can serve as robust predictors of vegetation associations. Metzger et al. proposed using EnS to create a consistent, national to global biodiversity observation network (Metzger et al. 2013a, 2013b), based on a methodology involving maximum likelihood analysis and clustering (Zomer et al. 2014, 2015). Metzger et al.'s global bioclimatic map has been used to develop regional maps of climate associations and also as a reference for comparison with regional data (Zomer et al. 2014). The MRC and CCAI maintains basinwide climate and landcover data for the LMB that includes dominant vegetation types; these include, for example, coniferous, deciduous, and evergreen forests that are the necessary inputs for the EnS methodology (Fig. 1).

1.4 Biodiversity and Protected Areas in the LMB

Effective management of LMB ecosystems and their services requires a baseline assessment of status and ongoing monitoring and evaluation to assess the outcomes of decisions. This provides the necessary information for EbA—a key EBM strategy for navigating an uncertain future. As an example, forest ecosystems provide services related to wild foods and non-timber forest products (NTFPs). Households with access to productive forestlands have adaptation strategies not available to the population living in urban areas when food becomes scarce or expensive. Livelihoods are also supported from NTFP harvesting that brings income. The overall climate change adaptation strategy is one of decentralization to provide greater access and opportunity to the population. This is particularly important to communities in rural areas that have less resources and may have fewer options compared to those living in cities. The implication is that food security should consider the contribution of forest ecosystems to unconventional food supplies and subsistence economic opportunities. This is reinforced by a recent survey of biodiversity in Lao PDR: "The plants and animals harvested in the Mekong and its surrounding habitats provide an important source of food and income for the people who live along its banks" (IUCN 2013). Others have gone further to state that shifting agricultural practices in forest ecosystems are inherently more sustainable because they are less intensive and co-exist with forest ecosystems (Yokoyama et al. 2006). Numerous international experts recommend that "ecosystem-based approaches can contribute to climate change mitigation and adaptation and to sustainable development more broadly. Spatial planning for ecosystem services at international, national and local levels will be an important component of ecosystem-based approaches" (Leadley et al. 2010). The initial focus was on evaluating landscape change basin-wide and the use of these data to characterize selected ecosystem services for water provisioning and nutrient and sediment retention (Trisurat et al. 2018). Although protected areas and biodiversity had been evaluated at the provincial level, e.g., Nan Province, Thailand, it did not include climate impacts on bioclimatic zones that support terrestrial ecosystems (Trisurat et al. 2019). Changes in rainfall and temperature will influence vegetation patterns, potentially shifting species ranges to higher elevations and favoring new dominant and co-dominant plants compared with current communities (Zhang et al. 2014). Protected areas that are unchanging will likely fail to provide the range of habitats and corridors necessary for species to adapt to climate change (Monzón et al. 2011).

2 Study Area and Methods

The Mekong River in Southeast Asia drains an area of 795,000 km^2 that is referred to as the Lower Mekong Basin (LMB, Fig. 1). Annual mean precipitation ranges from approximately 1000 mm in northeast Thailand to over 3,500 mm in northern Lao PDR (Trisurat et al. 2018). This vast aquatic ecosystem flows over 4,800 km from its source in the Tibetan Plateau to the South China Sea below Ho Chi Minh. Of the approximately 475 km^3 annual average discharge, 12% or 60 billion m^3 are withdrawn for agricultural, industrial and other consumptive use. Hydropower energy potential of the LMB is estimated at 30,000 MW with only 10% developed; however, 26 hydropower projects are underway, with 12 mainstream and 30 tributary

Table 1 Protected areas in the Lower Mekong Basin

Lower Mekong Basin				
	Protected area		Area	
IUCN category	No. of PAs	% of PA	km^2	%
Ia	20	11.4	10,054	5.5
II	56	31.8	35,683	19.5
III	1	0.6	130	0.1
IV	27	15.3	36,524	20.0
V	8	4.5	6,431	3.5
VI	22	12.5	37,267	20.4
Not Applicable	10	5.7	39,069	21.3
Not Reported	33	18.8	17,914	9.8
Grand Total	*176*	*100.0*	*183,071*	*100.0*

IUCN categories provided by number (No.) of protected areas (PA), % of total PA, area in km^2 and % of total area for each category
Source: UNEP-WCMC (2019)

dams planned over the next 20 years (MRC 2011a). Ecosystem services provide food, water, and energy security to the extent they can be maintained in the face of a growing population and increasing demand for food and energy (MEA 2005). This includes services for habitat for protected species, with landcover and protected area data provided by the MRC (MRC 2011c). According to the World Database on Protected Areas (UNEP-WCMC 2019), there are 176 protected areas within the Lower Mekong Basin (LMB), comprising 183,000 km^2, almost 30% of the total LMB (Table 1). Excluding IUCN (International Union for Conservation of Nature) categories "Not Applicable" and "Not Reported", protected areas still account for over 20% of the LMB.

Baseline climate was derived from data averaged from 1980 through 2010, while 2030 represented a 30-year average of the period 2015 to 2045, and 2060 represented the period 2045 to 2075 for terrestrial ecosystems using ESM ensemble climate forecasts from the SimCLIM dataset (Trisurat et al. 2018). The EnS methodology used the SimCLIM set of selected ESM and emissions scenario combinations, which included a multi-model ensemble ($n = 13$) of Coupled Model Intercomparison Project 5 (CMIP5, https://esgf-node.llnl.gov/projects/cmip5/) ESMs applied across three representative concentration pathways (RCP 2.6, 4.5, 8.5), to assess climate change impacts on vegetation growing conditions. Consistent with the Fourth National Climate Assessment (USGCRP 2018), the RCPs span a range of lower to higher greenhouse gas (GHG) concentration, corresponding to increased radiative forcing of 2.6, 4.5, and 8.5 watts per square meter (W/m^2). In other words, scenario RCP8.5 involves annual GHG emissions increasing throughout the twenty first century, leveling off by 2100, while lesser RCPs represent decreased GHG emissions (USGCRP 2018). Most global climate modeling experts agree that the biosphere is already experiencing a trajectory consistent with RCP8.5, unless substantial GHG emission mitigation steps are taken immediately, worldwide. Climate-envelope models identify constraints on species geographic ranges related to aspects of temperature, precipitation, relative humidity, including the minimum and maximum values (Lassiter et al. 2000). Bioclimatic envelopes characterize to the extent feasible the complete set of conditions conducive to ecosystems, species, and biological interactions, and are therefore used to gauge the impacts of climate change across the LMB protected areas.

3 Results

Five bioclimatic zones comprised of 25 bioclimatic strata were found within the LMB (Fig. 2), with zones ranging from extremely hot and xeric at the lower elevations to warm temperate and mesic at higher elevations. Mean annual temperatures are typically inversely correlated with average elevation. Additionally, 92.5% of the bioclimatic zones in the LMB are classified as extremely hot at baseline, and this percentage is expected to increase in 2030 under RCP8.5 (Table 2). Of particular note in 2030 is the reduction of the extent of the warm temperate and mesic zone from 1.5% to 0.2%. This represents a substantial, near-term loss of an already rare bioclimatic zone. There is also an increase in extremely hot and xeric (dry) from 11.4% to 17.1% (Table 2). By the year 2060 the zones appear to be limited to two primary types, extremely hot and mesic and extremely hot and moist, comprising 93.2% of the total area. Baseline LMB climate generally provides moist to mesic conditions conducive to forested ecosystems and rainfed agriculture; however, the warm temperate and mesic zone within protected areas diminishes under all scenarios and nearly disappears under RCP 8.5 by 2060. Bioclimatic zones for baseline and future climates in 2030 and 2060 across all emissions scenarios are illustrated (Fig. 3) and summarized for year 2060 across all emissions scenarios and compared to baseline and the entire LMB (Fig. 4).

The mean elevation of the bioclimatic zones within protected areas shifted upwards within mountainous terrain and across large plains or plateaus with elevation gradients under higher emission scenarios (Table 3). The extremely hot and mesic zone decreased in elevation; however, this is an artifact of the dramatic increase in extent of this zone across the LMB. The lower elevation, extremely hot and xeric zone shifted only marginally upslope in the highest emissions scenario by 2060, reflecting the small extent of this bioclimatic zone.

To emphasize the magnitude of bioclimatic change accompanying climate change, the percentage of the total area of each protected area that shifts to a different bioclimatic zone is shown (Fig. 5). Percent shifts in bioclimatic zone and strata are also summarized for the LMB and across all protected areas for both 2030 and 2060 and across all emissions scenarios (Table 4). Greater than 11% to almost 38% of the LMB is projected to shift to a different bioclimatic zone by 2030, and 9% to almost 88% will shift to a new bioclimatic zone by 2060 under RCP 8.5 (Table 4). Basinwide results were slightly greater than compared to protected areas, with 9% to 29% projected to shift to a different bioclimatic zone by 2030 and from 7% to more than 77% by 2060.

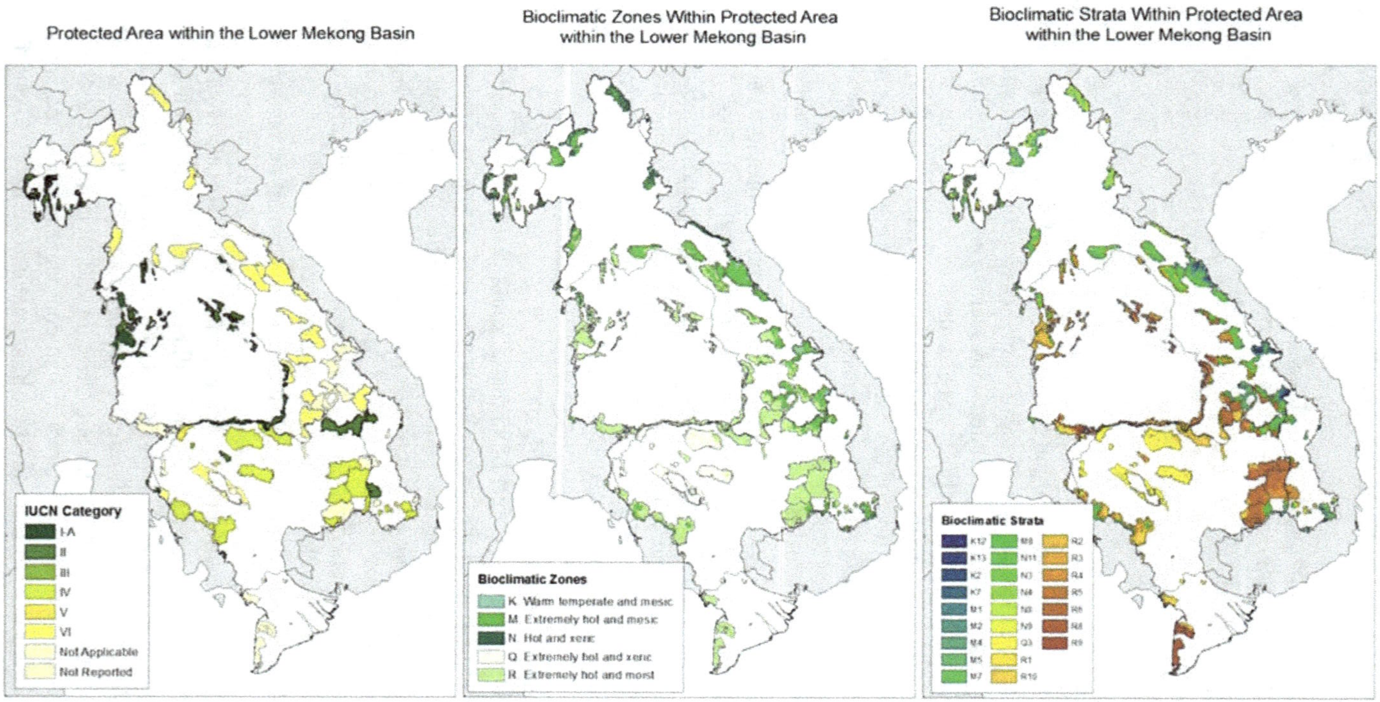

Fig. 2 Protected areas within the Lower Mekong Basin, also bioclimatic zones and strata within protected areas. (Source: UNEP-WCMC 2019)

Table 2 Areal extent (in km^2) and percent of total area (bottom) of bioclimatic zones within Lower Mekong Basin protected areas

Bioclimatic zone	Zone	1995	2030			2060		
			RCP_2.6	RCP_4.5	RCP_8.5	RCP_2.6	RCP_4.5	RCP_8.5
Area of bioclimatic zones within protected area within the Lower Mekong Basin								
Warm Temperate and Mesic	K	1,695	1,131	734	246	1,265	222	12
Extremely Hot and Mesic	M	29,136	26,615	25,145	27,596	27,189	29,487	79,688
Hot and Xeric	N	6,471	4,686	3,527	2,072	5,062	1,974	203
Extremely Hot and Xeric	Q	12,507	15,579	17,934	18,758	14,726	17,787	7,165
Extremely Hot and Moist	R	60,021	61,819	62,490	61,158	61,158	60,360	22,782
Percental of all protected area within the Lower Mekong Basin within bioclimatic zones								
Warm Temperate and Mesic	K	1.5	1.0	0.7	0.2	1.2	0.2	0.0
Extremely Hot and Mesic	M	26.5	24.2	22.9	25.1	24.8	26.8	72.5
Hot and Xeric	N	5.9	4.3	3.2	1.9	4.6	1.8	0.2
Extremely Hot and Xeric	Q	11.4	14.2	16.3	17.1	13.4	16.2	6.5
Extremely Hot and Moist	R	54.6	56.3	56.9	55.7	56.1	55.0	20.7

Source: UNEP-WCMC (2019)

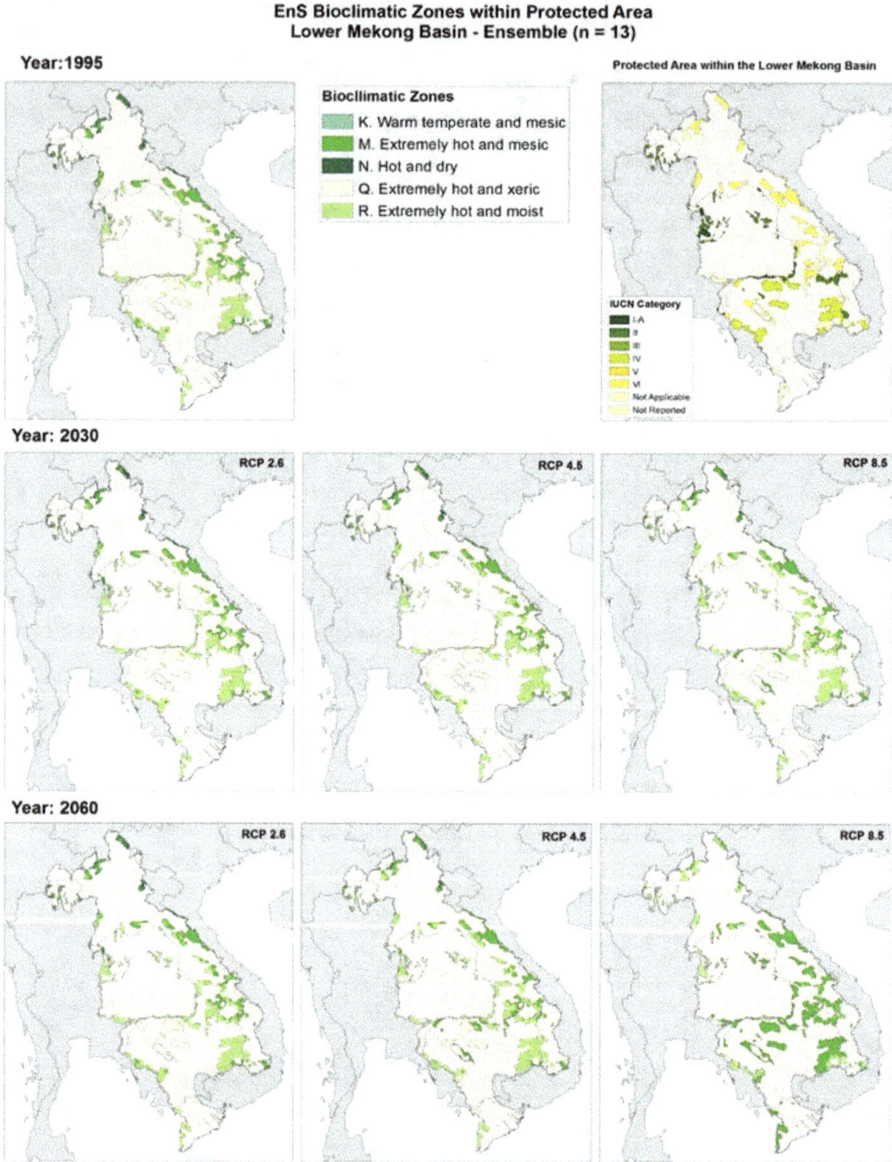

Fig. 3 Environmental Stratification (EnS) bioclimatic zones within protected areas in the Lower Mekong Basin, for baseline (1995), and 2030 and 2060 under emission scenarios RCP 2.6, 4.5, and 8.5

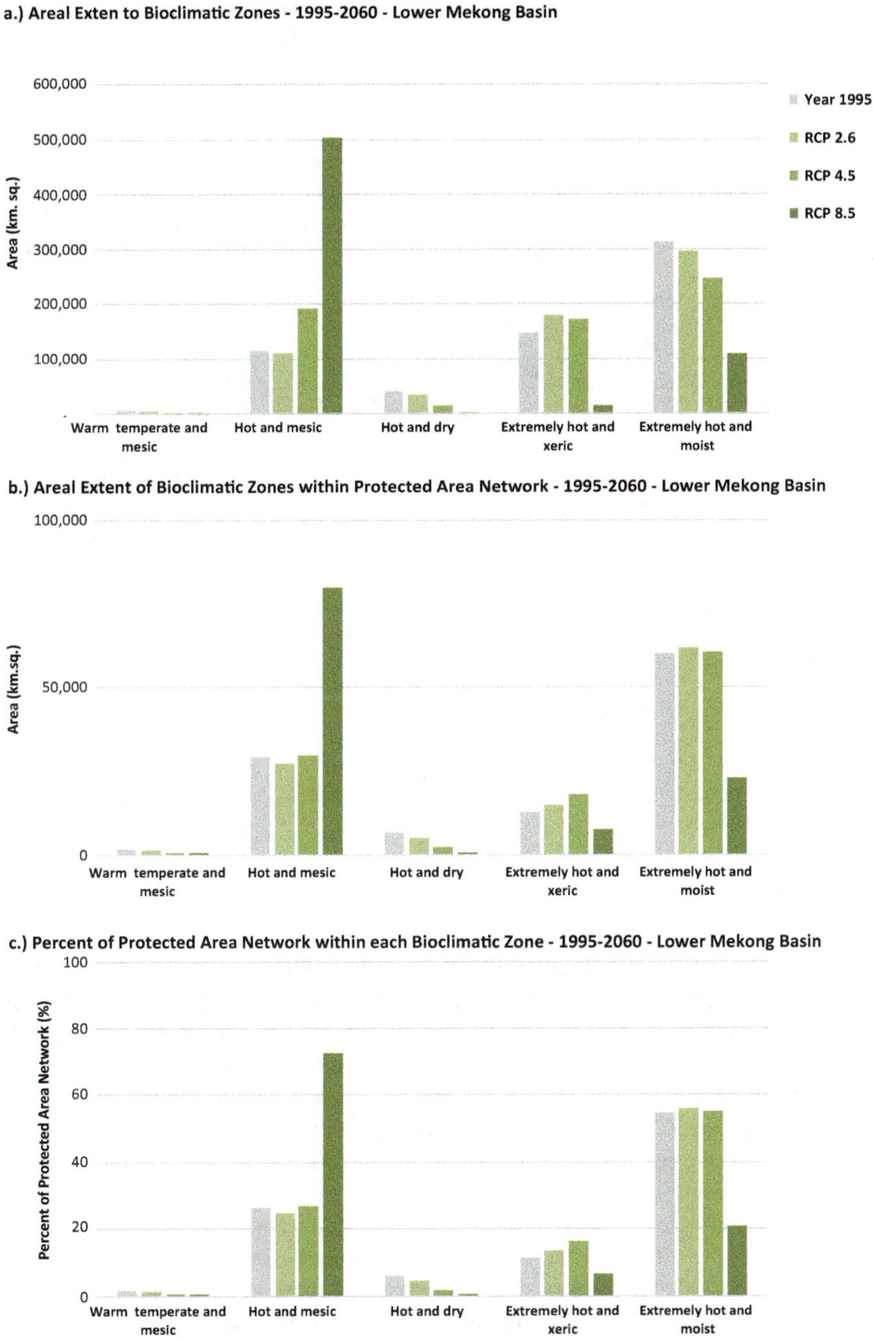

Fig. 4 The extent of bioclimatic zones in the Lower Mekong Basin (**a**) basinwide, (**b**) within protected areas, and (**c**) as percent of the total protected area in bioclimatic zone, from baseline (1995) to 2060 under three emission scenarios RCP 2.6, 4.5, and 8.5

Table 3 Projected change in the mean elevation above sea level (m asl) of bioclimatic zones in protected areas within the Lower Mekong Basin by 2030 and 2060, under three emission scenarios RCP 2.6, 4.5, and 8.5

		Elevation				Elevation shift		
		1995	RCP 2.6	RCP 4.5	RCP 8.5	RCP 2.6	RCP 4.5	RCP 8.5
Bioclimatic zone	Zone	m asl				m		
Projected change in mean elevation of bioclimatic zones and their upward shift by 2030—LMB								
Warm temperate and Mesic	K	1,535	1,601	1,666	1,823	66	131	288
Extremely hot and Mesic	M	721	783	824	564	62	103	**(157)**
Hot and Xeric	N	1,130	1,206	1,263	1,364	76	133	235
Extremely Hot and Xeric	Q	96	106	112	129	10	16	32
Extremely Hot and Moist	R	219	252	279	332	33	59	113
Projected change in mean elevation of bioclimatic zones and their upward shift by 2060—LMB								
Warm Temperate and Mesic	K	1,535	1,584	1,837	2,211	50	303	676
Extremely Hot and Mesic	M	721	769	496	268	48	**(225)**	**(453)**
Hot and Xeric	N	1,130	1,188	1,373	1,694	58	243	564
Extremely Hot and Xeric	Q	96	104	134	504	8	38	408
Extremely Hot and Moist	R	219	244	344	733	24	125	513

Numbers in bold indicate a negative shift

4 Discussion

As indicated by the roughly one-third (year 2030, RCP8.5) and two-thirds (2060, RCP8.5) of protected areas shifting to a new bioclimatic zone from baseline, climate change will have a substantial impact on protected areas in the LMB. Bioclimatic zones correspond to major vegetation assemblages (i.e., dominant and co-dominants) and translate broadly to landcover types such as coniferous, deciduous and evergreen forest. Zonal transitions are more profound than strata regarding ecosystem change, because a shift to a different bioclimatic zone indicates novel bioclimatic conditions with direct consequences for the biota and ecosystem function of that protected area. Changes in bioclimatic strata correspond to the level of vegetation species and can be investigated further if foundation species are of interest. Understanding the nature and direction of these changes provides crucial information for adaptation planning and management. In the near term (year 2030, RCP8.5) protected areas in central and northern LMB require attention to adaptation planning (i.e., average zonal shifts of 60% or greater). By 2060, under emissions scenario RCP8.5, protected areas all across the LMB are at risk. Only a few protected

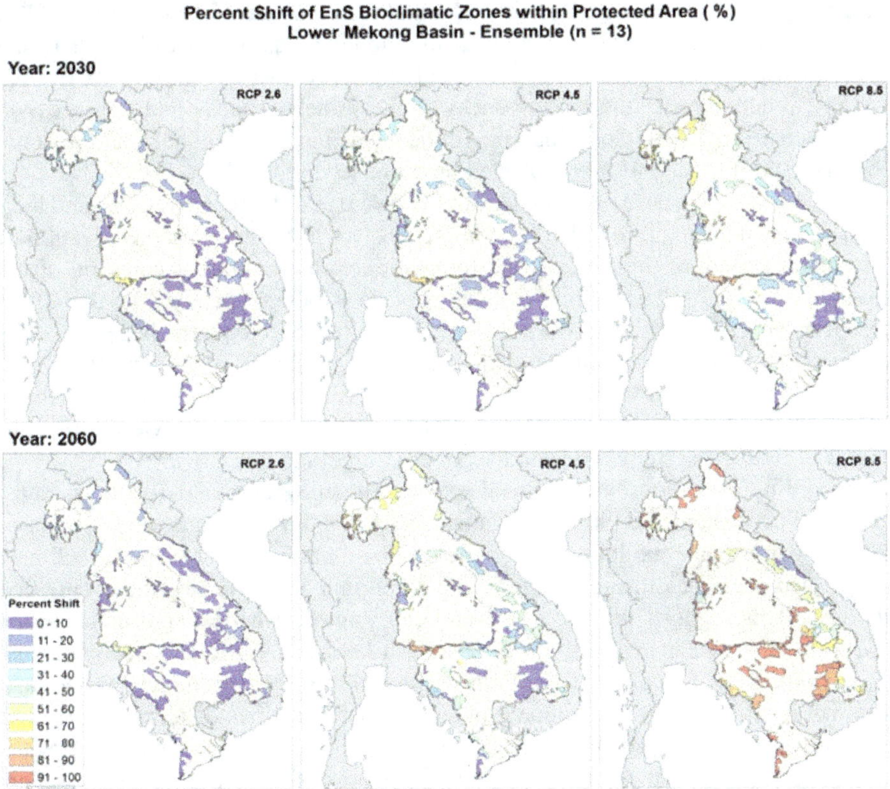

Fig. 5 Average percent of protected areas within the Lower Mekong Basin that shifted to a different bioclimatic zone from baseline (1995) as projected for 2030 and 2060 across three emission scenarios RCP 2.6, 4.5, 8.5

Table 4 Percent of total area for the Lower Mekong Basin (LMB) and protected areas that shifted to a different bioclimatic zone or strata by 2030 and 2060, under three emission scenarios RCP 2.6, 4.5, and 8.5

Percent of area shifting to a different bioclimatic zone or strata						
	2030			2060		
	RCP 2.6	RCP 4.5	RCP 8.5	RCP 2.6	RCP 4.5	RCP 8.5
Bioclimatic zones						
LMB-all	11.8	19.6	37.4	9.2	42.2	87.7
LMB-Protected area	9.3	16	29.1	7.1	31.7	77.6
Bioclimatic strata						
LMB-all	24.6	38.2	57.8	19.4	62.5	94.8
LMB-Protected area	32.1	50.8	68.2	24.9	70.4	90.2

areas are expected to experience minimal zonal shifts (i.e., average shift of 0–20%), and these areas are mainly located along the mountainous border of Lao PDR and Vietnam. Moreover, climate change is a landscape-level disturbance that is expected to substantially impact the effectiveness of the protected areas and conservation efforts across the LMB. Coordination will be needed across provincial and national borders to compensate for the projected bioclimatic shifts.

The need to synthesize existing data and knowledge of LMB ecosystems and their services is fundamental to EBM. A comprehensive, basin-wide characterization of the natural capital of the LMB is the foundation of an accurate accounting of the costs and benefits of sustainable development decisions to society: "To date there has been no comprehensive overview of the value of ecosystem benefits in the region, leading to a serious undervaluing by both politicians and even many local communities. A full review of Mekong ecosystem services is urgently overdue" (WWF 2013). Ideally, this should also include a comprehensive review of beneficiaries of final ecosystem goods and services (Ringold et al. 2010).

According to various international experts (Leadley et al. 2010), the greatest drivers of ecosystem and biodiversity change are "land use change, modification of river flow, freshwater pollution, and exploitation of marine resources" with "climate change and ocean acidification increasingly important drivers during the twenty first century." This suggests that the freshwater ecosystems, fisheries, and forests of the LMB are most threatened in the near term. Furthermore, "if greenhouse gas emissions continue along current trajectories, several Earth System models project that this will result in far greater climate-induced transformations of terrestrial biomes and marine biota than projected in earlier global biodiversity assessments." There is reason for optimism though, as the authors point out that "new socio-economic scenarios point to plausible development pathways of low greenhouse gas emissions and low land conversion that could lead to much lower biodiversity impacts." These scenarios of future development may be optimistic, but they are consistent with sustainable development goals and "require fundamental changes in development paradigms, but are coherent with known constraints on economics, resource use and human development goals" (Leadley et al. 2010).

There are challenges though to a complete characterization of LMB ecosystems. There is no single authoritative source, and there are gaps in the data which require further research for both freshwater and forest ecosystems, including habitat quality (WWF 2013). An exception is the ecological study of the Mekong upper floodplain and wetlands above Vientiane (IUCN 2013). This study is also exceptional for its companion socioeconomic study of community livelihoods along the river and surrounding forest that describe human uses of natural capital for both food and household income (Singer 2013). The Critical Ecosystem Partnership Fund provided a comprehensive review of hotspots of conservation potential in the larger Indo-Burma region that includes the LMB (CEPF 2011). Their strategic review addressed not only biodiversity but also the important social, legal, and political dimensions that must be considered when allocating resources for conservation of natural capital. However, the scope of their review is broad and provides little information for ecoregion characterization.

The World Wildlife Fund, with an ecosystem services emphasis, conducted a study of forest cover change in the Greater Mekong and compared their results to other data sources (WWF 2013). They provided maps contrasting two alternative scenarios that spanned a green energy, conservation-balanced future with business-as-usual exploitation of natural capital. They also addressed freshwater ecosystems, identifying 13 distinct but connected aquatic systems along the Mekong. The authors caution that information on ecosystem quality is particularly lacking. For example, remotely-sensed landcover data products typically do not discriminate between primary forest and disturbed, fragmented forest cover, although primary forest is known to be scarce in Lao PDR and Thailand (WWF 2013).

Overall, our understanding of social, economic and environmental system interactions and their dynamics is in its infancy, and experts argue in favor of a precautionary approach. The combined effects of time lags between drivers of ecosystem response and the measured indicators, thresholds of system change that result in nonlinear, accelerated response, and tipping points of irreversible ecosystem change, mean that "the impacts of global change on biodiversity are hard to predict, difficult to control once they begin, and slow and expensive to reverse once they have occurred" (Leadley et al. 2010). Fortunately, EBM and EbA provide useful, practical strategies for dealing with these uncertainties.

Acknowledgements This chapter was reviewed in accordance with the policy of the Center for Environmental Measurement & Modeling, U.S. Environmental Protection Agency, and approved for publication. However, the research presented was not performed or funded by EPA and was not subject to EPA's quality system requirements. Approval for publication does not signify that the contents necessarily reflect the view and policies of the Agency, nor does mention of trade names or commercial products constitute endorsement or recommendation for use. The authors express sincere thanks to the Mekong River Commission (MRC) Climate Change and Adaptation Initiative (CCAI) for funding this research and providing GIS and climate data. Aekkapol Aekakkararungroj (ADPC SERVIR-Mekong) is credited with providing Figure 1. The US State Department Embassy Science Fellows Program is acknowledged for sponsoring John M. Johnston to collaborate with the MRC on this study.

Disclaimer This chapter has been subjected to Agency review and has been approved for publication. The views expressed in this paper are those of the author(s) and do not necessarily reflect the views or policies of the U.S. Environmental Protection Agency.

References

Bagstad, K. J., Semmens, D. J., Waage, S., & Winthrop, R. (2013a). A comparative assessment of decision-support tools for ecosystem services quantification and valuation. *Ecosystem Services, 5*(0), 27–39. https://doi.org/10.1016/j.ecoser.2013.07.004.

Bagstad, K. J., Semmens, D. J., & Winthrop, R. (2013b). Comparing approaches to spatially explicit ecosystem service modeling: A case study from the San Pedro River, Arizona. *Ecosystem Services, 5*(0), 40–50. https://doi.org/10.1016/j.ecoser.2013.07.007.

Bai, Y., Zhuang, C., Ouyang, Z., Zheng, H., & Jiang, B. (2011). Spatial characteristics between biodiversity and ecosystem services in a human-dominated watershed. *Ecological Complexity, 8* (2), 177–183. https://doi.org/10.1016/j.ecocom.2011.01.007.

CEPF. (2011). *Ecosystem profile: Indo-Burma biodiversity hotspot*. Retrieved from https://www.cepf.net/sites/default/files/final.indoburma_indochina.ep_.pdf

Chiang, L.-C., Lin, Y.-P., Huang, T., Schmeller, D. S., Verburg, P. H., Liu, Y.-L., & Ding, T.-S. (2014). Simulation of ecosystem service responses to multiple disturbances from an earthquake and several typhoons. *Landscape and Urban Planning, 122*(0), 41–55. https://doi.org/10.1016/j.landurbplan.2013.10.007.

Daily, G. C., Polasky, S., Goldstein, J., Kareiva, P. M., Mooney, H. A., Pejchar, L., et al. (2009). Ecosystem services in decision making: Time to deliver. *Frontiers in Ecology and the Environment, 7*(1), 21–28. https://doi.org/10.1890/080025.

Delphin, S., Escobedo, F. J., Abd-Elrahman, A., & Cropper, W., Jr. (2013). Mapping potential carbon and timber losses from hurricanes using a decision tree and ecosystem services driver model. *Journal of Environmental Management, 129*(0), 599–607. https://doi.org/10.1016/j.jenvman.2013.08.029.

Feng, M., Liu, S., Euliss, N. H., Jr., Young, C., & Mushet, D. M. (2011). Prototyping an online wetland ecosystem services model using open model sharing standards. *Environmental Modelling & Software, 26*(4), 458–468. https://doi.org/10.1016/j.envsoft.2010.10.008.

Fu, B., Wang, Y. K., Xu, P., Yan, K., & Li, M. (2014). Value of ecosystem hydropower service and its impact on the payment for ecosystem services. *Science of The Total Environment, 472*(0), 338–346. https://doi.org/10.1016/j.scitotenv.2013.11.015.

Harmáčková, Z. V., & Vačkář, D. (2015). Modelling regulating ecosystem services trade-offs across landscape scenarios in Třeboňsko Wetlands Biosphere Reserve, Czech Republic. *Ecological Modelling, 295*(0), 207–215. https://doi.org/10.1016/j.ecolmodel.2014.10.003.

Hoyer, R., & Chang, H. (2014). Assessment of freshwater ecosystem services in the Tualatin and Yamhill basins under climate change and urbanization. *Applied Geography, 53*(0), 402–416. https://doi.org/10.1016/j.apgeog.2014.06.023.

ISPONRE. (2013). *Operational framework for ecosystem-based adaptation to climate change for Vietnam: A policy supporting document.* Retrieved from Hanoi, Viet Nam: http://awsassets.panda.org/downloads/wwf_vietnam_eba_operational_framework_2013.pdf.

IUCN. (2013). *Ecological survey of the Mekong River between Louangphabang and Vientiane Cities, Lao PDR, 2011–2012.* Retrieved from Vientiane, Lao PDR: https://cmsdata.iucn.org/downloads/ecologicalsurveymekong__iucn_cepf.pdf.

Jackson, B., Pagella, T., Sinclair, F., Orellana, B., Henshaw, A., Reynolds, B., et al. (2013). Polyscape: A GIS mapping framework providing efficient and spatially explicit landscape-scale valuation of multiple ecosystem services. *Landscape and Urban Planning, 112*(0), 74–88. https://doi.org/10.1016/j.landurbplan.2012.12.014.

Lassiter, R. R., Box, E. O., Wiegert, R. G., Johnston, J. M., Bergengren, J., & Suárez, L. A. (2000). Vulnerability of ecosystems of the mid-Atlantic region, USA, to climate change. *Environmental Toxicology and Chemistry, 19*(4), 1153–1160. https://doi.org/10.1002/etc.5620190448.

Leadley, P., Pereira, H. M., Alkemade, R., Fernandez-Manjarrés, J. F., Proença, V., Scharlemann, J. P. W., & Walpole, M. J. (2010). *Biodiversity scenarios: Projections of 21st century change in biodiversity and associated ecosystem services.* Retrieved from Montreal, Canada: https://www.cbd.int/doc/publications/cbd-ts-50-en.pdf.

Leh, M. D. K., Matlock, M. D., Cummings, E. C., & Nalley, L. L. (2013). Quantifying and mapping multiple ecosystem services change in West Africa. *Agriculture, Ecosystems & Environment, 165*(0), 6–18. https://doi.org/10.1016/j.agee.2012.12.001.

Linke, P. (2014). On the development of strategies for water and energy management in the context of the water-energy-food nexus. In J. D. S. Mario R. Eden & P. T. Gavin (Eds.), *Computer aided chemical engineering* (34, pp. 196–201). Amsterdam: Elsevier.

Marquès, M., Bangash, R. F., Kumar, V., Sharp, R., & Schuhmacher, M. (2013). The impact of climate change on water provision under a low flow regime: A case study of the ecosystems services in the Francoli river basin. *Journal of Hazardous Materials, 263*(Part 1(0)), 224–232. https://doi.org/10.1016/j.jhazmat.2013.07.049.

MEA. (2005). Ecosystems and human health: Some findings from the Millenium ecosystem assessment. In *Ecosystems and human well-being: Health synthesis.* Geneva, Switzerland: World Health Organization.

Metzger, M. J., Brus, D. J., Bunce, R. G. H., Carey, P. D., Gonçalves, J., Honrado, J. P., et al. (2013a). Environmental stratifications as the basis for national, European and global ecological monitoring. *Ecological Indicators, 33*, 26–35. https://doi.org/10.1016/j.ecolind.2012.11.009.

Metzger, M. J., Bunce, R. G. H., Jongman, R. H. G., Sayre, R., Trabucco, A., & Zomer, R. (2013b). A high-resolution bioclimate map of the world: A unifying framework for global biodiversity research and monitoring. *Global Ecology and Biogeography, 22*(5), 630–638. https://doi.org/10.1111/geb.12022.

MONRE. (2013). *Strengthening community and ecosystem resilience against climate change impacts: Lao PDR case study from field testing an operational framework for ecosystem-based adaptation*. Retrieved from Vientiane, Lao PDR: http://www.monre.gov.la/home/.

Monzón, J., Moyer-Horner, L., & Palamar, M. B. (2011). Climate change and species range dynamics in protected areas. *Bioscience, 61*(10), 752–761. https://doi.org/10.1525/bio.2011.61.10.5.

MRC. (2010). *State of the Basin report 2010*. Retrieved from Vientiane, Lao PDR: http://www.mrcmekong.org/assets/Publications/basin-reports/MRC-SOB-report-2010full-report.pdf.

MRC. (2011a). *Assessment of basin-wide development scenarios: Main report*. Vientiane, Laos: Mekong River Commission. http://www.mrcmekong.org/assets/Publications/basin-reports/BDP-Assessment-of-Basin-wide-Dev-Scenarios-2011.pdf.

MRC. (2011b). *Climate change adaptation initiative: 2011–2015 programme document*. Retrieved from Vientiane, Lao PDR: http://www.mrcmekong.org/assets/CCAI-2011-2015-documentFinal.pdf.

MRC. (2011c). *Planning atlas of the lower Mekong River basin*. Retrieved from http://www.mrcmekong.org/assets/Publications/basin-reports/BDP-Atlas-Final-2011.pdf

Munang, R., Thiaw, I., Alverson, K., Mumba, M., Liu, J., & Rivington, M. (2013). Climate change and ecosystem-based adaptation: A new pragmatic approach to buffering climate change impacts. *Current Opinion in Environmental Sustainability, 5*(1), 67–71. https://doi.org/10.1016/j.cosust.2012.12.001.

Ranganathan, J., Raudsepp-Hearne, C., Lucas, N., Irwin, F., Zurek, M., Bennett, K., et al. (2008). *Ecosystem services: A guide for decision makers*. Retrieved from http://pdf.wri.org/ecosystem_services_guide_for_decisionmakers.pdf

Rasul, G. (2014). Food, water, and energy security in South Asia: A nexus perspective from the Hindu Kush Himalayan region. *Environmental Science & Policy, 39*(0), 35–48. https://doi.org/10.1016/j.envsci.2014.01.010.

Ringold, P. L., Nahlik, A. M., Boyd, J., & Bernard, D. (2010). *Report from the workshop on indicators of final ecosystem goods and services for wetlands and estuaries*. Retrieved from http://www.epa.gov/nheerl/arm/streameco/

Rosenthal, A., Arkema, K., Verutes, G., Bood, N., Cantor, D., Fish, M., et al. (2013). *Identification and valuation of adaptation options in coastal-marine ecosystems: Test case from Placencia, Belize*. Retrieved from Stanford University: https://www.mpaaction.org/sites/default/files/Rosenthal%20et%20al_2013_Identification%20and%20Valuation%20of%20Adaptation%20Options%20in%20Coastal-Marine%20Ecosystems.pdf.

Sánchez-Canales, M., López Benito, A., Passuello, A., Terrado, M., Ziv, G., Acuña, V., et al. (2012). Sensitivity analysis of ecosystem service valuation in a Mediterranean watershed. *Science of The Total Environment, 440*(0), 140–153. https://doi.org/10.1016/j.scitotenv.2012.07.071.

Sierra-Correa, P. C., & Cantera Kintz, J. R. (2015). Ecosystem-based adaptation for improving coastal planning for sea-level rise: A systematic review for mangrove coasts. *Marine Policy, 51*(0), 385–393. https://doi.org/10.1016/j.marpol.2014.09.013.

Singer, U. (2013). *Livelihoods and resource management survey on the Mekong between LouangPhabang and Vientiane cities, Lao PDR*. Retrieved from Vientiane, Lao PDR: https://portals.iucn.org/library/sites/library/files/documents/2013-054.pdf.

Su, C., & Fu, B. (2013). Evolution of ecosystem services in the Chinese Loess Plateau under climatic and land use changes. *Global and Planetary Change, 101*(0), 119–128. https://doi.org/10.1016/j.gloplacha.2012.12.014.

Trisurat, Y., Aekakkararungroj, A., Ma, H.-o., & Johnston, J. M. (2018). Basin-wide impacts of climate change on ecosystem services in the Lower Mekong Basin. *Ecological Research, 33*(1), 73–86. https://doi.org/10.1007/s11284-017-1510-z.

Trisurat, Y., Shirakawa, H., & Johnston, J. M. (2019). Land-use/land-cover change from socio-economic drivers and their impact on biodiversity in Nan Province, Thailand. *Sustainability, 11* (3), 649.

UNEP-WCMC. (2019). *The world database on protected areas (WDPA)*. Retrieved from www.protectedplanet.net.

USGCRP. (2018). *Impacts, risks, and adaptation in the United States: Fourth national climate assessment, Volume II*. Retrieved from Washington, DC, USA: https://nca2018.globalchange.gov/.

Vigerstol, K. L., & Aukema, J. E. (2011). A comparison of tools for modeling freshwater ecosystem services. *Journal of Environmental Management, 92*(10), 2403–2409. https://doi.org/10.1016/j.jenvman.2011.06.040.

Vignola, R., McDaniels, T. L., & Scholz, R. W. (2013). Governance structures for ecosystem-based adaptation: Using policy-network analysis to identify key organizations for bridging information across scales and policy areas. *Environmental Science & Policy, 31*(0), 71–84. https://doi.org/10.1016/j.envsci.2013.03.004.

Wamsler, C., Luederitz, C., & Brink, E. (2014). Local levers for change: Mainstreaming ecosystem-based adaptation into municipal planning to foster sustainability transitions. *Global Environmental Change, 29*(0), 189–201. https://doi.org/10.1016/j.gloenvcha.2014.09.008.

WWF. (2013). *Ecosystems in the Greater Mekong: Past trends, current status, possible futures*. Retrieved from wwf.panda.org/greatermekong.

Yokoyama, S., Tanaka, K., & Phalakhone, K. (2006). *Forest policy and Swidden agriculture in Laos*. Paper presented at the NIE-SEAGA: Sustainability and Southeast Asia, Singapore.

Zhang, M.-G., Zhou, Z.-K., Chen, W.-Y., Cannon, C. H., Raes, N., & Slik, J. W. F. (2014). Major declines of woody plant species ranges under climate change in Yunnan, China. *Diversity and Distributions, 20*(4), 405–415. https://doi.org/10.1111/ddi.12165.

Zomer, R. J., Trabucco, A., Metzger, M. J., Wang, M., Oli, K. P., & Xu, J. (2014). Projected climate change impacts on spatial distribution of bioclimatic zones and ecoregions within the Kailash Sacred Landscape of China, India, Nepal. *Climatic Change, 125*(3), 445–460. https://doi.org/10.1007/s10584-014-1176-2.

Zomer, R. J., Xu, J., Wang, M., Trabucco, A., & Li, Z. (2015). Projected impact of climate change on the effectiveness of the existing protected area network for biodiversity conservation within Yunnan Province, China. *Biological Conservation, 184*, 335–345. https://doi.org/10.1016/j.biocon.2015.01.031.

9

Advancing Aquatic Ecosystem-Based Management with Full Consideration of the Social-Ecological System

Gerjan Piet, Gonzalo Delacámara, Marloes Kraan, Christine Röckmann, and Manuel Lago

Abstract In this study we present an integrated Ecosystem-Based Management (EBM) approach that attempts to reconcile several concepts including integrated ecosystem assessment (IEA), marine spatial planning, resilience thinking, and complex adaptive systems. The approach builds on the IEA process but enhances it by explicitly considering the full social-ecological system (SES) and the creation of a generic framework for assessment of ecosystem status and management strategy evaluation.

Lessons Learned

- This approach reconciles many existing concepts that describe the ecological system, the social (or socio-economic) system and EBM into a unifying approach
- It consists of concrete steps which identify issues for the practitioner to consider, gives examples that provide the basis for a common framework,
- It provides guidance on how to make the framework (more) operational and is applicable to any aquatic ecosystem

G. Piet (✉)
Wageningen Marine Research, IJmuiden, The Netherlands
e-mail: gerjan.piet@wur.nl

G. Delacámara
IMDEA Water Institute, Alcalá de Henares, Madrid, Spain

M. Kraan
Wageningen Marine Research, IJmuiden, The Netherlands

Environmental Policy Group, Wageningen University, Wageningen, Netherlands

C. Röckmann
Wageningen Marine Research, IJmuiden, The Netherlands

Wageningen Economic Research, The Hague, The Netherlands

M. Lago
Ecologic Institute, Berlin, Germany

- It allows the incorporation and synthesis of interdisciplinary information on SES into practical and useful linkage frameworks for EBM plan development and implementation

Needs to Advance EBM

- Where consideration of the full SES can be overwhelmingly complex leading to inaction, we propose to work with a subsection of the SES (called subSES). This subSES can then be the starting point for building the knowledge base for EBM decision making. We provide practical guidance how to construct such a knowledge base for both the ecological and the social system.
- The subSES, in conjunction with the available knowledge base, then drives the development of the knowledge base and determines the type of risk assessment(s) that can be applied. Science should then inform the process to translate the high-level societal goals into operational objectives, identify the main barriers that prevent achievement of these objectives, and guide the relevant authorities that develop an EBM plan.
- A novel component of this approach is that the EBM plan distinguishes between management measures (interacting with the ecological system) and policy instruments (interacting with social processes) that together harness the knowledge base of the subSES.

1 Introduction

Conventional management of aquatic resources, based on a specific policy or directed to a sub-sector or flagship species, has often failed to deliver on the policy objectives (Long et al. 2015). As an alternative, Ecosystem-Based Management (EBM), which has a more holistic understanding of the ecosystem and its linkages, is widely accepted as the key concept to guide contemporary decision-making (Börgstrom et al. 2015; Cormier et al. 2017). The conventional management perspectives have assumed, either implicitly or explicitly, that effective policy making is hindered by lacking or inadequate knowledge of ecological processes, functions, and services (Ruckelshaus et al. 2009). Such an ecological focus has failed to produce the full picture of best available knowledge for effective decision-making (Christie 2011). Cormier et al. (2017) distinguish decision-making, which is essentially a specific choice among alternatives, from policy-making, which is a process of identifying a problem and setting societal goals and objectives. In practice, the policy is implemented through a management plan expected to 'carry into effect' the policy objective. An effective policy cycle requires the incorporation of social and institutional processes, such as the involvement of various institutional actors (Röckmann et al. 2015) and an understanding of the governance context (Bissix and Rees 2001). The need for a more holistic approach that incorporates ecological and socio-economic factors (Bianchi 2008), however, can result in in-action due to overwhelming complexity (DeFries and Nagendra 2017).

As a paradigm, EBM addresses uncertainty and complexity, it is an interdisciplinary visioning of multiple objectives, and as such, EBM can be categorised as a 'wicked problem' (Berkes 2012). 'Wicked problems' have no definitive formulation, no clear stopping rule and no objectively right or wrong solutions and no final resolution (Rittel and Webber 1973); (Ludwig 2001). DeFries and Nagendra (2017) suggest the following solutions for addressing wicked problems: multi-sectoral decision-making; institutions that enable management to span across administrative boundaries; adaptive management; markets that incorporate natural capital; and collaborative processes to engage diverse stakeholders and address inequalities. Integrating environmental and socio-economic processes (including institutional, ethical and cultural) requires a single conceptual framework (Christie 2011). The concept of social-ecological systems (SES) can provide such a framework (De Lange et al. 2010). This SES can be split into smaller social-ecological sub-systems (subSES) to address the issue of complexity.

Relevant information and processes can be considered in individual compartments that form the SES. Applying the concept of SES in aquatic EBM implies that any distinction between social and natural systems is artificial and arbitrary by definition, because they are connected (Berkes 2012).

Socio-Ecological Systems can be considered complex adaptive systems (CAS) (De Lange et al. 2010) as the structure, functions, and dynamics of CAS emerge from the interaction and connectedness of the system's constituent parts and with other systems (Hagstrom and Levin 2017). By acknowledging that SES are CAS, management can overcome the drawbacks of conventional approaches. Instead of searching for optimal solutions, linear dynamics, or marginal changes under complete information, a shift towards a more dynamic management approach can be made, where non-linear changes, uncertainty, and surprise are intrinsic characteristics of the system. In addition to the known unknowns, e.g., lack of historic data of species, CAS come with new uncertainties that cannot be tackled through standard sensitivity analysis (Polasky et al. 2011).

Sustainability is among the ultimate objectives of EBM (Long et al. 2015). Ostrom (2009) identified four core subsystems of SES sustainability: (1) resource systems; consisting of (2) resource units; (3) governance systems; and (4) users. When modelling a SES, it may be worth distinguishing these subsystems. In the face of ongoing changes and their uncertain consequences as well as exposure to uncertain shocks, the key to sustainability is enhancing the resilience of a SES (Folke et al. 2005; Nelson et al. 2006; Biggs et al. 2015). Enhancing resilience in terms of persistence, adaptability and transformability (Folke et al. 2010) means preserving the SES's adaptive capacity in order to remain within a certain range of conditions that meet the sustainability goals. Resilience thinking promotes governance frameworks that are able to reconcile the conflicting interests and visions of different stakeholders in a transparent and accountable way so as to foster cooperation among them and enhance stakeholders' ability to commit to legitimate and transparent policy objectives (Dietz et al. 2003). In addition, these governance frameworks should also pave the way to achieve collectively agreed goals through robust institutions, with stakeholders who are able to regularly adjust to changes in the ecological and the social-economic systems (Nelson et al. 2006).

Linkage frameworks such as DPSIR, Driver-Pressure-State-Impact-Response (OECD 1994; EEA 1995; Elliott 2002), are commonly used in the context of environmental management to describe how human activities impact the state of the ecosystem (Halpern et al. 2008; Knights et al. 2013) and hence the supply of ecosystem services to human well-being (Elliott et al. 2017). Linkage frameworks rely on accurate descriptions of linkages (e.g., stressor-receptor or pressure-state relationships) and can be informed by qualitative, quantitative, or expert judgement-based assessments, or any combination of these (Knights et al. 2014). Such linkage frameworks have been applied in an EBM context to guide the selection of management measures and their evaluation using risk-based approaches (Knights et al. 2015; Piet et al. 2015; Borgwardt et al. 2019; Teixeira et al. 2019; Culhane et al. 2019).

Few, if any, examples of EBM planning initiatives informed by advanced science have been implemented across multiple sectors (Katsanevakis et al. 2011; Cormier et al. 2017). In order to enhance the use of salient science (Röckmann et al. 2015) into the policy-making process, Cormier et al. (2017) proposes four steps: strategic goal setting; tactical objectives; management measures; and adaptive management. Integrated ecosystem assessments (IEAs) have also been proposed as a tool to operationalise EBM as they provide a framework for organizing science in order to inform decisions in marine EBM at multiple scales and across sectors (Levin et al. 2009a; Walther and Möllmann 2014; Tallis et al. 2010; Harvey et al. 2017). It is therefore not surprising that the IEA process described by (Levin et al. 2009b) is closely aligned to the EBM processes described by Börgstrom et al. (2015) or Ansong et al. (2017). The (further) development and operationalization of an IEA/EBM approach is however hindered by the lack of a systematic, critical appraisal.

In this study, we merge the existing IEA and EBM processes into a single generic approach that can be applied in all aquatic systems. This approach is then enhanced by explicitly considering the full social-ecological system (SES), which incorporates both environmental and socio-economic processes (Christie 2011). In order to advance this process, we provide the systematic criteria that allow critical appraisal of its progress based on the aspects of EBM, identified in the review by Long et al. (2015) together with additional practical guidance that emerges from concepts such as resilience thinking and CAS. Further, we provide operational guidance on the development of relevant subSES to inform this approach.

2 Advancing EBM

EBM should be able to guide decision-makers by identifying trade-offs between societal goals (Walther and Möllmann 2014). These trade-offs may involve the choice between the conservation goals of specific ecological components (Aanesen et al. 2014), between ecological and socio-economic objectives (ICES 2017), between different ecosystem services (Turkelboom 2017; Dick 2017), or the conflicting interests of specific stakeholders. While decision-making is often primarily aimed at effectively achieving specific conservation goals it ultimately involves socio-economic considerations including: (1) the sharing of costs and benefits among stakeholders; (2) the balance between short- and longer-term benefits; (3) the need to forgo current rents in exchange for future security; and (4) local opportunity costs and regional and global benefits.

EBM should be considered an incremental process as opposed to a single, giant leap away from traditional management (DeFries and Nagendra 2017; Borgstrom et al. 2015). As EBM revolves around a cyclical process, advancements can be made with every iteration of the adaptive management cycle. In this study, we adopt the IEA process and enhance earlier frameworks by incorporating the SES concept

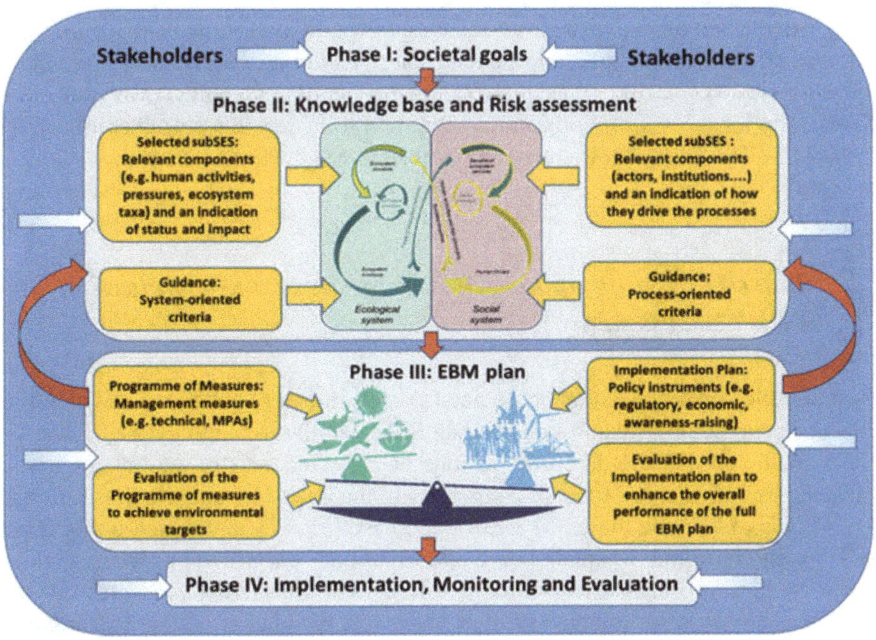

Fig. 1 One cycle of the adaptive, cyclical ecosystem-based management approach built around a balanced representation of the selected social-ecological sub-system (subSES). The figure depicts the phases occurring in the science domain (identified in Fig. 2) but identifies where in each phase cross-domain interaction occurs (i.e., stakeholder involvement) with the wider society. In each phase, the main contributions of this study to advance EBM are indicated

(Fig. 1) to shape the knowledge base and by providing practical guidance that allows an appraisal of progress. We assess progress in developing the knowledge base by looking at the key principles of EBM according to Long et al. (2015) and their alignment to relevant concepts such as IEA (Levin et al. 2009b; Samhouri et al. 2014), wicked problems (DeFries and Nagendra 2017), EBM phases (Borgstrom et al. 2015), policy-making processes (Cormier et al. 2017), ecosystem-based marine spatial planning (Ansong et al. 2017), and resilience thinking (Folke et al. 2010). We build a common framework around these concepts and organise them into four operational phases (identification of societal goals; developing the knowledge base and risk assessments; EBM plan development; implementation, monitoring, and evaluation), which differ in the scientific expertise required.

2.1 Phase I: Identification of Societal Goals

Phase I involves the scoping of societal goals, policy objectives, and perceived threats form the starting point of the IEA (Levin et al. 2009b, 2014) and EBM (Ansong et al. 2017; Cormier et al. 2017). It includes the analysis of policy synergies, conflicts, and an understanding of opportunities and challenges for developing EBM alternatives as well as stakeholder participation to help identify and prioritise among them. A possible issue is that policy objectives, mostly applying to global or regional scales, often refer to conditions of the ecological system only while at the local level, the objectives often aim to restore the sustainability of the whole SES. This may require reconciling objectives at different scales.

2.2 Phase II: Setting up the Knowledge Base and Conducting a Risk Assessment

Phase II builds on the inventory of societal goals in the previous phase and identifies relevant social-ecological sub-systems (subSES), similar to the 'focal SES' (Ostrom 2009). Such subSES should consist of one or more linkages in the linkage framework tied to one or more societal goals and are the basis to elaborate those subSES into what is to become the EBM knowledge base (Fig. 2), and which may contain both qualitative and/or quantitative information. The linkage framework is constructed using consistent typologies of the activities, their pressures and the ecosystem components affected by them (see e.g. Knights et al. 2015; Borgwardt et al. 2019) and with categories that resonate with stakeholders. If needed hierarchical typologies can be applied. For example the activity 'fishing' can be divided into specific types of fishing (e.g. demersal or pelagic) or an ecosystem component can be divided into functional groups or even species. The application of the linkage framework should ascertain that those subSES cover all the elements that matter to

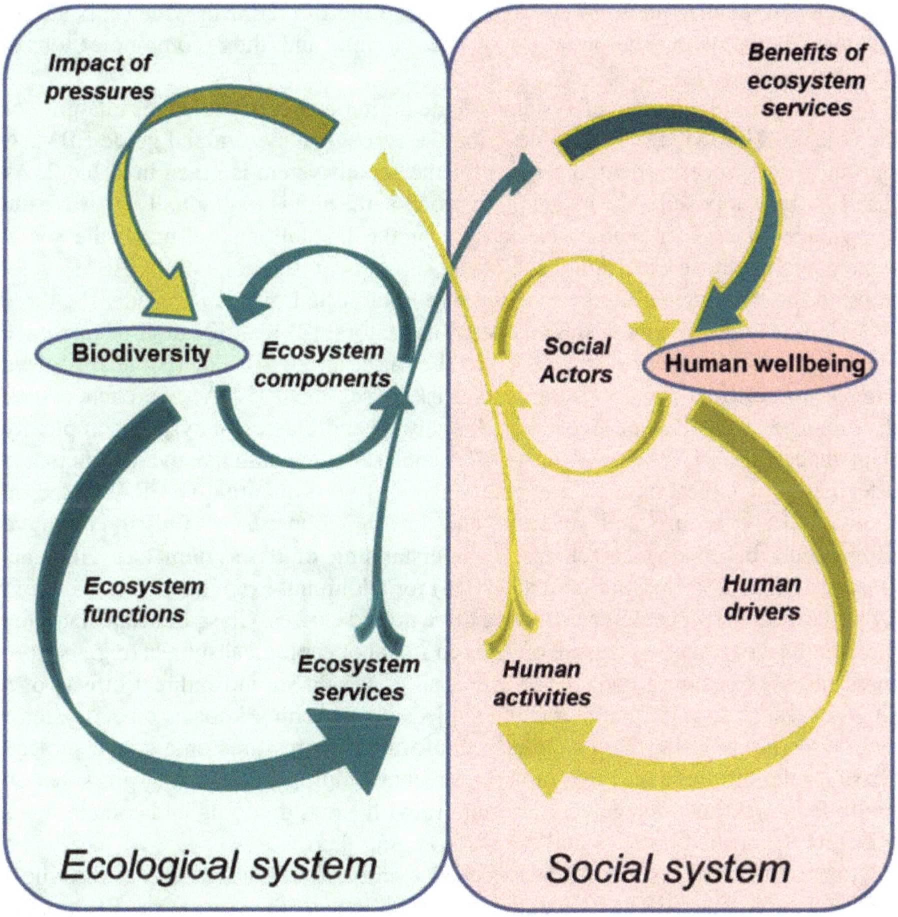

Fig. 2 The social-ecological system (adopted and modified from (Gómez et al. 2016)) consisting of an ecological system-based supply-side where the flow of ecosystem services into the social system contributes to human well-being and a social-system-based demand-side where the human activities and their pressures impact the ecosystem components and their functioning

the problem identified in Phase I, as well as any additional issues that have emerged through stakeholder consultation. Note, however, that with the introduction of more detailed categories, the complexity of the linkage framework (i.e. number of linkages) and thus information demand increases.

Different types of data, generally covering different subSES, are to be used in EBM phases II and III. Information on SES complexity is a requirement for diagnosing why some SES are sustainable while others are not (Ostrom 2009). Therefore, the inherent complexity of the SES and subSES should be harnessed, rather than eliminated. In practice, this implies that different subSES can be analysed at the appropriate level of detail without being hindered by the complexities and requirements of the full SES. At the same time the full SES, in which this subset operates, helps with interpretation of the outputs and their communication to stakeholders.

Table 1 provides a suite of system-oriented criteria to determine the suitability of the (sub)SES knowledge base to describe the ecological system and guide EBM. A similar list of process-oriented criteria for the social system is given in Table 2. As IEA/EBM is supposed to be an adaptive process, the aim is to gradually improve the knowledge base of the ecological system or the institutional set-up of the social system against those criteria in each of the subsequent iterations of the EBM cycle.

With the structure of the knowledge base established, we can consider the three IEA steps: (1) development of ecosystem indicators; (2) identification of reference levels; and (3) conducting risk analyses (Levin et al. 2014). For risk analysis, we suggest to consider the Environmental Risk Assessment (ERA) approach, which distinguishes the different levels of risk analyses and classes of system complexity (Holsman et al. 2017). Level 1 of the ERA consists of a qualitative evaluation that is often based on expert opinion. Level 2 consists of semi-quantitative ERAs based on estimates of exposure and severity, and Level 3 consists of fully quantitative assessments based on a mechanistic understanding of the system (see Holsman et al. (2017) and Stelzenmüller et al. (2015) for additional examples). The complexity of the system can be described using three nested classes. Class 1 encapsulates the direct impact of a single pressure on a given social or ecological subject (e.g., bottom trawl fisheries catching cod). Class 2 measures the direct and indirect effects of a single pressure on multiple interacting subjects (e.g., fishing impacts on ecosystems as in (Hobday et al. 2011)) or the effects of multiple pressures on a single subject. Class 3 refers to the direct and indirect effects of multiple interacting pressures on multiple interacting subjects (e.g., bottom trawl fishing, dredging and contaminants affecting the seabed habitats and the fish foraging there).

It is important to distinguish between the overall, less detailed ERA of the whole SES and the detailed ERAs conducted on the subSES. Within the overall ERA of the whole SES, the qualitative or semi-quantitative ERAs of the full SES (i.e., Class 3) can be used to identify the linkages that introduce the greatest risk to the ecosystem (aligned to the threats in Phase I). These Level 1 or 2 ERAs are primarily aimed at guiding decision makers on which sector-specific management measures to focus (Cormier et al. 2017).

The levels of risk analyses and classes of system complexity described above determine the type of ERA required. Formal description of the SES, through a linkage framework, can provide further characterisation of the classes. This linkage framework can then also be used to assess the quality of the ERA for each level of the analytical tools. Ultimately, the aim is to advance from a qualitative ERA towards an increasingly quantitative ERA that is more elaborate and realistic in terms of the 'reciprocal and cumulative interactions among multiple (interacting) pressures and multiple interacting subjects' (Holsman et al. 2017) that make up the full SES. Distinguishing the subSES and the full SES, each with complementary ERAs can reduce the level of complexity in order to ensure salient information into the policy cycle (Folke et al. 2005; Biggs et al. 2015).

Table 1 Ecological EBM criteria based on the EBM principles in Long et al. (2015) for assessing the knowledge base of the ecological part of the social-ecological system (SES) and hence the core sustainability SES subsystems: resource systems and resource units (Ostrom 2009). Guidance is provided to assess to what extent the knowledge base has advanced in order to support EBM. Links to other criteria in this table or Table 2 are in *italics*

EBM criteria	Generic assessment guidance and considerations
Ecological integrity and biodiversity	This can be achieved by defining and conserving a diversity of species traits or functional groups that support the integrity of the ecosystem, or check the three aspects: variety; balance; and disparity. Are these explicitly considered in the knowledge base? The ecological structural components determine the functioning of the ecological system. Hence the link to the '*Consider ecosystem connections*' criterion: a knowledge base that covers more relevant components or detail is better.
Consider ecosystem connections	This is determined by the ecological part of the SES (e.g., by mapping critical connections) and is linked to the '*Ecological integrity and biodiversity*' criterion as more components and/or detail increase this aspect (e.g., in terms of taxa considered in the food web) this can be improved with an indication of the importance of a connection (e.g., pressure-state relationships, predator-prey relationships). Knowledge on the ecological functioning of the ecological structural components determines the provisioning of ecosystem services which contribute to human well-being and as such can be incorporated into (economic) markets.
Account for dynamic nature of ecosystems	Variation in the ecological part of the SES (e.g., due to perturbations) should be considered. Longer time-series are better. Question the assumption of perfect foresight. Include exogenous scenarios of socio-economic drivers or environmental change (e.g., climate scenarios). Strengthen feedbacks that maintain desired regimes, break or disturb feedbacks that maintain undesired regimes; look for non-linearity in the system as these are often the cause for the dynamic nature.
Acknowledge uncertainty	This requires transparency on the quality of the knowledge base which could be reflected, for example, through the assessment of uncertainties, reporting of crucial (model) assumptions and confidence intervals in the output. Uncertainty is inherent to complex adaptive systems (such as the marine) and their management. In addition to the known unknowns (e.g., lack of historic data of species), complex systems come with new uncertainties that cannot be tackled through standard sensitivity analysis (Polasky et al. 2011). The social EBM criteria '*Adaptive management*' and '*stakeholder participation*' (see Table 2) become increasingly important if uncertainty in the current knowledge base is high.
Appropriate spatial and temporal scales	What are the appropriate spatial and temporal scales of the (eco) system? For example, resolution of spatial grid and temporal units (e.g., years, months). Which scales to consider? Not just spatial/temporal but also in different domains (e.g., ecological, jurisdictional, administrative or political). Use a systems framework to address relevant scales and how they interact. Assessment should occur at the ecosystem scale. If other scales are relevant and do not match with the ecosystem scale this needs to be identified.

(continued)

Table 1 (continued)

EBM criteria	Generic assessment guidance and considerations
Distinct boundaries	Acknowledge boundaries and thus the fluxes and influences from outside of the boundaries of the ecosystem. Consider both jurisdictional boundaries as well as ecosystem boundaries (see *Appropriate Spatial and Temporal Scales*). Are transboundary issues considered? For example: Terrestrial run-off into rivers and lakes or inflow of rivers into the coastal/marine ecosystem. The definition of boundaries should allow the adaptation of institutions in a good social-ecological fit (see *recognise coupled SES*).
Recognise coupled SES	Are all relevant flows considered between the social and the ecological system that make up the SES? How many linkages, or how much of the activities, pressures, ecosystem components and the ecosystem services they provide is covered in the subSES (used in the management strategy evaluation) compared to the full SES. Is it understood how these link to the actors that drive the relevant social processes (see Table 2)?
Consider cumulative impacts	Apply an integrated perspective, including all relevant activities and their pressures acting on the ecosystem (see *recognise coupled SES*). Consider whether synergistic or antagonistic cumulative effects apply (Crain et al. 2008).

2.3 *Phase III: Planning of EBM*

This phase should result in a comprehensive EBM plan that can mitigate the threats and achieve the policy objectives identified in Phase I. In building a comprehensive EBM plan, the Phase II knowledge base is used to first guide the design of the EBM plan, and then evaluate the plan's potential performance before implementation in Phase IV.

Most examples of management plans have focused on ecological outcomes (e.g., Rademeyer et al. 2007; Ansong et al. 2017; Samhouri et al. 2014). Since EBM is concerned with the management of SES, the EBM plan that is developed and tested in this phase should cover both the ecological and social components of the system. To that end, the EBM plan consists of two interconnected, structured, yet differentiated sets of decisions, management measures, and policy instruments, each primarily addressing a specific aspect of the SES (see Fig. 3):

- Management measures are integrated into a Programme of Measures, a combined set of actions aimed at achieving environmental objectives and thus to enhance and protect the ecological system. Potential management measures can be classified according to the three ISO 31,000 risk management categories: prevention; mitigation; and recovery controls (Cormier et al. 2013, 2018) that can be aligned to the EBM management measure typology outlined in Piet et al. (2015), depending on where in the linkage framework the measure intervenes.
    - Prevention controls manage the causes of the risk and are aimed at the human activity and/or the pressure. Examples are input control (e.g., scrapping schemes to reduce the capacity of the fishing fleet), output controls that

Table 2 Social EBM criteria for assessing the knowledge base of the social part of the social-ecological system (SES) and hence the core sustainability SES subsystems: governance systems and users (Ostrom 2009). These criteria are based on the key EBM principles identified by Long et al. (2015) and link to the relevant governance actors. Guidance is provided to assess to what extent the institutional set-up and its governance processes can support EBM. Links to other criteria in this table or Table 1 are in *italics*

EBM criteria	Actors	Generic assessment guidance and considerations
Use of scientific knowledge	Science	This includes the scientific use of all types of knowledge including local knowledge, traditional knowledge or citizen science. Has the knowledge been produced according to the scientific standards? Is the methodology appropriate? Are procedures transparent? Peer-reviewed? Is there consensus on the quality of the available (scientific) knowledge? This requires both the interaction between scientists and decision-makers to foster salience in scientific input as well as the interaction between scientists and other actors to foster credibility in knowledge production (see Röckmann et al. 2015). Ultimately the outcome of the process should be perceived as evidence-based.
Inter-disciplinarity		Was the appropriate expertise in terms of relevant disciplines applied when producing the knowledge? Can stakeholder knowledge be integrated? The aim is to progress from multi- to inter- to transdisciplinary science. See *Stakeholder Involvement.*
Stakeholder involvement		Science could benefit from knowledge available with other stakeholders, notably the business sector. Stakeholders can play a role in providing knowledge (see '*Use of Scientific Knowledge*') collecting data (monitoring; cooperative research). The feedback of stakeholders on making choices under uncertainty is also important. Reed (2008) identifies eight best practices that improve the quality and effectiveness of stakeholder participation.
Integrated management	Management	In this context integrated can be interpreted as cross-sectoral, inter-disciplinary and/or holistic (i.e., encompassing the whole SES). Which of these (or other) perspectives are incorporated into the management process? Compliance of the SES aspect *Human activities and their pressures* is a requirement. The Decision-making across administrative boundaries is tightly linked to the *Distinct boundaries* criterion where jurisdictional boundaries may be different from ecosystem boundaries. Building resilience requires a governance capable of balancing heterogeneity, redundancy, modularity and connectivity at *Appropriate temporal and spatial scales* (Elmhirst et al. 2009; Levin et al. 2013)
Adaptive management		The management should be adaptive as it needs to deal with the inherent uncertainty of EBM. Learning-by-doing is needed when outcomes of decisions are uncertain because of complex system dynamics. This is linked to ecological EBM criteria: *Acknowledge uncertainty, Account for dynamic nature of ecosystems* and

(continued)

Table 2 (continued)

EBM criteria	Actors	Generic assessment guidance and considerations
		Appropriate monitoring. Rather than choosing optimal paths and decision rules in a deterministic framework, facing current risks and considerable uncertainties requires governance frameworks able to adapt to the multiple circumstances that may prevail in the foreseeable future.
Apply the precautionary approach		Does the institutional set-up allow the application of the precautionary approach? This requires compliance to the SES aspect *Changes and Uncertainty*.
Stakeholder involvement		Managers depend on the input from science but could benefit from knowledge available with other stakeholders, notably the business sector. Also, the feedback of stakeholders on making choices (or co-decision making) under uncertainty is important. As compliance of the SES aspect is a requirement; stakeholder involvement in policy implementation can be instrumental.
Appropriate monitoring		A requirement of *Adaptive management* is adequate monitoring. The quality of the monitoring is reflected in the proportion of the relevant components of the SES for which sufficient data is collected at *appropriate spatio-temporal scale* and the level of *uncertainty* to allow *scientific knowledge to guide informed decision-making*. Monitoring programs can be developed in collaboration with the other stakeholders, i.e. multi-sector actor resulting in cooperative research. The monitoring data should be transformed into salient and legitimate scientific knowledge to guide informed decision-making. The degree to which that actually occurs needs to increase in order to advance EBM.
Decisions reflect societal choice	Policy making	Specifying clear goals increases efficiency and efficacy of the MSP process and allows the identification of potential trade-offs of proposed management strategies. Specify what trade-offs should be considered, e.g. amongst stakeholders, or between short- and longer-term benefits. '*Stakeholder involvement*' is required specifically in order to make decision-making inclusive and reflect societal choice (i.e., legitimacy).
Stakeholder involvement		For rationale see Röckmann et al. (2015). Check if the 'typology of eight levels of participation' (Reed 2008) is applied. The degree of stakeholder interaction should be appropriate for the specific context. Also, the feedback of stakeholders on making choices (or co-decision making) under uncertainty is important. Is there information on compliance? Stakeholder involvement in decision making can be instrumental to ascertain compliance.
Sustainability		All three pillars of sustainability (i.e., ecological, economic and social) should be considered in the trade-offs informing the decision-making process.
Stakeholder involvement	All resource users	The participation and involvement of all the resource users is the backbone of a successful EBM process. This may be

(continued)

Table 2 (continued)

EBM criteria	Actors	Generic assessment guidance and considerations
		through a top-down process initiated by the government or a bottom-up process where the users self-organize. In case of a top-down process stakeholder participation should reflect and be based on all sectors which are affected by the plan, local community actors and environmental non-governmental organizations (NGOs) in addition to the stakeholder groups specifically mentioned (i.e., science, policy makers and managers). This to ascertain all relevant societal claims, values and relevant aspects and impacts can be considered in the process and involved at each stage and that implementation and monitoring of strategies are effectively done. The likelihood of self-organization of the users to achieve a sustainable SES was found to depend on several aspects of the SES covered by the other EBM criteria, see (Ostrom 2009).

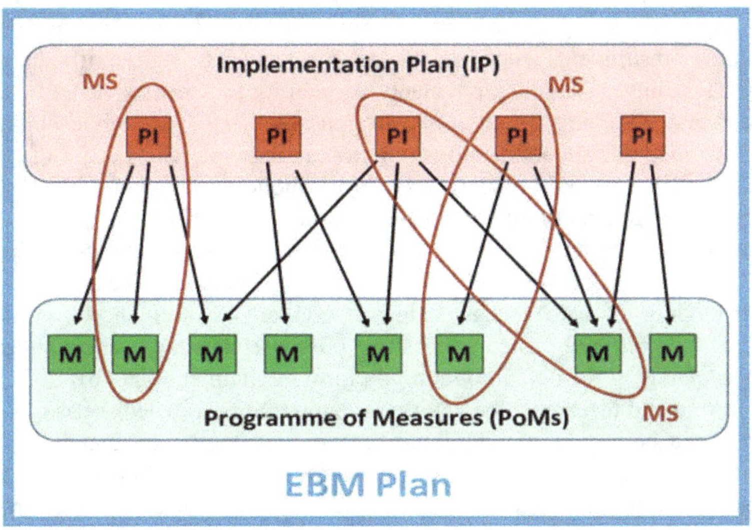

Fig. 3 Diagram explaining the elements that make up an ecosystem-based management plan. PI = Policy Instruments, M = Management Measure, MS = Management Strategy

prevent the pressure from entering the system (e.g., catch controls in fisheries) or spatial- and/or temporal distribution controls (e.g., marine protected areas or real-time closures).

- Mitigation and recovery controls are implemented to reduce the likelihood and magnitude of each consequence as a result of the risk event occurring. We distinguish mitigation controls that target the pressure once it is present in the system (e.g., beach cleaning after oil spills) and recovery controls targeting the ecosystem component (state) that is impacted (e.g., habitat restoration or stocking programs). For both mitigation controls there may be a lag in the response depending on the persistence of the pressure or the resilience of the species. For recovery controls, the lag is only determined by the resilience. The choice of the type of management measure thus determines the time horizon when results can be expected.

- Policy instruments are integrated into an Implementation Plan. This consists of all the arrangements or reforms that are required in the governing system (as part of the social system) for the implementation of the Programme of Measures and the overall performance of the full EBM plan. The following types of policy instruments can be distinguished (see Lago et al. 2015; Frelih-Larsen et al. 2016):

- Legislative instruments, including various (inter)national conservation laws or regulations
- Regulatory instruments, including the setting of targets or standards aimed at maintaining a certain level of environmental quality, prohibits (i.e., bans) or allows (i.e., permits) an individual or business to perform certain acts, or to have a certain portion (or amount) of a product
- Economic instruments such as pricing mechanisms (e.g., tariffs, taxes and charges, trading of permits), payments, or liability schemes
- Instruments involving information, awareness-raising, and public engagement such as training and qualifications (e.g., obtaining certificates or proof of qualification) related to environmental protection, public information programs, stakeholder and public participation, or innovation groups that aim to build capacity and knowledge about a particular environmental, economic, or practical issue
- 'Nudges' are possible alternatives to the instruments mentioned above, whereby the behaviour of actors (e.g., industry, civilians) is influenced (or nudged) towards the preferred choice via positive reinforcement and indirect suggestions such as changing the default option in a form (Valatin et al. 2016)
- Monitoring and research aimed at improving the SES knowledge base. This may include the collecting of additional data or developing an understanding on specific gaps that may correspond to specific nodes or linkages in the linkage framework that hampered this EBM cycle but may be useful in the next EBM cycle.

The design step of the EBM plan commences with the selection of appropriate candidate management measures. This is guided by the ERA, which identifies the linkages contributing most to environmental impact risk indicating the potential management measures following the approach developed by Piet et al. (2015). Those candidate management measures should cover all the major threats and are considered most likely to reduce the environmental impact risk. A pre-screening exercise using the '10-tenets of adaptive management' (Barnard and Elliott 2015) may be conducted to ascertain *a priori* that all possible issues are considered that may prevent the success of management measures or policy instruments. Failure to comply to any one of the tenets should be reason to re-consider a particular management measures or policy instrument. By applying these tenets as screening criteria for the list of potential issues, those issues that do not tie directly to a tenet, should be removed from further testing in the evaluation step.

In this evaluation step, the future performance of an alternative EBM plan is compared to that of the existing (e.g. baseline, business-as-usual) management plan. Both the alternative and baseline management plans may result in different outcomes depending on exogenous drivers (e.g., socio-economic or climate scenarios). For the evaluation of the alternative EBM plan(s) against baseline, we propose three outcome-oriented criteria:

- *Effectiveness: Does the plan achieve the pre-determined target?* This is what usually constitutes effective evaluation of a management strategy (see Punt et al. 2016 for best practices) and involves the simulation of specific indicator trajectories with their error distributions relative to policy targets (e.g. fishing mortality indicator relative to the target of Maximum Sustainable Yield). Effectiveness of an individual measure, or of a programme of measures, along with its implementation plan, is defined by the contribution it makes to bridge the information gap between understanding baseline conditions and target conditions.

- *Efficiency: Is the plan conducive to enhance human wellbeing?* This refers to the capacity of citizens and social institutions to take advantage of existing opportunities (as determined by technology, resource endowments and actual availability, physical and human capital, etc.) to improve human wellbeing in a sustainable way. It is a concept that applies to the users of a particular service (i.e., those who may have the opportunity to utilise the service without making anyone else worse off), the stakeholders in a particular decision context (i.e., who may have the option to cooperate in the preservation of a resource and share the benefits amongst them), or the government (i.e., who may have the possibility of improving the environment without worsening opportunities in terms of economic activities). This criterion is ultimately an assessment of sustainable development, where each generation should aim at improving its wellbeing within the available opportunities as long as this does not compromise the options available to future generations. The benefits and costs are defined as any positive or negative impacts on human wellbeing, irrespective of whether the affected individuals are aware of them, or whether they can be valued through market prices or any other ad-hoc valuation exercise. When comparing benefits and costs, the issue is that costs are often monetised and are relatively certain, whereas benefits may be difficult to monetise and are definitely more uncertain (though mostly just as real).

- *Equity: Are the benefits being shared in a socially just way?* The distribution of benefits and costs across stakeholders must be perceived as fair. Besides the contribution of the EBM plan, to social equity, the legitimacy of the EBM plan requires the perception that its consequences are fairly distributed among the affected parties both in the present as well as the future (i.e., intergenerational equity).

2.4 Phase IV: Implementation, Monitoring and Evaluation

This is the phase where EBM becomes operational based on the planning in the previous phases. According to Cormier et al. (2017), it is the competent authorities of specific sectors that are accountable to implement the measures that are designed to manage their specific operations. The role of science is to: (1) inform those sector-specific authorities on the detail of the measures before implementation; as well as (2) design the monitoring programs; and (3) conduct and inform the subsequent evaluation of the performance of those measures after implementation. The performance is evaluated testing indicators against a benchmark as a measure of achieving an objective. When the benchmark is not met, the goals and objectives needs to be re-examined and/or the management regime re-assessed (Behn 2003; Poister et al. 2010) in phases I and III of the next EBM cycle.

While a monitoring program is primarily intended to assess the status of the SES and the performance of the EBM plan, it can also feed relevant information into the knowledge base and hence need to be aligned with system-oriented criteria of Table 1. The implementation of any future alternative EBM plan is determined by the governance context and the institutional processes captured in the Phase II knowledge base with the process-oriented criteria in Table 2. The performance of the EBM plan can then be used to guide the planning (Phase III) in the next EBM cycle, which builds on the previous EBM plan by adopting those management measures or policy instruments that performed well, and modifying or replacing those that failed. Incorporating the feedbacks from Phase IV and integrating it into the next EBM cycle is crucial to a successful EBM process.

3 Discussion and Conclusions

The EBM approach presented here is an attempt to combine or reconcile many existing concepts that describe the ecological system, the social (or socio-economic) system and EBM into a unifying approach with guidance on how to make it (more) operational in any aquatic ecosystem based on a diverse existing literature incorporating concepts from IEA, Marine Spatial Planning, ERA, resilience and CAS (Long et al. 2015; Cormier et al. 2017; Levin et al. 2009b; Samhouri et al. 2014; Levin et al. 2009b; Samhouri et al. 2014; Ansong et al. 2017; Folke et al. 2010; Hagstrom and Levin 2017). We have attempted to translate these concepts into concrete steps, identify issues for the practitioner to consider, give examples that provide the basis for a common framework, and suggest practical guidance for the incorporation and synthesis of interdisciplinary information on SES into practical and useful linkage frameworks for EBM plan development and implementation. One example of a practical application of this framework can be found in (Piet et al. 2019).

Integrated ERAs have been included in environmental impact assessments for many decades but it was not until the work of Halpern et al. (2008) that the cumulative effects of multiple stressors received much attention. Since then, many other such integrated assessments have taken place in marine waters (e.g., Coll et al. 2012; Knights et al. 2013; Korpinen et al. 2012) and are now also covering inland and transitional waters (e.g., Borgwardt et al. 2019). These integrated assessments all apply different methodologies which may differ on their ability to inform EBM and associated monitoring requirements (Borja et al. 2016). Even within a specific integrated assessment there are methodological choices to be made depending on the chosen subSES and/or the application of the ERA (Piet et al. 2017).

The EBM process consisting of four phases (I–IV) has been built around the SES concept that brings together the natural and social scientific disciplines involved in EBM. The contribution of this study to advance EBM occurs primarily in Phase II and Phase III. Advancements in Phase II consists of: (1) approaches to reduce a complex suite of SES to one or more focused subSES to avoid inaction from overwhelming complexity, a common problem when resolving wicked problems such as EBM; and (2) practical guidance based on key EBM principles (Long et al. 2015) involving both the ecological system as well as the social system. Advancements in Phase III consists of the organizing the structure and typology of an EBM plan that is explicitly linked to both the ecological system as well as the social system. Even though the findings of this study are primarily relevant for the science domain, it explicitly acknowledges the interaction with wider society. This approach structurally incorporates consultation with other stakeholder groups in order to enhance the credibility in knowledge production and ascertain salient scientific input in the domain of policy-makers, decision-makers, and managers (Röckmann et al. 2015).

Despite the many issues that are still unresolved, this study provides the theoretical and conceptual basis to apply some of the methodological studies in this volume (Borgwardt et al. 2019; Teixeira et al. 2019) in order to advance the implementation of EBM toward and achievement of policy objectives in support of the societal goals for our aquatic systems (see Piet et al. 2019).

Acknowledgement This project is part of the AQUACROSS project (Knowledge, Assessment, and Management for AQUAtic Biodiversity and Ecosystem Services aCROSS EU policies) funded by the *European Union's Horizon 2020 research and innovation programme* under grant agreement No 642317.

References

Aanesen, M., Armstrong, C. W., Bloomfield, H. J., & Röckmann, C. (2014). What does stakeholder involvement mean for fisheries management? *Ecology and Society, 19*(4), 35.

Ansong, J., Gissi, E., & Calado, H. (2017). An approach to ecosystem-based management in maritime spatial planning process. *Ocean & Coastal Management, 141*, 65–81.

Barnard, S., & Elliott, M. (2015). The 10-tenets of adaptive management and sustainability: An holistic framework for understanding and managing the socio-ecological system. *Environmental Science & Policy, 51*, 181–191.

Behn, R. D. (2003). Why measure performance? Different purposes require different measures. *Public Administration Review, 63*, 586–606.

Berkes, F. (2012). Implementing ecosystem-based management: Evolution or revolution? *Fish and Fisheries, 13*, 465–476.

Bianchi, G. (2008). The concept of the ecosystem approach to fisheries. In G. Bianchi (Ed.), *FAO the ecosystem approach to fisheries* (pp. 20–38, Chap. 2). Oxfordshire: CABI.

Biggs, R., Schlüter, M., & Schoon, M. L. (2015). *Principles for building resilience: Sustaining ecosystem services in social-ecological systems* (pp. 1–290).

Bissix, G., & Rees, J. A. (2001). Can strategic ecosystem management succeed in multiagency environments? *Ecological Applications, 11*, 570–583.

Borgström, S., Bodin, Ö., Sandström, A., & Crona, B. (2015). Developing an analytical framework for assessing progress toward ecosystem-based management. *Ambio, 44*, 357–369.

Borgwardt, F., Robinson, L., Trauner, D., Teixeira, H., Nogueira, A. J., Lillebø, A., Piet, G., Kuemmerlen, M., O'Higgins, T., McDonald, H., Arevalo-Torres, J., Barbosa, A. L., Iglesias-Campos, A., Hein, T., & Culhane, F. (2019). Exploring variability in environmental impact risk from human activities across aquatic realms. *Science of the Total Environment, 652*, 1396–1408.

Borja, A., Elliott, M., Andersen, J. H., Berg, T., Carstensen, J., Halpern, B. S., Heiskanen, A.-S., et al. (2016). Overview of integrative assessment of marine systems: The ecosystem approach in practice. *Frontiers in Marine Science, 3*, 20.

Christie, P. (2011). Creating space for interdisciplinary marine and coastal research: Five dilemmas and suggested resolutions. *Environmental Conservation, 38*, 172–186.

Coll, M., Piroddi, C., Albouy, C., Lasram, F. B., Cheung, W. W. L., Christensen, V., Karpouzi, V. S., et al. (2012). The Mediterranean Sea under siege: Spatial overlap between marine biodiversity, cumulative threats and marine reserves. *Global Ecology and Biogeography, 21*, 465–480.

Cormier, R., Kannen, A., Elliott, M., Hall, P., & Davies, I. M. (2013). *Marine and coastal ecosystem-based risk management handbook.* ICES Cooperative Research Report No. 317. 60 pp.

Cormier, R., Kelble, C. R., Anderson, M. R., Allen, J. I., Grehan, A., & Gregersen, O. (2017). Moving from ecosystem-based policy objectives to operational implementation of ecosystem-based management measures. *ICES Journal of Marine Science, 74*, 406–413.

Cormier, R., Elliott, M., & Kannen, A. (2018). IEC/ISO Bow-tie analysis of marine legislation: A case study of the marine strategy framework directive. ICES Cooperative Research.

Crain, C. M., Kroeker, K., & Halpern, B. S. (2008). Interactive and cumulative effects of multiple human stressors in marine systems. *Ecology Letters, 11*, 1304–1315.

Culhane, F., Teixeira, H., Nogueira, A. J., Borgwardt, F., Trauner, D., Lillebø, A., Piet, G., Kuemmerlen, M., McDonald, H., O'Higgins, T., & Barbosa, A. L. (2019). Risk to the supply of ecosystem services across aquatic ecosystems. *Science of the Total Environment, 660*, 611–621.

DeFries, R., & Nagendra, H. (2017). Ecosystem management as a wicked problem. *Science, 356*, 265.

De Lange, H. J., Sala, S., Vighi, M., & Faber, J. H. (2010). Ecological vulnerability in risk assessment – A review and perspectives. *Science of the Total Environment, 408*, 3871–3879.

Dick, J., Turkelboom, F., et al. (2017). *Stakeholders' perspectives on the operationalisation of the ecosystem service concept: Results from 27 case studies.* Ecosystem Services.

Dietz, T., Ostrom, E., & Stern, P. C. (2003). The struggle to govern the commons. *Science, 302*, 1907–1912.

EEA. (1995). *Europe's environment: The Dobris assessment.* European Environmental Agency, 8 pp.

Elliott, M. (2002). The role of the DPSIR approach and conceptual models in marine environmental management: An example for offshore wind power. *Marine Pollution Bulletin, 44*: III–VII.

Elliott, M., Burdon, D., Atkins, J. P., Borja, A., Cormier, R., de Jonge, V. N., & Turner, R. K. (2017). "And DPSIR begat DAPSI(W)R(M) !" – A unifying framework for marine environmental management. *Marine Pollution Bulletin, 118*, 27–40.

Elmhirst, T., Connolly, S. R., & Hughes, T. P. (2009). Connectivity, regime shifts and the resilience of coral reefs. *Coral Reefs, 28*, 949–957.

Folke, C., Hahn, T., Olsson, P., & Norberg, J. (2005). Adaptive governance of social-ecological systems. *Annual Review of Environment and Resources, 30*, 441–473.

Folke, C., Carpenter, S. R., Walker, B., Scheffer, M., Chapin, T., & Rockström, J. (2010). Resilience thinking: Integrating resilience, adaptability and transformability. *Ecology and Society, 15*(4). https://doi.org/10.5751/ES-03610-150420.

Frelih-Larsen, A., Bowyer, C., Albrecht, S., Keenleyside, C., Kemper, M., Nanni, S., et al. (2016). *Updated inventory and assessment of soil protection policy instruments in EU member states.* Final Report to DG Environment. Berlin: Ecologic Institute.

Gómez, C. M., Delacámara, G., Arevalo-Torres, J., Barbiere, J., Barbosa, A.L., & Iglesias-Campos, A. (2016). The AQUACROSS innovative concept. Deliverable 3.1, European Union's Horizon 2020 Framework Programme for Research and Innovation Grant Agreement No. 642317.

Hagstrom, G. I., & Levin, S. A. (2017). Marine ecosystems as complex adaptive systems: Emergent patterns, critical transitions, and public goods. *Ecosystems, 20*, 458–476.

Halpern, B. S., Walbridge, S., Selkoe, K. A., Kappel, C. V., Micheli, F., D'Agrosa, C., Bruno, J. F., et al. (2008). A global map of human impact on marine ecosystems. *Science, 319*, 948–952.

Harvey, C. J., Kelble, C. R., & Schwing, F. B. (2017). Implementing "the IEA": Using integrated ecosystem assessment frameworks, programs, and applications in support of operationalizing ecosystem-based management. *ICES Journal of Marine Science, 74*, 398–405.

Hobday, A. J., Smith, A. D. M., Stobutzki, I. C., Bulman, C., Daley, R., Dambacher, J. M., Deng, R. A., et al. (2011). Ecological risk assessment for the effects of fishing. *Fisheries Research, 108*, 372–384.

Holsman, K., Samhouri, J., Cook, G., Hazen, E., Olsen, E., Dillard, M., Kasperski, S., Gaichas, S., Kelble, C. R., Fogarty, M., & Andrews, K. (2017). An ecosystem-based approach to marine risk assessment. *Ecosystem Health and Sustainability, 3*(1), e01256. https://doi.org/10.1002/ehs2.1256.

ICES. (2017). WKBESIO: Report of the WorKshop on balancing economic, social, and institutional objectives in integrated assessments.

Katsanevakis, S., Stelzenmuller, V., South, A., Sorensen, T. K., Jones, P. J. S., Kerr, S., Badalamenti, F., et al. (2011). Ecosystem based marine spatial management: Review of concepts, policies, tools, and critical issues. *Ocean & Coastal Management, 54*, 807–820.

Knights, A. M., Koss, R. S., & Robinson, L. A. (2013). Identifying common pressure pathways from a complex network of human activities to support ecosystem-based management. *Ecological Applications, 23*, 755–765.

Knights, A. M., Culhane, F., Hussain, S. S., Papadopoulou, K. N., Piet, G. J., Raakaer, J., Rogers, S. I., et al. (2014). A step-wise process of decision-making under uncertainty when implementing environmental policy. *Environmental Science & Policy, 39*, 56–64.

Knights, A. M., Piet, G. J., Jongbloed, R. H., Tamis, J. E., White, L., Akoglu, E., Boicenco, L., et al. (2015). An exposure-effect approach for evaluating ecosystem-wide risks from human activities. *ICES Journal of Marine Science, 72*, 1105–1115.

Korpinen, S., Meski, L., Andersen, J. H., & Laamanen, M. (2012). Human pressures and their potential impact on the Baltic Sea ecosystem. *Ecological Indicators, 15*, 105–114.

Lago, M., Mysiak, J., Gomez, C. M., Delacamara, G., & Maziotis A. (Eds.). (2015, October). *Use of economic instruments in water management – Insights from international experience.* Cham: Springer.

Levin, P. S., Fogarty, M. J., Murawski, S. A., & Fluharty, D. (2009a). Integrated ecosystem assessments: Developing the scientific basis for ecosystem-based management of the ocean. *PLoS Biology, 7*, 23–28.

Levin, P. S., Fogarty, M. J., Murawski, S. A., & Fluharty, D. (2009b). Integrated ecosystem assessments: Developing the scientific basis for ecosystem-based management of the ocean (perspective). *PLoS Biology, 7*, e1000014.

Levin, S., Xepapadeas, T., Crépin, A. S., Norberg, J., De Zeeuw, A., Folke, C., Hughes, T., et al. (2013). Social-ecological systems as complex adaptive systems: Modeling and policy implications. *Environment and Development Economics, 18*, 111–132.

Levin, P. S., Kelble, C. R., Shuford, R. L., Ainsworth, C., deReynier, Y., Dunsmore, R., Fogarty, M. J., et al. (2014). Guidance for implementation of integrated ecosystem assessments: A US perspective. *ICES Journal of Marine Science, 71*, 1198–1204.

Long, R. D., Charles, A., & Stephenson, R. L. (2015). Key principles of marine ecosystem-based management. *Marine Policy, 57*, 53–60.

Ludwig, D. (2001). The era of management is over. *Ecosystems, 4*, 758–764.

Nelson, G. C., Bennett, E., Berhe, A. A., Cassman, K., DeFries, R., Dietz, T., Dobermann, A., et al. (2006). Anthropogenic drivers of ecosystem change: An overview. *Ecology and Society, 11*, –29.

OECD. (1994). *Environmental indicators: OECD Core set*. Paris: Organisation for Economic Co-operation and Development.

Ostrom, E. (2009). A general framework for analyzing sustainability of social-ecological systems. *Science, 325*, 419–422.

Piet, G. J., Jongbloed, R. H., Knights, A. M., Tamis, J. E., Paijmans, A. J., van der Sluis, M. T., de Vries, P., et al. (2015). Evaluation of ecosystem-based marine management strategies based on risk assessment. *Biological Conservation, 186*, 158–166.

Piet, G. J., Knights, A. M., Jongbloed, R. H., Tamis, J. E., de Vries, P., & Robinson, L. A. (2017). Ecological risk assessments to guide decision-making: Methodology matters. *Environmental Science & Policy, 68*, 1–9.

Piet, G., Culhane, F., Jongbloed, R., Robinson, L., Rumes, B., & Tamis, J. (2019). An integrated risk-based assessment of the North Sea to guide ecosystem-based management. *Science of the Total Environment, 654*, 694–704.

Poister, T. H., Pitts, D. W., & Edwards, L. H. (2010). Strategic management research in the public sector: A review, synthesis, and future directions. *American Review of Public Administration, 40*, 522–545.

Polasky, S., de Zeeuw, A., & Wagener, F. (2011). Optimal management with potential regime shifts. *Journal of Environmental Economics and Management, 62*, 229–240.

Punt, A. E., Butterworth, D. S., de Moor, C. L., de Oliveira, J. A. A., & Haddon, M. (2016). Management strategy evaluation: Best practices. *Fish and Fisheries, 17*, 303–334.

Rademeyer, R. A., Plaganyi, E. E., & Butterworth, D. S. (2007). Tips and tricks in designing management procedures. *ICES Journal of Marine Science, 64*, 618–625.

Reed, M. S. (2008). Stakeholder participation for environmental management: A literature review. *Biological Conservation, 141*, 2417–2431.

Rittel, H. W. J., & Webber, M. M. (1973). Dilemmas in a general theory of planning. *Policy Sciences, 4*(2), 155–169.

Röckmann, C., van Leeuwen, J., Goldsborough, D., Kraan, M., & Piet, G. (2015). The interaction triangle as a tool for understanding stakeholder interactions in marine ecosystem based management. *Marine Policy, 52*, 155–162.

Ruckelshaus, M., Essington, T., & Levin, P. (2009). Puget Sound, Washington, USA. In K. M. H. Leslie (Ed.), *Ecosystem-based management for the oceans* (pp. 201–226). Washington, DC: Island Press.

Samhouri, J. F., Haupt, A. J., Levin, P. S., Link, J. S., & Shuford, R. (2014). Lessons learned from developing integrated ecosystem assessments to inform marine ecosystem-based management in the USA. *ICES Journal of Marine Science, 71*, 1205–1215.

Stelzenmüller, V., Fock, H. O., Gimpel, A., Rambo, H., Diekmann, R., Probst, W. N., Callies, U., et al. (2015). Quantitative environmental risk assessments in the context of marine spatial management: Current approaches and some perspectives. *ICES Journal of Marine Science, 72*, 1022–1042.

Tallis, H., Levin, P. S., Ruckelshaus, M., Lester, S. E., McLeod, K. L., Fluharty, D. L., & Halpern, B. S. (2010). The many faces of ecosystem-based management: Making the process work today in real places. *Marine Policy, 34*, 340–348.

Teixeira, H., Lillebø, A. I., Culhane, F., Robinson, L., Trauner, D., Borgwardt, F., Kummerlen, M., Barbosa, A., McDonald, H., Funk, A., & O'Higgins, T. (2019). Linking biodiversity to ecosystem services supply: Patterns across aquatic ecosystems. *Science of the Total Environment, 657*, 517–524.

Turkelboom, F., Leone, M., et al. (2017). *When we cannot have it all: Ecosystem services trade-offs in the context of spatial planning*. Ecosystem Services.

Valatin, G., Moseley, D., & Dandy, N. (2016). Insights from behavioural economics for forest economics and environmental policy: Potential nudges to encourage woodland creation for climate change mitigation and adaptation? *Forest Policy and Economics, 72*, 27–36.

Walther, Y. M., & Möllmann, C. (2014). Bringing integrated ecosystem assessments to real life: A scientific framework for ICES. *ICES Journal of Marine Science, 71*, 1183–1186.

10

Unravelling the Relationship between Ecosystem-Based Management, Integrated Coastal Zone Management and Marine Spatial Planning

Martin Le Tissier

Abstract Coastal zones are among the most productive areas in the world, offering a wide variety of valuable habitats and ecosystems services that have attracted humans and human activities over millennia. But equally coastal zones are also among the most vulnerable areas to climate change, natural hazards and other anthropogenic perturbations. The impacts of coastal change are far reaching and are already changing the wellbeing of coastal communities. It is essential to make use of long-term management tools to enhance the conservation of coastal resources whilst increasing the sustainability of their uses. Integrated Coastal Management (ICM) and Marine Spatial Planning (MSP) are both tools that attempt to override traditional sectoral approaches that lead to disconnected decisions and missed opportunities for more sustainable coastal development. Ecosystem-Based Management (EBM) describes the comprehensive integrated management of human activities based on the best available scientific knowledge to achieving sustainable use of ecosystem goods and services and maintenance of ecosystem integrity. However, there is a degree of contradiction regarding the juxtaposition of EBM to ICM and MSP—does it underpin and coordinate the implementation of them or does ICM and MSP coordinate the application of EBM principles to management practices and goals or does the difference in terminology detract from the real challenge of achieving sustainability of the world's coastal and marine areas? This chapter provides insights into the juxtaposition of these concepts and suggests a promising future approach founded on Biodiversity Portfolio Analysis (BPA).

Lessons Learned

- Adherence to terminology rather than end goals can blur the emphasis and principles of processes needed to address environmental challenges in coastal and marine areas.

M. Le Tissier (✉)
MaREI Center for Marine and Renewable Energy, Beaufort Building, Environmental Research Institute, University College Cork, Co. Cork, Ireland
e-mail: martin.letissier@ucc.ie

- There is a need to recognise the coastal and marine environment as a portfolio of constituent elements that need to be managed as a cohesive whole.
- Principles and processes of EBM need to be juxtaposed with other methods and tools for understanding environmental challenges faced by society now and in the future.

Needs to Advance EBM

- Stronger inculcation of societal elements and their explicit inclusion in EBM principles and processes is a necessary requirement.
- Developing a coherent approach to link EBM with existing policy processes and outcomes is important.

1 Introduction

Scientific and policy communities have increasingly recognised that complex environmental problems are an existential threat to humanity that require integrative, interdisciplinary approaches to understand and manage the interaction between social and ecological systems (Binder et al. 2015; Defries and Nagendra 2017). Concurrently, there is also a growing recognition that environmental change is experienced at many inter-related scales from local to global with cause and impact also ranging from near to distant (Adger and Brown 2010). Whilst the sciences have quantified and documented environmental change and its consequences [op. cit.], research has also shown that human activities are largely behind the driving forces that are the proximate causes of environmental change (Stern et al. 1992).

Formal concepts of ecosystem management initially arose in the 1970s aligned to wildlife management and were first transferred to the marine environment as part of the United Nations Conference on the Law of the Sea (UNCLOS) (Forst 2009; Long et al. 2015). Environmental policies increasingly advocate a holistic approach to coastal and marine resource management (Buhl-Mortensen et al. 2017) that address an increasing degree of anthropogenic pressures on coastal and marine environments as well as conflicts between multiple users competing for space and resources. The Ecosystem Approach (EA) is a strategy that underpins the objectives of the Convention on Biological Diversity as "a strategy for the integrated management of land, water and living resources that promotes conservation and sustainable use in an equitable way" (CBD 2000), and is widely referred to as Ecosystem-based Management or EBM (Long et al. 2015). EA is embedded in the concept of sustainable development, which requires that the needs of future generations are not compromised by the actions of people today, and puts emphasis on a management regime that maintains the health of the ecosystem alongside appropriate human use of the marine environment, for the benefit of current and future generations (ICES 2005; Defries and Nagendra 2017). Key features of EBM centre around notions that management should be holistic and not focused on single sectors or species [c.f., Garcia et al. 2003] and directed to managing human activities and their impacts on ecosystems rather than the ecosystem itself (Leslie and McLeod 2007). However, the reality has been that both scientific research and management programmes have applied concepts of EBM with very diverse variations in emphasis, definition of terms and application of principles (Yaffee 1999; Arkema et al. 2006; Sardà et al. 2014; Long et al. 2015). In large part this is because there has been a mismatch in institutional arrangements (Alexander and Haward 2019) where Integrated Coastal Management (ICM) and Marine Spatial Planning (MSP) have been employed to redress a traditional sectoral focus of marine environmental and resource management (Ehler and Douvere 2009; Freestone et al. 2010; UNEP 2010; Smith et al. 2017) and have stronger policy and governance as well as management components (Javier 2015). ICM and MSP have had greater policy presence because they are seen as providing a means to improve decision making by providing a framework to analyse competing human activities and managing their impact on the marine environment (Buhl-Mortensen et al. 2017).

2 Unravelling EBM, ICM and MSP

Degradation of coastal and marine ecosystems is an often cited impact of environmental change (e.g., Stern et al. 1992; Arkema et al. 2006; Alexander and Haward 2019), and is a consequence of post-1900 industrialisation and post-World War II economic and population growth (Yasuhara et al. 2012), which have increasingly seen both human migration to the world's coastal zones and exploitation of coastal and marine space and resources as demand has outstripped availability of land-based space and resources (Caddy and Grithiths 1995; Long et al. 2015). EBM has been advocated as a key pillar in the sustainable management of coastal zones and marine areas in conjunction with ICM and MSP respectively (UN Environment 2018; Langlet and Rayfuse 2019).

There is an ever-growing, largely academic, literature that debates the relationship between the definitions and roles of EBM with ICM and MSP (see for example, Haines-Young and Potschin 2011; NOAA 2011; Aswani et al. 2012; Celliers et al. 2016) in the context of whether one is subsumed within the other (Golitsyn 2010) and the legal basis and definitions of these terms in different jurisdictions are discussed in detail in this volume (See Enright and Boteler 2020; O'Hagan 2020; Harwell 2020). The aims of ICM and MSP whilst principle-based, in common with EBM, are primarily place-based in their implementation in order to:

- Reduce conflicts, and enhance synergies, between sectors and their activities, and
- Protect and conserve the environment and its resources upon which those activities are dependent.

ICM and MSP seek to achieve these objectives by:

- Integrating between levels of government and other management authorities (including across administrative boundaries),
- Integrating between disciplines, and
- Integrating across spatial and temporal scales.

Although EBM essentially includes much of these aims and objectives it differs in its sense of 'place', which is focussed on ecosystem units rather than sectoral activities and, whilst acknowledging that humans are part, are dependent on and occur with the ecosystem, places a different emphasis of management objectives. The consequence is that EBM can place an unequal weighting to the three tenets of sustainable development—environmental, economic and social equity—and lead to the view of human activity as impacts to ecosystem long-term viability and longevity. In contrast, MSP and ICM are essentially planning processes that seek to overcome the fragmentation in single-sector management approaches and analyse and organise human activities in coastal and marine areas to achieve economic and social objectives whilst safeguarding ecological integrity (Sandersen et al. 2013). Thus ICM and MSP have evolved with a greater emphasis towards designing rules and procedures for how to govern an area (Rodriguez 2017). This can lead to decisions that emphasises a 'balance' between protection of the environment and the maintenance and development of coastal and marine dependent economies: such a perspective is inherent to many coastal and marine strategies such as the EU integrated maritime policy (European Commission 2007). Therefore, whilst EBM has evolved to include many of the same principles as ICM and MSP (Haines-Young and Potschin 2011) it has a greater emphasis on conserving ecosystems and ecosystem services, which brings in additional management challenges over and above merging sectoral approaches to exploitation of coastal and marine space and resources (Golitsyn 2010). What this has meant is that, regardless of the terminology used, the practices of EBM by implementing authorities have tended to focus on applying an EBM approach to inform strategies and plans that seek to promote the sustainable growth of coastal and maritime economies, the sustainable development of marine areas and the sustainable use of marine resources. For instance the EU Marine Strategy Framework Directive as the environmental pillar of the EU's cross-cutting Integrated Maritime Policy, and aspects of the National Environmental Policy Act (NEPA) in the USA.

3 Current Coastal and Marine Management Regimes

The coastal and marine environment and its resources have been managed through fragmented and sectoral approaches (Altvater and Passarello 2018; Kelly et al. 2018) that fail to incorporate the complexity and interconnection of marine ecosystems and the cumulative pressures that different human activities have on species and habitats (see O'Hagan 2020 for discussion on the dominance of the Common Agricultural Policy and the Common Fisheries Policy over activities within the environment for

Europe). EBM is described as an integral component of management regimes, such as ICM and MSP (Kittinger et al. 2014; Robinson et al. 2014; Javier 2015), but has developed in a number of parallel approaches leading to a plethora of terminology and variation in the detail of principles depending on the ultimate management interest. Essentially the recognition of the importance of EBM within any management approaches is centred on acknowledging the relationship of ecosystem services to human welfare and emphasising the need for tools that encourage coordination and cooperation, participation, transparency, public interest, etc., to achieve the governance of these spaces. At the same time, the increasing interest in developing new uses and activities bring an important concern about its environmental consequences (NOAA 2011). A further benefit of EBM is its focus on natural boundaries and interconnections rather than the 'un-natural' administrative boundaries that characterise social and economic organisation (Paxinos et al. 2008; Environmental Law Institute 2009a; UNEP 2012). Therefore, EBM may be viewed as a process that implements the concepts of an Ecosystem Approach into planning management regimes for coastal and marine areas (Douvere 2008; Ansong et al. 2017).

The expectation that EBM with ICM and MSP can close the gaps between societal objectives and the state of the environment is great (Karlsson 2019). Many nations and regions have management regimes that seek to harmonise laws, policy, plans and strategies within prevailing institutional arrangements to achieve sustainable development of coastal and ocean space that balance different uses of the space and resources (Balgos et al. 2005). Increasingly, EBM has been applied to coastal and marine areas to incorporate both environmental and non-environmental factors into management regimes in order to inculcate human systems as components of the natural environment (Domínguez-Tejo et al. 2016). These initiatives seek to embed EBM as a procedure to ensure that management processes achieve an equitable and sustainable balance between conservation of the environment and persistence of ecosystem services into the future with the multiple demands on coastal and marine space and resources (Rodriguez 2017). The practical application of EBM within management regimes has recognised that there is a divergence from theory depending on political and socio-economic priorities (Jones et al. 2016) but that EBM provides opportunities to address challenges of functional metrics and indicators, spatial and temporal measures to address multiple and cumulative uses of resources, integration across sectors and adaptive management (Rodriguez 2017).

There is a growing awareness that management regimes, and the way they are informed by EBM approaches, need to be updated to account for contradictory environmental/societal and economic/stakeholder goals that can incentivise an economic and environmental perspective to act against each other (Breen et al. 2012). Furthermore, EBM, ICM and MSP need to evolve methodologies that better account for the consequences of current and future scenarios of development and uses of coastal and marine areas.

4 The Future

The academic debate over the supremacy and juxtaposition of EBM, ICM and MSP terms (and their variants) of one above the other perhaps misses the point that they all purport to have a primary objective oriented around concepts of sustainable development, which is principally about meeting the needs of the present without compromising those of the future (Brundtland 1987; Holden et al. 2014). To deliver 'sustainability', and develop an organisational methodology for the integrated management of natural resources requires approaches that complements but go beyond 'classical' conservation concepts such as endangered species or habitat protection and various types of protected area designations (English Nature 2003). Such considerations lie behind concepts that seek to extend ideas of humans as an integral component of the ecosystem (Millennium Ecosystem Assessment 2005) to those of humanity existing within planetary boundaries (Raworth 2012; Dearing et al. 2014) and ecosystem services. However, assessing the diversity of social relationships with coasts and marine space can prove difficult for scientists and practitioners in order to protect and conserve the services and benefits they provide (Kittinger et al. 2014). In particular outstanding challenges exist in regard to the spatial distribution of social, environmental and economic values; contrasting across different types of uses and their cumulative effects; addressing risk and uncertainty; and the juxtaposition of administrative and jurisdictional boundaries versus ecological boundaries (Domínguez-Tejo et al. 2016). This presents barriers to incorporating social dimensions of marine ecosystems into ecosystem-based management, which can in turn affect the success of planning and management initiatives encompassed by ICM and MSP.

Sustainable development is often addressed as a finite utopian vision of some 'perfect' future. However, sustainable development is increasingly viewed as a process—or transformative pathway—towards a largely unknown (or uncertain) future that can change over time and space whereby concepts of adaptation and resilience hold greater importance (Denton et al. 2014; Feola 2015; Romero-Lankao et al. 2016). Management of coastal and marine space and resources often focuses on "islands" of high value ecosystems, in terms of economic value or conservation (Hills et al. 2009). However, integrated management requires land and seascape-levels analysis of all ecosystem values from both monetary and non-monetary perspectives. In viewing sustainable development as a pathway the opportunity for the balance between different elements—environmental, economic and social—can vary over time and space. In actuality coastal and marine spaces are a mosaic of different interests vying for the 'best' return on activities whether they are conflictory or complementary, exploitative or conservationist. Furthermore, relative returns can vary over time and space. In practice, this means that successful EBM needs to incorporate analyses of risk and return to ensure that management outcomes lead to greater security across all elements—human and environment—of coastal and marine systems over short and long timescales.

The Biodiversity portfolio Analysis (BPA) method is derived from the logic used in share (equity) portfolio management in terms of balancing within a portfolio the returns with the risks (Figge 2001; Breen et al. 2012). Optimising the returns from a share portfolio, or a suite of ecosystems in a landscape, is dependent on the relationship between the units in terms of risk and return in time and space. This leads to an approach that more approximates with portfolio management that provides flexibility to adapt and adjust over time and space as situations change (Hills et al. 2009; Breen and Hynes 2014)—especially given the uncertainty associated with all three pillars of sustainability. It has been suggested that BPA as part of a holistic management approach could lead to strategies that 'favour' conservation as a longer term strategy maintaining future options in comparison with current management approaches that can be 'weighted' towards sustaining the status quo of existing coastal and marine use with less emphasis on future potential development and opportunities (Figge 2001; Hills et al. 2009; Breen et al. 2012). In practice this means that drawing distinctions between pillars of sustainability and assigning a hierarchy of one over the other is likely to be counterproductive. ICM and MSP influence the spatial and temporal distribution of human activities and, to attain sustainable development, can only be effective if EBM, with its focus on the costs and benefits that ecosystem services provide, runs concurrently (Douvere and Ehler 2006; Environmental Law Institute 2009a, b; Forst 2009; Soma et al. 2015; Hummel et al. 2017; Altvater and Passarello 2018). In this way sustainable development of coastal and marine spaces can be achieved through:

- Addressing the heterogeneity of coastal and marine areas to reflect the interdependency of ecology and human activities upon each other.
- Influencing the behaviour of humans and their activities over time in a way that respects ecological limitations and boundaries.
- Addressing the conflicts and compatibility issues that arise from different human activities targeting the same ecological resources and/or different ecological resources from the same space.
- Steering single-sector management to become integrated across multiple sector decision making.

The complexity of managing coastal and marine areas is ever increasing as growing populations identify more uses for the services and benefits provided by the world's coastal and marine areas, and build the capacity to exploit opportunities (Francis et al. 2019). As the number and variety of both possible uses of coastal and marine spaces, and interest groups pursuing uses, escalates it is important to be able to assess the effect of management regimes not only in the short term but over medium and longer term planning horizons. A portfolio tactic allows an assessment to be made of a range of possible scenarios the outcomes from an EBM approach to achieve a sustainable future, whilst minimising the risk to that objective.

References

Adger, W. N., & Brown, K. (2010). Progress in global environmental change. *Global Environmental Change, 20*, 547–549. https://doi.org/10.1016/j.gloenvcha.2010.07.007.

Alexander, K. A. A., & Haward, M. (2019). The human side of marine ecosystem-based management (EBM): 'Sectoral interplay' as a challenge to implementing EBM. *Marine Policy, 101*, 33–38. https://doi.org/10.1016/j.marpol.2018.12.019.

Altvater, S., & Passarello, C. (2018). Policy brief implementing the ecosystem-based approach in maritime spatial planning. Available at: https://www.msp-platform.eu/sites/default/files/20181025_ebainmsp_policybrief_mspplatform.pdf. Accessed 11 May, 2020.

Ansong, J., Gissi, E., & Calado, H. (2017). An approach to ecosystem-based management in maritime spatial planning process. *Ocean & Coastal Management, 141*, 65–81. https://doi.org/10.1016/j.ocecoaman.2017.03.005.

Arkema, K. K., Abramson, S. C., & Dewsbury, B. M. (2006). Marine ecosystem-based management: From characterization to implementation. *Frontiers in Ecology and the Environment, 4*, 525–532. https://doi.org/10.1890/1540-9295(2006)4[525,MEMFCT]2.0.CO;2.

Aswani, S., Christie, P., Muthiga, N. A., et al. (2012). The way forward with ecosystem-based management in tropical contexts: Reconciling with existing management systems. *Marine Policy, 36*, 1–10. https://doi.org/10.1016/j.marpol.2011.02.014.

Balgos, M. C., Ricci, N., Walker, L., et al. (2005). Compilation of summaries of National and Regional Ocean Policies. The Nippon Foundation Research Task Force on National Ocean Policies.

Binder, C. R., Pahl-wostl, C., & Knieper, C. (2015). Frameworks for analyzing social-ecological systems. Available at: https://www.tias-web.info/wp-content/uploads/2015/07/Frameworks-for-SES-analysis_CB-small.pdf. Accessed 11 May, 2020.

Breen, B., & Hynes, S. (2014). Shortcomings in the European principles of Integrated Coastal Zone Management (ICZM): Assessing the implications for locally orientated coastal management using Biome Portfolio Analysis (BPA). *Marine Policy, 44*, 406–418. https://doi.org/10.1016/j.marpol.2013.10.002.

Breen, P., Robinson, L. A., Rogers, S. I., et al. (2012). An environmental assessment of risk in achieving good environmental status to support regional prioritisation of management in Europe. *Marine Policy, 36*, 1033–1043. https://doi.org/10.1016/j.marpol.2012.02.003.

Brundtland, G. H. (1987). Report of the World Commission on environment and development: Our common future. United Nations.

Buhl-Mortensen, L., Galparsoro, I., Vega Fernández, T., et al. (2017). Maritime ecosystem-based management in practice: Lessons learned from the application of a generic spatial planning framework in Europe. *Marine Policy, 75*, 174–186. https://doi.org/10.1016/j.marpol.2016.01.024.

Caddy, J. F., & Grithiths, R. C. (1995). *Living marine resources and their sustainable development: Some environmental and institutional perspectives*. FAO Fisheries Technical Paper. No. 353. FAO, Rome.

CBD. (2000). Report of the Fifth meeting of the conference of the parties to the convention on biological diversity.

Celliers, L.,Taljaard, S., van Niekerk, L. (2016). MSP, ICM, ABM, and EBM - the alphabet of confusion? Presented at the 13th Biodiversity Planning Forum, George, South Africa 9 June 2016.

Dearing, J. A., Wang, R., Zhang, K., et al. (2014). Safe and just operating spaces for regional social-ecological systems. *Global Environmental Change, 28*, 227–238. https://doi.org/10.1016/j.gloenvcha.2014.06.012.

Defries, R., & Nagendra, H. (2017). Ecosystem management as a wicked problem. *Science (80-), 270*, 265–270. https://doi.org/10.1126/science.aal1950.

Denton, F., Wilbanks, T. J., Abeysinghe, A. C., et al. (2014). *Climate-resilient pathways: Adaptation, mitigation, and sustainable development*. Cambridge, UK: Cambridge University Press.

Domínguez-Tejo, E., Metternicht, G., Johnston, E., & Hedge, L. (2016). Marine spatial planning advancing the ecosystem-based approach to coastal zone management: A review. *Marine Policy, 72*, 115–130. https://doi.org/10.1016/j.marpol.2016.06.023.

Douvere, F. (2008). The importance of marine spatial planning in advancing ecosystem-based sea use management. *Marine Policy, 32*, 762–771. https://doi.org/10.1016/j.marpol.2008.03.021.

Douvere, F., & Ehler, C. (2006). Issues and prospects. Ecosystem-based marine spatial management: An evolving paradigm for the management of coastal and marine places. *Ocean Yearbook, 23*, 1–26.

Ehler, C., & Douvere, F. (2009). *Marine spatial planning: A step-by-step approach toward ecosystem-based management.* Intergovernmental Oceanographic Commission and Man and the Biosphere Programme. IOC Manual and Guides No. 53, ICAM Dossier No. 6. Paris: UNESCO.

English Nature. (2003). *Adopting an ecosystem approach for improved stewardship of the maritime environment: Some overarching issues.* English Nature Research Reports Number 538.

Enright, S.R., & Boetler, B. (2020). The ecosystem approach in international law. In T. O'Higgins, M. Lago, & T. H. DeWitt (Eds.), *Ecosystem-based management, ecosystem services and aquatic biodiversity: Theory, tools and applications* (pp. 333–352). Amsterdam: Springer.

Environmental Law Institute. (2009a). *Expanding the use of ecosystem-based management in the Coastal Zone Management Act.* Environmental Law Institute (ELI).

Environmental Law Institute. (2009b). *Ocean and coastal ecosystem based management. Implementation handbook.* Washington, DC: Environmental Law Institute®.

European Commission. (2007). *SEC(2007) 1280. An Integrated Maritime Policy for the European Union. Impact assessment summary.* European Commission.

Feola, G. (2015). Societal transformation in response to global environmental change: A review of emerging concepts. *Ambio, 44*, 376–390. https://doi.org/10.1007/s13280-014-0582-z.

Figge, F. (2001). *Managing biodiversity correctly—efficient portfolio management as an effective way of protecting species.* Centre for Sustainability Management, University of Luneburg.

Forst, M. F. (2009). The convergence of Integrated Coastal Zone Management and the ecosystems approach. *Ocean & Coastal Management, 52*, 294–306. https://doi.org/10.1016/j.ocecoaman.2009.03.007.

Francis, T. B., Levin, P. S., Punt, A. E., et al. (2019). Linking knowledge to action in ocean ecosystem management: Elementa. https://doi.org/10.1525/elementa.338

Freestone, D., Cicin-Sain, B., Hewawasam, I., & Hamon, G. (2010). *Draft policy brief on improving governance: Achieving integrated, ecosystem-based ocean and coastal management.* 5th Global Conference on Oceans, Coasts and Islands, p 22.

Garcia, S. M., Zerbi, A., & Aliaume, C., et al. (2003). *The ecosystem approach to fisheries. Issues, terminology, principles, institutional foundations, implementation and outlook.* FAO Fisheries Technical Paper. No. 443. Rome: FAO.

Golitsyn, V. (2010). Major challenges of globalisation for seas and oceans: Legal aspects. In D. Vidas (Ed.), *Law, technology and science for oceans in globalisation* (pp. 59–73). Brill Academic Publisher.

Gopnik, M. (2008). *Integrated marine spatial planning in U.S. waters: The path forward. The Marine conservation initiative.* Gordon and Betty Moore Foundation.

Haines-Young, R., & Potschin, M. (2011). *Integrated coastal zone management and the ecosystem approach.* Deliverable D2.1, PEGASO Grant agreement no: 244170. CEM Working Paper No 7.

Harwell, D. R. (2020). Ecosystem services in U.S. environmental law and governance for the ecosystem-based management practitioner. In T. O'Higgins, M. Lago, & T. H. DeWitt (Eds.), *Ecosystem-based management, ecosystem services and aquatic biodiversity: Theory, tools, and applications* (pp. 373–402). Amsterdam: Springer.

Hills, J., Carlisle, M., Le Tissier, M., et al. (2009). Landscape-scale analysis of ecosystem risk and returns: A new tool for ICZM. *Marine Policy, 33*, 887–900. https://doi.org/10.1016/j.marpol.2009.04.017.

Holden, E., Linnerud, K., & Banister, D. (2014). Sustainable development: Our common future revisited. *Global Environmental Change, 26*, 130–139. https://doi.org/10.1016/j.gloenvcha.2014.04.006.

Hummel, C., Provenzale, A., Van Der Meer, J., et al. (2017). Ecosystem services in European protected areas: Ambiguity in the views of scientists and managers? *PLoS One, 12*, e0187143. https://doi.org/10.1371/journal.pone.0187143.

ICES. (2005). *Guidance on the application of the ecosystem approach to management of human activities in the European Marine Environment*. ICES Cooperative Research Report No. 273.

Javier, G. S. (2015). The approach of integrated coastal zone management: From technical to political point of view. *Journal of Coastal Zone Management, 18*, 2–4. https://doi.org/10.4172/2473-3350.1000e111.

Jones, P. J. S., Lieberknecht, L. M., & Qiu, W. (2016). Marine spatial planning in reality: Introduction to case studies and discussion of findings. *Marine Policy, 71*, 256–264. https://doi.org/10.1016/j.marpol.2016.04.026.

Karlsson, M. (2019). Closing marine governance gaps? Sweden's marine spatial planning, the ecosystem approach to management and stakeholders' views. *Ocean & Coastal Management, 179*, 104833. https://doi.org/10.1016/j.ocecoaman.2019.104833.

Kelly, C., Ellis, G., & Flannery, W. (2018). Conceptualising change in marine governance: Learning from transition management. *Marine Policy, 95*, 24–35. https://doi.org/10.1016/j.marpol.2018.06.023.

Kittinger, J. N., Koehn, J. Z., Le Cornu, E., et al. (2014). A practical approach for putting people in ecosystem-based ocean planning. *Frontiers in Ecology and the Environment, 12*, 448–456. https://doi.org/10.1890/130267.

Langlet, D., & Rayfuse, R. (2019). *The ecosystem approach in ocean planning and governance: Perspectives from Europe and beyond*. Leiden: Brill Nijhoff.

Leslie, H. M., & McLeod, K. L. (2007). Confronting the challenges of implementing marine ecosystem-based management. *Frontiers in Ecology and the Environment, 5*, 540–548. https://doi.org/10.1890/060093.

Long, R. D., Charles, A., & Stephenson, R. L. (2015). Key principles of marine ecosystem-based management. *Marine Policy, 57*, 53–60. https://doi.org/10.1016/j.marpol.2015.01.013.

Millennium Ecosystem Assessment. (2005). *Ecosystems and human well-being: Synthesis. A report of the Millennium Ecosystem Assessment*. World Resources Institute.

NOAA. (2011). *Clarifying the relationships among ecosystem based management; integrated ecosystem assessments; and, coastal and marine spatial planning*. NOAA response to SAB/ESMWG letter of April 5, 2010. NOAA.

O'Hagan, A. M. (2020). Ecosystem-based management (EBM) and ecosystem services in EU law, policy and governance. In T. O'Higgins, M. Lago, & T. H. DeWitt (Eds.), *Ecosystem-based management, ecosystem services and aquatic biodiversity: Theory, tools and applications* (pp. 353–372). Amsterdam: Springer.

Paxinos, R., Wright, A., Day, V., et al. (2008). Marine spatial planning: Ecosystem-based zoning methodology for marine management in South Australia. *Journal of Conservation Planning, 4*, 37–59.

Raworth, K. (2012). A safe and just space for humanity: Can we live within the doughnut? Oxfam discussion papers.

Robinson, L. A., Culhane, F. E., Baulcomb, C., et al. (2014). *Towards delivering ecosystem-based marine management: The ODEMM approach*. ODEMM.

Rodriguez, N. J. I. (2017). A comparative analysis of holistic marine management regimes and ecosystem approach in marine spatial planning in developed countries. *Ocean & Coastal Management, 137*, 185–197. https://doi.org/10.1016/j.ocecoaman.2016.12.023.

Romero-Lankao, P., Gnatz, D. M., Wilhelmi, O., & Hayden, M. (2016). Urban sustainability and resilience: From theory to practice. *Sustain, 8*, 1–19. https://doi.org/10.3390/su8121224.

Sandersen, H. T., Mikkelsen, E., Moksness, E., & Vølstad, J. H. (2013). Knowledge issues in ICZM and EBM applied on small geographic scales: Lessons from a case study in Risør, Norway. In E. Moksness, E. Dahl, & J. Støttrup (Eds.), *Global challenges in integrated coastal zone management*. Wiley.

Sardà, R., O'Higgins, T., Cormier, R., et al. (2014). A proposed ecosystem-based management system for marine waters: Linking the theory of environmental policy to the practice of environmental management. *Ecology and Society, 19*, 51. https://doi.org/10.5751/ES-07055-190451.

Smith, D. C., Fulton, E. A., Apfel, P., et al. (2017). Implementing marine ecosystem-based management: Lessons from Australia. *ICES Journal of Marine Science, 74* (7), 1990–2003. https://doi.org/10.1093/icesjms/fsx113.

Soma, K., van Tatenhove, J., & van Leeuwen, J. (2015). Marine governance in a European context: Regionalization, integration and cooperation for ecosystem-based management. *Ocean & Coastal Management, 117*, 4–13. https://doi.org/10.1016/j.ocecoaman.2015.03.010.

Stern, P. C., Young, O. R., & Druckman, D. (1992). Global environmental change: Understanding the human dimensions. Washington, DC: The National Academies Press. https://doi.org/10.17226/1792

UN Environment. (2018). *Conceptual guidelines for the application of marine spatial planning and integrated coastal zone management approaches to support the achievement of sustainable development goal targets 14.1 and 14.2*. UN Regional Seas Reports and Studies No. 207.

UNEP. (2010). *Marine and coastal ecosystem-based management. An introductory guide to managing oceans and coasts better*. UNEP.

UNEP. (2012). *Synthesis document on the experience and use of marine spatial planning*. UNEP/CBD/SBSTTA/16/INF/18.

Yaffee, S. L. (1999). Three faces of ecosystem management. *Conservation Biology, 13*, 713–725.

Yasuhara, M., Hunt, G., Breitburg, D., et al. (2012). Human-induced marine ecological degradation: Micropaleontological perspectives. *Ecology and Evolution, 2*, 3242–3268. https://doi.org/10.1002/ece3.425.

11

Mitigating Negative Unintended Impacts on Biodiversity in the Natura 2000 Vouga Estuary

Ana I. Lillebø, Heliana Teixeira, Javier Martínez-López, Ana Genua-Olmedo, Asya Marhubi, Gonzalo Delacámara, Verena Mattheiß, Pierre Strosser, Timothy G. O'Higgins, and António A. J. Nogueira

Abstract This chapter presents the co-development of the Ecosystem-Based Management (EBM) planning process in the Vouga estuary for the mitigation of unintended impacts on biodiversity resulting from the 2019/2020 management plan. This estuary, part of Ria de Aveiro coastal lagoon located on the north-west coast of Portugal (40°38′N, 08°45′W), connects the Vouga river catchment area to the Atlantic Ocean. Ria de Aveiro, part of the Natura 2000 network, is characterised by high biodiversity and a wide range of ecosystem services. However, it is also a vulnerable territory that requires a management plan in practice for environmental protection, targeting threatened species and habitats, but also to enable socio-economic welfare. Framed by EBM principles, the stepwise planning approach aimed at identifying the governance boundaries and institutions, the policy objectives, synergies, and gaps relevant to managing biodiversity, and to promote participatory actions with local stakeholders and policy-makers to understand their objectives. These three first steps enabled us to understand the social-ecological system and to co-develop relevant EBM solutions. In the final step, the proposed

A. I. Lillebø (✉) · H. Teixeira · A. Genua-Olmedo · A. A. J. Nogueira
CESAM & Department of Biology, University of Aveiro, Aveiro, Portugal
e-mail: lillebo@ua.pt; heliana.teixeira@ua.pt; ana.genua@ua.pt; antonio.nogueira@ua.pt

J. Martínez-López
BC3 – Basque Centre for Climate Change, Leioa, Spain
e-mail: jmartinez@cebas.csic.es

A. Marhubi · G. Delacámara
IMDEA Water Institute, Parque Científico Tecnológico de la Universidad de Alcalá, Alcalá de Henares, Madrid, Spain
e-mail: asya.marhubi@imdea.org; gonzalo.delacamara@imdea.org

V. Mattheiß · P. Strosser
AcTeon - 5 Place Sainte-Catherine, Colmar, France
e-mail: v.mattheiss@acteon-environment.eu; p.strosser@acteon-environment.eu

T. G. O'Higgins
MaREI, Environmental Research Institute, Cork, Ireland
e-mail: tim.ohiggins@ucc.ie

EBM solutions were evaluated for effectiveness, efficiency, equity and fairness, and then compared to the present condition. The co-developed solutions target science, policy and stakeholders interfaces. Namely, scientific knowledge applied to restore saltmarshes and seagrasses, policy objectives harmonising monitoring across EU Directives and integrate territorial management instruments, and management process involving stakeholders throughout.

Lessons Learned The co-created EBM plan for the Vouga estuary Natura 2000 site is foreseen to support the further development of the Vouga Estuary Management Plan. To this end, it is also foreseen to support actions for a more comprehensive understanding of the social-economic implications of the provided ecosystem services in line with the Centro Portugal region strategy for smart specialisation (Portugal RIS3 Centro).

These are:

- *Continue to increase stakeholder participation*: stakeholders want to contribute to management and actively participate in the co-creation of adaptive management solutions;
- *Integrate and coordinate policies*: proceed with the development of the Vouga estuary management plan considering connectivity across water domains, landowners and users;
- *Promote adaptive management and acknowledge unintended impacts*: harmonise existing mandatory monitoring programmes to support regular evaluation and enable adaptive management involving stakeholders to respond to future management needs and challenges.

Needs to Advance EBM

- At the scale of Natura 2000 Vouga estuary, EBM plans should be co-created with input from local stakeholders and policy-makers. To protect biodiversity, managers should consider climate change projections and acknowledge uncertainty. For the successful implementation of the identified water and nature policies in places like the Vouga estuary, any actions need to ensure involvement of users and landowners.
- At a global scale, and particularly at European Union scale, it has been acknowledged that biodiversity protection is still deficient and that, at current trends, the EU Strategy for 2020 will fail to achieve its goal of halting loss of biodiversity. To this end, EBM, that encompasses any management or policy options intended to restore, enhance or protect the resilience of the ecosystem, appears as a valuable approach in support of EU Strategy beyond 2020.

1 Introduction

The United Nations (UN) 2011 declaration of the 2011–2020 Decade on Biodiversity brought to the forefront news the urgent need to halt the loss of biodiversity as well as its overall vision for 2050 of *"living in harmony with nature"*. It also made clear the fact that ecosystem functioning and the provision of ecosystem services essential for human well-being are supported by biological diversity. Within the UN Environment Programme, the Convention on Biological Diversity (CBD) developed The Strategic Plan for Biodiversity with a shared vision, mission, and set of strategic goals: the Aichi Biodiversity Targets.[1]

To this end, the European Union (EU) Biodiversity Strategy to 2020 aims at ensuring the existence and conservation of biodiversity and ecosystem services and at halting the loss of global biodiversity (European Commission 2011). Its main objective is to fulfil the implementation of nature protection legislation, with special emphasis in Natura 2000 sites with high biodiversity value. This strategy includes six targets focused on: better protection and restoration of ecosystems and their associated services; establishment of green infrastructure; development of sustainable agriculture and fisheries; control of invasive alien species; and an EU contribution to stop global biodiversity loss.

Action 5 is based on improving knowledge on ecosystem services. The use of maps helps to achieve this action by characterizing the spatial heterogeneity of ecosystems and services they supply, and the associated pressures and impacts. They also help to translate scientific evidence into information that is understandable for policy and decision making (Maes et al. 2016). Thus, mainstreaming values of biodiversity and ecosystem services into decision-making is expected to help increase awareness about the implications of further degradation and loss of natural ecosystems on human well-being (Teixeira et al. 2018, 2019).

An Ecosystem-Based Management (EBM) of aquatic ecosystems is more likely to support a timely achievement of the EU 2020 Biodiversity Strategy targets than isolated sectorial management initiatives (Piet et al. 2017; Martínez-López et al. 2019a). Such an integrative approach to ecological, social and governance principles sets an adequate context to apply socio-ecological concepts such as ecosystem services in practical management initiatives (Lillebø et al. 2019; Martínez-López et al. 2019b). The EBM planning process involves the coordination of policies, institutions and practices (Drakou et al. 2017; Piet et al. 2017; Rouillard et al. 2018), representing a holistic approach that aims to balance multiple interrelated dimensions of ecological integrity and human well-being (Gómez et al. 2016, 2017; Langhans et al. 2019).

Following Rouillard et al. (2018), the proposed approach aiming at mitigating negative unintended impacts on biodiversity in the Natura 2000 Vouga estuary

[1] Aichi Biodiversity Targets: https://www.cbd.int/doc/strategic-plan/2011-2020/Aichi-Targets-EN.pdf.

considers the following principles (Curtin and Parker 2014; Gómez et al. 2016, 2017; Martin et al. 2018):

- *EBM considers ecological integrity, biodiversity, resilience and ecosystem services;*
- *EBM is carried out at appropriate spatial scales;*
- *EBM develops and uses multi-disciplinary knowledge;*
- *EBM builds on social–ecological interactions, stakeholder participation and transparency;*
- *EBM supports policy coordination;*
- *EBM incorporates adaptive management.*

1.1 Study Site

Ria de Aveiro is a shallow coastal lagoon located on the north-west coast of Portugal (40°38′N, 08°45′W). The adjacent coast experiences strong seasonal upwelling, the designated North Atlantic Upwelling that supports high levels of productivity especially in summer (Lopes et al. 2014). The lagoon establishes the aquatic continuum between the upstream catchment area (3500 km^2) of the Vouga river that contributes with circa 80% of the freshwater inflow, and the Atlantic Ocean through a single connection (1.3 km length, 350 m wide and 20 m depth) (e.g., Lillebø et al. (eds) 2015; Stefanova et al. 2015; Sousa et al. 2016; Lopes et al. 2017). These hydrographical settings determine that the Vouga river estuary is located within the boundaries of Ria de Aveiro coastal lagoon.

Due to its valuable natural capital, listed under both the Birds Directive and the Habitats Directive, Ria de Aveiro is a classified site under the Natura 2000 network, entailing a Special Protection Area (SPA) that includes extensive saltmarsh habitats and the adjacent marine area. Since 2011, the lagoon is also an International Long-Term Ecosystem Research (ILTER) site. Within the lagoon watershed Aveiro city represents the major urban settlement with circa 60,000 inhabitants. Like other social-ecological systems, the Vouga estuary is subject to co-competing land and water uses (Lillebø et al. (eds) 2015). Previous trans-disciplinary studies acknowledged the importance of the Vouga estuary's geographic location combined with its natural capital, which has enabled the development of a wide variety of economic, cultural and recreational activities (Lillebø et al. (eds) 2015; Dolbeth et al. 2016; Lillebø et al. 2016; Sousa 2017; O'Higgins et al. 2019). However, this area often requires human intervention for protection, or to enable economic activities, due to anthropogenic pressures impacting the hydro-morphological conditions of the lagoon, the Vouga estuary, and the adjacent Baixo Vouga Lagunar freshwater section of the Vouga river, and natural pressures like ocean storm surges, coastal erosion, and torrential rain and flood events (Pereira and Coelho 2013; Lillebø et al. (eds) 2015; Dolbeth et al. 2016; Lopes et al. 2017; Luís et al. 2018).

1.2 *Biodiversity Challenge in the Natura 2000 Vouga Estuary*

Two management interventions, occurring during 2019/2020, will likely have negative unintended impacts on biodiversity (Lillebø et al. 2019; Martínez-López et al. 2019b):

- Dredging programme to enable hydrodynamic equilibrium and navigability in Ria de Aveiro coastal lagoon (APA 2018);
- Extension of a flood bank to disable surface saltwater intrusion into Baixo Vouga Lagunar agricultural areas, named '*Sistema de Defesa Primária do Baixo Vouga Lagunar*' (DGADR 2017).

The dredging programme's ultimate goal is to improve lagoon navigability and is expected to allocate part of its dredged sediments to reinforce the banks at lower elevation zones, threatened by surface saltwater intrusion from inundation, for the protection of infrastructures and goods. Additional dredged sand will be used for beach replenishment. The extension of the flood bank is expected to improve accessibilities, foster agricultural and livestock activities, and protect wildlife and other economic activities, namely ecotourism with bird watching tours, angling, and recreational activities at the upstream area of the flood bank. These two management options will cause negative, unintended impacts on biodiversity, including changes of the system's eco-hydrodynamics, including water current velocity, turbidity, and tidal prism that will impact seagrasses and saltmarshes (Lillebø et al. (eds) 2015, 2019; Dolbeth et al. 2016). Additionally, downstream saltmarshes will be subdued to "coastal squeeze" as the combined effect of the physical flood bank barrier with the increase tidal prism will result in longer submersion periods that saltmarsh species are not adapted to (Martínez-López et al. 2019b).

The goals of our EBM approach are to:

- Contribute to operationalising an EBM planning process in response to foreseen unintended impacts resulting from the present management options;
- Mitigate unintended impacts from a major dredging programme targeting the hydrodynamic equilibrium (APA 2018);
- Mitigate unintended impacts from the extension of a flood bank targeting surface saltwater intrusion into agricultural areas (DGADR 2017);
- Make use of the best available information in a trans-disciplinary context.

To reach these goals, the overarching policies, programmes, key governance institutions, and objectives relevant to managing biodiversity were identified. Stakeholders were engaged throughout the process in order to co-define the baseline, co-develop management scenarios, and co-create the EBM plan.

2 The EBM Planning Approach

To address the governance challenges in the frame of the Vouga estuary, the EBM planning approach followed a stepwise procedure.

2.1 *Step One: Setting the Governance Boundary*

In order to be policy relevant the governance boundaries for this EBM approach were set at 500 m from the aquatic realm boundary (Fig. 1) following the Vouga Estuary Management Plan under development by the Portuguese Environmental Agency.[2] This Plan links public administration and private sectors, and provides the basis to effectively manage Ria de Aveiro's natural capital, ecosystem services and associated socio-economic activities. It encompasses an integrated land-use management plan with appropriate measures to protect all wetlands habitats, e.g., transitional waters, mud and sand flats, seagrasses and salt marshes, as well social, economic and cultural development. As shown by O'Higgins et al. (2019), the 500 m boundary is well aligned with ecosystem service production and consumption boundaries of this resource system. In this way, the Vouga Estuary Management Plan policy boundary overlaps with the production and consumption of relevant ecosystem services. Furthermore, the Plan foresees the articulation of territorial management instruments, plans and programmes at different scales, from local to regional (Centro Portugal region) and national, covering appropriate spatial scales for the EBM approach.

2.2 *Step Two: Identify Policy Objectives, Synergies, and Gaps*

The most relevant national policy plans and programmes (and institutions responsible for implementation of the policy instruments), objectives for the planning process of EBM responses, as well as linkages to EU Policies, are presented in Table 1. These initiatives cover aspects from nature conservation, to water quality and management, to climate change adaptation and tourism. The later are key drivers for sustainable economic growth of the Centro Portugal region (Dolbeth et al. 2016). At local/regional scales, it is also important to consider sectoral plans and programmes that integrate operations, enable collaborative work among institutions, and promote articulation of environmental, economic and social factors. The most relevant national and regional institutions to be considered in the planning process of EBM responses are presented in Table 2.

---

[2]Agência Portuguesa do Ambiente, I.P.—APA, https://www.apambiente.pt/.

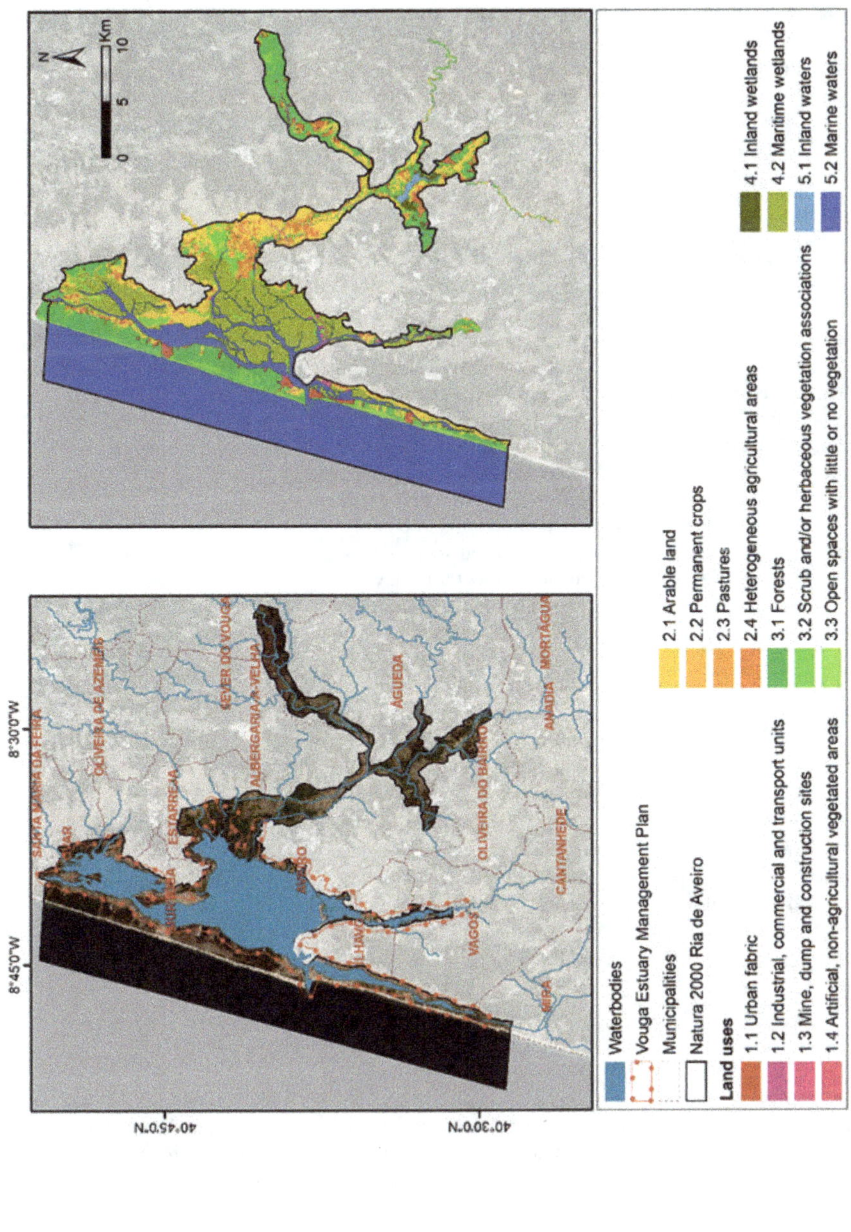

Fig. 1 The Ria de Aveiro coastal lagoon and boundaries of the Vouga Estuary Management Plan. Land uses are shown in the lower panel

Table 1 Identification of national relevant policy plans and programmes, and objectives for the EBM planning process at Ria de Aveiro: (a) link to EU policies and (b) link to regional/local policies

Policy plans and programmes	Objectives	(a) Link to EU policies
Sectoral Plan for Natura 2000 Network (PSRN2000) Institute for Nature Conservation and Forests (ICNF)	Territorial management tool for implementation of the national policy for conservation of biological diversity, aiming at safeguarding and enhancement of sites and SPAs of the continental territory, as well as maintenance of species and habitats in a favourable conservation status in these areas	Birds Directive (2009/147/EC); Habitats Directive (92/43/EEC)
National Water Plan (Decreto-Lei no. 76/2016) *Inter-ministerial Commission for Water management:* APA/ARHC; ICNF; Regional Directorate for Agriculture and Fisheries (DRAP); Directorate-General for Marine Resources (DGRM)	Governmental cross-sectoral management for the next 10 years: Increase water productivity and promoting rational use, with maximum respect for territorial integrity of the river basins; Protection, conservation and rehabilitation of water resources and associated ecosystems; Meeting needs of the population and country's economic and social development; Respect for relevant national and Community legislation and satisfaction of international commitments assumed by the Portuguese State; Access to information and participation of citizens in management of water resources	Water Frame Directive (WFD) (2000/60/EC) Floods Directive (2007/60/EC) Marine Strategy Framework Directive (2008/56/EC)
River Basin Management Plan (PGBH—RH4) Portuguese Environment Agency (APA/ARHC)	Outlines water planning for the tri-basin region of Vouga, Mondego and Lis, in accordance with WFD	WFD (2000/60/EC)
National Strategic Plan for climate change adaptation (ENAAC) Ministry of Environment	Establishes the need for adaptation. Contains the National adaptation strategy, and associated action plan, including reducing vulnerability and increasing response capacity.	EU Strategy on Adaptation to Climate Change (COM (2013) 216
National Strategic Plan for Tourism (PENT) Ministry of Economy and Innovation	Serves as basis for implementation of a series of initiatives aimed at fostering sustained growth of national tourism over the coming 10 years, and guiding activities of Portugal	EU strategy for a smart, sustainable and inclusive growth (COM (2014) 85 final, 2014/0044)

(continued)

Table 1 (continued)

Policy plans and programmes	Objectives	(a) Link to EU policies
	National Tourism Authority, as the key public body for the sector.	
		(b) Link to regional/local policies
Polis Litoral Ria de Aveiro APA/ARHC; ICNF	Integrated Operations of Rehabilitation and Recovery of Coastal Areas. Strong collaborative work between central administration and the Ria de the Aveiro Region Inter-municipal Community (CIRA)	Address the regional policy instruments Contribute to the Vouga estuary management plan Contribute to the Regional strategy for smart specialization (RIS3 Centro)
Coastal Zone Management Programme Ovar—Marinha Grande (POC-OMG) APA/ARHC	Reconcile the various conflicts of uses of the coastal zone, promoting articulation of environmental, economic and social factors related to coastal management.	
River Basin Management Plan (PGBH—RH4) Portuguese Environment Agency (APA/ARHC)	Outlines water planning for the tri-basin region of Vouga, Mondego and Lis, in accordance with WFD	

Source: Lillebø et al. (2019)

Each policy main objective (Table 1) identifies regional policy instruments contributing to the Vouga Estuary Management Plan, which aims at contributing to the Centro Portugal region strategy for smart specialisation (Portugal RIS3 Centro). Within this strategy, sea-related economic activities were selected as a strategic priority together with agriculture, forestry, tourism, information and communication technologies, materials, biotechnology, and health and wellness. The boundary for the Vouga Estuary Management Plan is presented and discussed in Fidélis and Carvalho (2015), and is considered as management boundary in O'Higgins et al. (2019) and in the proposed EBM approach (see Sect. 2.1). The Vouga Estuary Management Plan requires coordination with:

- Sectoral Plan for Natura 2000 Network (Institute for Nature Conservation and Forests; ICNF, I.P.), the territorial management tool for implementation of the national policy for the conservation of biological diversity;
- National Strategic Plan for climate change adaptation, following climate change projections, and containing the National adaptation strategy and associated action plan relevant in Centro Portugal coastal area.

The proposed EBM approach requires monitoring the policy impact of unintended pressures resulting from present management options. Although most of the information is reported in the frame of these EU water-related and Nature Directives, data sets are not harmonised. Therefore, the main gap identified concerns

Table 2 Identification of main institutions and policy domains for the EBM planning process

Institution	Policy domain	Additional information
Portuguese Environmental Agency (APA, I.P.) through the Regional Hydrographic Administrations (ARH Centro)	River Basin Management Plan (WFD) and Flood Risk Management Plan (Floods Directive) for hydrographic Region 4 (RH4) that includes Vouga, Mondego and Lis Rivers, and the foreseen estuary land use and management plans.	APA/ARH Centro is responsible for: water resources management; spatial planning of water resources, uses (including the economic analysis) and demands, and law enforcement; and for strategic and integrated planning of the coastal zone.
Institute for Nature Conservation and Forests (ICNF, I.P.)	Sectoral Plan for Natura 2000 Network (Habitats, Birds Directives; Biodiversity Strategy)	ICNF, I.P. is the national authority for nature conservation, biodiversity and forests; articulates and promotes integration of forest policy and conservation of nature and biodiversity in policies to combat desertification; to mitigate climate change and its effects; and to reduce country's energy dependence.
Centro Region Department of Agriculture and Fisheries (DRAPC)	Common Agricultural Policy (CAP) and Common Fisheries Policy (CFP)	DRAPC is a service of the Ministry of Agriculture, Forestry and Rural Development, whose mission is to participate in formulation and implementation of policies in agriculture, rural development and fisheries, as well as collaborate in policies in areas of forests, food security and plant health, in liaison with relevant central bodies and services within the framework of the efficiency of local management of resources.
Directorate general for Natural Resources, Safety and Maritime Services (DGRM)	Marine Strategy Framework Directive (MSFD) and Maritime Spatial Planning (MSP)	DGRM is a government entity of the Ministry of the Sea, a central office of direct administration of State, with administrative autonomy with the mission, under maritime administration functions, to implement and execute policies concerning maritime safety and prevention of pollution by ships and ensure regulation, inspection, national coordination and control of activities developed under these policies.

(continued)

Table 2 (continued)

Institution	Policy domain	Additional information
The Centro Regional Coordination and Development Commission (CCDRC)	Promote an integrated and sustainable development of Portugal's Centro region (NUT II)	CCDRC is tasked with coordinating and promoting at the regional level governmental policies with regard to Regional Planning and Development, Environment, Land Management, Inter-Regional and Cross-Border Cooperation and also support local government and inter-municipal associations. CCDR-C's fields of intervention also encompass management of regional operational programmes funded by the EU, and other regional development financing instruments.

the lack of harmonised monitoring programmes for the Water Framework Directive (WFD, 2000/60/EC) and Habitats Directive (HD, 92/43/EEC). In this context, one of the main challenges concerns the use of best available information.

2.3 *Step Three: Understand Stakeholder Objectives*

Vouga estuary Natura 2000 site governance involves a multiplicity of institutions, organisations and stakeholders, and involves articulation of programs and plans of local, regional and national levels (Teles et al. 2014; Fidélis and Carvalho 2015; Lillebø et al. (eds) 2015; Sousa et al. 2016; Sousa 2017; Fidélis et al. 2019). The Portuguese Environment Agency, through its Regional Hydrographic Administration for Portugal Centro Region (APA/ARH Centro) was engaged at a very early stage of the project, helping identify key management questions. Other stakeholders engaged at the kick-off stage of the work and contributed to the datasets that support scenario development include the Institute for Nature Conservation and Forests (ICNF, I.P.), Regional Directorate for Agriculture and Fisheries (DRAP Centro), Hydrographic Institute (IH), and Directorate-General for Marine Resources (DGRM). Stakeholder's participatory moments included two workshops (WS I and WS II) and a final seminar, where the co-created EBM plan was presented. All participants received a non-technical, open-access book in Portuguese detailing the entire EBM process and the main results. Approximately 70 stakeholders representing the four major groups, namely policy/governance, public administration, business, and non-governmental organizations, were invited to participate (Table 3).

Table 3 Identification of key stakeholders for the planning process of EBM responses

Policy/Governance	
Environment	APA/ARH Centro—Portuguese Environmental Agency ICNF—Institute for Nature Conservation and Forests
Fisheries and agriculture	DRAPC—Centro Region Department of Agriculture and Fisheries
Marine	DGRM—Directorate-General for Natural Resources, Safety and Maritime Services
Public administration	
Regional administration	CCDRC—The Centro Regional Coordination and Development Commission CIRA—Inter-municipal Community of the Aveiro Region
Municipalities within the Natura 2000 classified area	Águeda, Albergaria-a-Velha, Anadia, Aveiro, Estarreja, Ílhavo, Mira Murtosa, Oliveira do Bairro, Ovar, Vagos.
Parishes within the Natura 2000 classified area	E.g., Angeja, Avanca, Beduído & Vieiros, Bunheiro, Cacia, Canelas & Fermelã, Esgueira, Espinhel, Fermentelos, Gafanha Da Boa Hora, Gafanha Da Encarnação, Gafanha Da Nazaré, Gafanha Do Carmo, Glória & Vera Cruz, Murtosa, Óis da Ribeira, Ouca, Ovar Union of parishes, Pardilhó, Requeixo, Salreu, Santo André De Vagos, São Jacinto, São Salvador, Sosa, Torreira, Vagos & Santo António De Vagos, Válega.
Business	
Industry	Portucel—The Navigator Company
Tourism	Incrível Odisseia—Moliceiros boat rides Sterna—Solar boat tours and bird watching
Agriculture	ABBVL—Association of Beneficiaries of Baixo Vouga Lagunar ACRM—Association of Breeders of Marinhoa Breed ALDA—Association of Agriculture of the District of Aveiro
Fisheries	APARA—Artisanal Fishing Association of the Region of Aveiro
Aquaculture	APA—Portuguese aquaculture association
Services	APA—Port of Aveiro Administration (APA)
Other	
Local associations	AVELA—Sailing club ADERAV—Association for the study and protection of the Natural and Cultural Heritage of Aveiro Region CCPAV—Hunting and Fishing Club of Aveiro/Vouga
Non-governmental organizations (NGO's)	FAPAS—Fund for the Protection of Wild Animals GEOTA—Study Group on Spatial Planning and Environment LPN—League for the Protection of Nature SPEA—The Portuguese Society for the Study of Birds ASPEA—Portuguese Association of Environmental Education

Source: Lillebø et al. (2019)

2.3.1 Stakeholders' Perception and Spatial Multi-Criteria Analysis

At the first workshop (WS I), 17 stakeholders representing the four major groups participated, signing an informed consent agreement form, and were asked to identify the relevance of ecosystem services in Ria de Aveiro for building alternative management scenarios (Fig. 2). Participants were invited to express their opinion regarding expected beneficial effects and persistent concerns related to the current management options and contribute to the spatial multi-criteria analysis through prioritization of ecosystem services (Lillebø et al. 2019; Martínez-López et al. 2019b). This prioritization reflected stakeholders' social preferences regarding ecosystem services in order to find optimal management actions (sensu Villa et al. 2002; Martínez-López et al. 2019b). The method adopted ensures transparency of the participatory process, which is of paramount importance as different sectoral interests, such as conservationists, local users and from the business sector, like tourism, may express different priorities in relation to a set of ecosystem services of interest. This is crucial to make the participatory valuation of ES an opportunity for a more comprehensive, fair and integrative perspective for EBM (Martínez-López et al. 2019b). This socio-ecological approach illustrates how planned and structured co-developed solutions can effectively contribute and support adaptive management and conservation of coastal ecosystems (Lillebø et al. 2019).

2.3.2 Recommendations for EBM Implementation

At the second workshop (WS II), 15 stakeholders representing the four major groups, which signed an informed consent agreement form, were asked to identify the relevant issues that should be included in the adaptation strategy as well as the opportunities and constraints of implementation.

Participants were invited to join round-table groups (Fig. 3) following a 'world café' methodology to discuss three topics:

- Environment and ecosystem services (spatial distribution of EBM measures, identification of areas for remediation of marshes, benefits and constraints);
- Institutions and equity (identification of institutions involved, process coordinators, benefits and constraints);

Fig. 2 Overview of WS I participatory moments: habitats spatial distribution maps; presentation of WS I objectives; spatial multi-criteria analysis for ES valuation

Fig. 3 Overview of WS II participatory moments: The three 'world café' sessions considered

- Operationalization and sustainable development (identification of existing activities supported by the benefits provided by marshes, business opportunities, benefits and constraints).

Stakeholders were also invited to answer the question "*In which way the EBM methodology can be better or not in relation to the management approaches used until now?*"

2.4 Step Four: Understand the Social-Ecological System

The assessment of Vouga estuary's current state included the identification of habitats, specific public and private primary human activities, and respective pressures in the entire Natura 2000 territory. To address the ecological perspective of the system, data sources from scientific publications, projects (e.g., LAGOONS EU FP7; ADAPT-MED EU FP7 ERA-NET; LTER-RAVE FCT, AQUACROSS EU H2020), national agencies (e.g., above mentioned), online platforms (e.g., Copernicus datasets) and from national/regional official reports were integrated. In order to harmonise habitats classification, all data sets (mainly following Annex I of EU Habitats Directive; Sousa et al. 2016) were converted into the EUNIS habitat classification, following the official correspondence table available at the European Environmental Agency (EEA) portal (http://eunis.eea.europa.eu/habitats.jsp). Data sets on ecosystem services (mainly following CICES, V4.3; Lillebø et al. (eds) 2015; Sousa et al. 2016) were updated and classified following the latest Common International Classification of Ecosystem Services (CICES, V5.1; https://cices.eu/) (Haines-Young and Potschin 2017; O'Higgins et al. 2019). The potential of a given habitat to supply ecosystem services was attained using a lookup table on the contribution of each EUNIS habitat compiled based on expert judgment (Teixeira et al. 2019). The identified ecosystem services were aggregated into eleven ecosystem services in order to enable spatial multi-criteria analysis by stakeholders (Table 4). This table includes the correspondence code from CICES v4.3 to 5.1, considering the identified services for the considered territory, as well as selected aggregation of services used in the scope of the participatory moments in order to optimize communication and active participation of stakeholders.

Table 4 Assessment of ecosystem services provided by Vouga estuary: (a) provisioning; (b) regulation and maintenance; (c) cultural

ES code	Group	Class	v4.3	v5.1	Subclass Ria de Aveiro
(a) CICES section: Provisioning					
ES1 Biotic based energy sources	Mechanical energy	Animals reared to provide energy (incl. mechanical)	1.3.2.1	1.1.3.3	Physical labour provided by cattle supporting agricultural activities
	Biomass-based energy sources	*Cultivated plants* (incl. fungi, algae) grown as a source of energy I *Wild plants* (terrestrial and aquatic, incl. fungi, algae) used as a source of energy	1.3.1.1	1.1.1.3 & 1.1.5.3	*Not applicable at the selected management area*
ES2 Abiotic energy sources	Renewable abiotic energy sources	Freshwater surface water used as an energy source I Coastal and marine water used as energy source	N/A	4.2.1.3 & 4.2.1.4	*Not applicable at the selected management area*
ES3 Biotic materials	Biomass	Fibres and other materials from wild *plants/animals* for direct use or processing (excl. genetic materials)	1.2.2.1 & 1.2.2.2	1.1.5.2 & 1.1.6.2	Reeds seasonally harvested Worms collected in intertidal mudflats and used as bait Macroalgae are collected for *in-situ* aquaculture Seagrasses and macroalgae ("moliço") harvesting Sea rush used as cattle bedding and then as a fertilizer
ES4 Abiotic materials	Non-metallic	Mineral substances used for material purposes	N/A	4.3.1.2	Sand extraction
	Water	*Surface water/ ground water* (and subsurface) used as a material (non-drinking purposes)	1.2.2.1 & 1.2.2.2	4.2.1.2 & 4.2.2.2	The lagoon provides surface water for salt production and forest-fire control, and ground water for inland aquaculture, agriculture, livestock, urban and industrial purposes

(continued)

Table 4 (continued)

ES code	Group	Class	v4.3	v5.1	Subclass Ria de Aveiro
ES5 Nutritional biotic substances	Biomass	Wild plants (terrestrial and aquatic, incl. fungi, algae) used for nutrition	1.1.1.3	1.1.5.1	Wild glasswort *Salicornia* sp. harvested and sold as a gourmet product
		Wild animals used for nutritional purposes	1.1.1.4	1.1.6.1	Fish and shellfish: lamprey *Petromyzon marinus*, European eel *Anguilla anguilla*, allis shad *Alosa alosa*, clams *Ruditapes decussatus* and *Venerupis corrugata,* cockle *Cerastoderma edule* cuttlefish *Sepia officinalis*
		Plants cultivated by *in-situ* aquaculture grown for nutritional purposes	1.1.1.5	1.1.2.1	Macroalgae farming
		Animals reared by *in-situ* aquaculture for nutritional purposes	1.1.1.6	1.1.4.1	Aquaculture farms of marine fish and bivalves
ES6 Nutritional abiotic substances	Mineral	Mineral substances used for nutritional purposes	N/A	4.3.1.1	Marine salt extraction (salt pans)
	Water	Surface water for drinking	1.1.2.1	4.2.1.1	*Not applicable at the selected management area*
(b) CICES section: Regulation and maintenance					
ES7 Mediation of flows	Mass flows	Control of erosion rates | Buffering and attenuation of mass movement	2.2.1.1 & 2.2.1.2	2.2.1.1 & 2.2.1.2	Overall, coastal dunes, salt marshes and seagrass meadows contribute to maintain the lagoon integrity
	Liquid flows	Hydrological cycle and water flow regulation (incl. flood control and coastal protection)	2.2.2.1 & 2.2.2.2	2.2.1.3	Seagrass meadows and salt marshes reduce sediment resuspension and turbidity in the water column, contributing to increase the light availability in the water column São Jacinto dunes, salt marshes and reeds provide resilience to extreme weather events and act as physical buffering of climate change

(continued)

Table 4 (continued)

ES code	Group	Class	v4.3	v5.1	Subclass Ria de Aveiro
ES8 Mediation of waste toxics and other nuisances	Mediation by biota | Mediation by ecosystems	Bio-remediation by micro-organisms, algae, plants, and animals	2.1.1.1 & 2.1.1.2	2.1.1.1	Bio-remediation by ecosystem components (e.g., halophytes); Decomposition/mineralisation processes of plant material mediated by microorganisms; Biological filtration by oysters, clams and mussels in aquaculture and by wild animals
		Filtration/sequestration/storage/accumulation by micro-organisms, algae, plants, and animals | Dilution by freshwater and marine ecosystems	2.1.1.2 & 2.1.2.1 & 2.1.2.2	2.1.1.2 & 5.1.1.1	Bio-physicochemical filtration/sequestration/storage/accumulation of pollutants by the lagoon habitats; Adsorption and binding of metals and organic compounds in ecosystems, as a result of combination of biotic and abiotic factors; Hydrodynamic dilution of pollutants (tidal action)
ES9 Maintenance of physical chemical biological conditions	Lifecycle maintenance, habitat, gene pool protection	Maintaining nursery populations and habitats (incl. gene pool protection)	2.3.1.2	2.2.2.3	Maintaining nursery habitat for fisheries species and commercial invertebrates; Seagrasses, salt marshes including extended areas of reeds, intertidal mudflats, sand flats and salt pans
	Pest, disease control	Pest control (incl. invasive species) | Disease control	2.3.2.1 & 2.3.2.2	2.2.3.1 & 2.2.3.2	Maintaining the system in a healthy status (e.g., from alien species or diseases)
	Soil formation, composition	Decomposition and fixing processes and their effect on soil quality	2.3.3.2	2.2.4.2	Decomposition of biological materials and their incorporation in sediments
	Water conditions	Regulation of the chemical condition of salt waters by living processes	2.3.4.2	2.2.5.2	Water purification by tidal wetlands, including seagrass meadows and salt marshes

(continued)

Table 4 (continued)

ES code	Group	Class	v4.3	v5.1	Subclass Ria de Aveiro
	Atmospheric composition and climate regulation I Gaseous/air flows	Regulation of chemical composition of atmosphere and oceans I Maintenance and regulation by inorganic natural chemical and physical processes	2.3.5.1	2.2.6.1 & 5.2.2.1	Global climate regulation by greenhouse gas/carbon sequestration by seagrass meadows and salt marshes, water columns and storage in sediments and their biota; Transport of carbon into oceans
		Regulation of temperature and humidity, including ventilation and transpiration	2.3.5.2 & 2.2.3.2	2.2.6.2	Micro and regional climate regulation by the Ria de Aveiro lagoon water body that includes the Vouga estuary
(cont.) (c) CICES section: Cultural					
ES10 Physical & intellectual interactions with biota, ecosystems, land & seascapes environmental settings	Physical and experiential interactions	Characteristics of living systems that enable activities promoting health, recuperation or enjoyment through passive or observational interactions	3.1.1.1	3.1.1.1 & 6.1.1.1	*In-situ* bird watching of: Resident birds, e.g., *Charadrius alexandrinus* Migratory birds, e.g., *Himantopus himantopus*
		Characteristics of living systems that that enable activities promoting health, recuperation or enjoyment through active or immersive interactions I Natural, abiotic characteristics of nature that enable active or passive physical and experiential interactions	3.1.1.2	3.1.1.2 & 6.1.1.1	Walking, diving, biking, sailing, boating, kite surfing, windsurfing, kayaking, swimming, leisure fishing (angling) and leisure hunting.
	Intellectual and representative interactions	Characteristics of living systems that enable scientific	3.1.2.1	3.1.2.1 & 6.1.2.1	Ria de Aveiro is subject matter for research

(continued)

Table 4 (continued)

ES code	Group	Class	v4.3	v5.1	Subclass Ria de Aveiro
		investigation or the creation of traditional ecological knowledge I Natural, abiotic characteristics of nature that enable intellectual interactions			
		Characteristics of living systems that enable education and training	3.1.2.2	3.1.2.2	Natural and cultural heritage of the lagoon are subject matter of education (e.g., guided boat tours in Ria, science activities in the summer with the support of the University of Aveiro, BioRia Environmental trails, Natural Reserve of São Jacinto Dunes, "Marinha da Troncalhada" salt pan ecomuseum, "Santiago da Fonte" salt pan (belongs Aveiro University), ship-museum "Santo André" (an extension of the Maritime Museum of Ílhavo); "Casa Gafanhoa" municipal museum (testimony of the rural livelihoods of Ílhavo municipality)
		Characteristics of living systems that are resonant in terms of culture or heritage	3.1.2.3	3.1.2.3	Archaeological sites (e.g., shipwrecks, ship hull, and other isolated findings); Traditional architecture (e.g., "Palheiros", "Casa Gafanhoa"); Traditional boats (e.g., "Moliceiro", "Bateira", "Mercantel"); Traditional activities (e.g., salt production)
		Characteristics of living systems that enable aesthetic experiences	3.1.2.5	3.1.2.4	Sense of place; Artistic representations of nature (e.g., ceramic tiles, painted shells); Inspiration for some painters and writers, interested in the history and heritage of the lagoon and its users

(continued)

Table 4 (continued)

ES code	Group	Class	v4.3	v5.1	Subclass Ria de Aveiro
ES11 Spiritual symbolic & other interactions with biota ecosystems & land seascapes environmental settings	Spiritual and/or emblematic	Elements of living systems used for entertainment or representation I Natural, abiotic characteristics of nature that enable spiritual, symbolic and other interactions	3.1.2.4	3.2.1.3 & 6.2.1.1	*Ex-situ* experiences through local festivals related with the lagoon's products and activities (e.g., "Festa da Ria" summer festival with traditional "moliceiro" boats race; Cod fish festival; Eel and "ovos moles" from Aveiro; food festival; International marine salt festival; "FARAV" handcraft festival, Lamprey festival, Allis shad (*Alosa alosa*) festival
	Other cultural outputs	Characteristics or features of living systems that have an existence value	3.2.2.1	3.2.2.1	Enjoyment provided by salt pans, salt marshes, seagrasses and wild species
		Characteristics or features of living systems that have an option or bequest value	3.2.2.2	3.2.2.2	Willingness to preserve salt pans, salt marshes, seagrasses and wild species for future generations

Equivalence of CICES classification (Group v4.3 & Class v5.1) to aggregated ecosystem services (ES code) used at the participatory moments in the scope of the EBM approach. Note: the assessment included the biologically mediated process and the abiotic outputs
Adapted from Martínez-López et al. (2019b) and O'Higgins et al. (2019)

Analysis of ecosystem services valuations (Martínez-López et al. 2019a) revealed two major stakeholder opinion groups, stating within group more similar preferences regarding ecosystem services, and whose composition was heterogeneous and not related to specific stakeholder groups identified. Weights in the spatial multi-criteria analysis in the selected Natura 2000 area took into account the mean of the ecosystem services scores given by individuals in the same opinion group. However, in the absence of strong and significant differentiation among the two opinion groups' valuations, a compromise map was generated, representing the average prioritization of ecosystem services by all participants.

Overall, the stakeholders' valuation clearly revealed the importance attributed to ecosystem services directly provided by water (freshwater, transitional, and coastal/ marine), with special emphasis on the lagoon ecosystem (Lillebø et al. 2019). However, different preference patterns may arise if the focus is set on smaller scales or in specific areas of the case study as also demonstrated by Martínez-López et al. (2019a).

The final ecosystem services valuation maps were then compared with areas that will be affected by the dredging programme and flood bank. The mapping process involved the combination of several lines of evidence. The spatial explicit information on habitat mapping, human activities, and identified ecosystem services (Teixeira et al. 2019) was considered along with stakeholder's ecosystem services prioritization, the ecosystem services provisioning risk assessment (Lillebø et al. 2018), and stakeholders' persisting concerns regarding the foreseen measures, crucial to effectively highlight the most critical areas for the implementation of EBM management measures.

2.5 Step Five: Specification of Relevant EBM Solutions (as Part of the EBM Cycle)

Specification of relevant EBM solutions requires a clear definition of each component of Drivers-Pressures-State relationships as well as their causal links (Gómez et al. 2017; Teixeira et al. 2018). The applied approach, after Gómez et al. (2017) links the socio-economic and the ecological systems by making a clear distinction between:

- *"the activities that benefit from the provision of natural goods and services for the production of final goods and services that are of direct concern for human welfare;*
- *the drivers of pressures affecting ecosystems, represented by the specific demands of naturally provided goods and services in the quantity, quality required at specific places and moments of time;*
- *the primary activities that (co-) produce goods and services provided by natural capital that are of direct concern to explain the pressures over ecosystems."*

The assessment of the current state included the identification of the specific primary activities and the respective pressures considering the identified habitats (see Sect. 3.4). The specification of the relevant EBM solution requested the following:

- Identify a baseline scenario, incorporating the considered management measures;
- Formulate objectives related to the unintended impacts on biodiversity;
- Screen measures and instruments to understand ecological and social components;
- Construct a narrative reflecting management measures, stakeholders' perception of ecosystem services valuation, and science-based knowledge generated, to support planning the EBM response;
- Evaluate proposed habitat restoration measures using EBM criteria, taking into consideration policies and feasibility, to show that compliance is achievable.

2.6 Step Six: Evaluate the EBM Solutions

After co-defining EBM management alternatives, the following step concerns the co-evaluation of proposed management alternatives, following Piet et al. (2017), by applying established pre-screening criteria: effectiveness (i.e., hitting the target); efficiency (making the most for human well-being); and equity and fairness (i.e., sharing the benefits). This allowed to determine the performance of the proposed EBM measures compared to a baseline situation "*in terms of environmental impacts, subsequent costs and benefits of human wellbeing at individual and collective levels, and the distribution of these impacts and costs throughout society*" (Piet et al. 2017).

3 The Co-created EBM Plan for the Vouga Estuary Natura 2000 Site

3.1 The Governance Boundary

As for other socio-ecological systems the Vouga estuary governance is complex, involving several institutions with multi-level and multi-spatial scales of governance (Lillebø et al. (eds) 2015; Sousa 2017; Fidélis et al. 2019), with different governance models applied for integrated water resource management (Teles et al. 2014; Fidélis et al. 2019). Fidélis et al. (2019) assessed alternative governance models for Ria de Aveiro considering "*the organizational settings established to accommodate the different policy priorities existing in an estuary, their decision-making tools and processes, responsibility boundaries, stakeholder involvement schemes, and the means to face the challenges of a dynamic and vulnerable system*". This analysis, built upon Teles et al. (2014), presents an in-depth discussion highlighting the need for a paradigm change that implies high levels of institutional reforms. The authors concluded that "*regardless of the model adopted, it is crucial to derive a stable collaborative framework of decision-making in order to integrate action plans and policies for integrated water resource management in estuarine areas*" which is in line with the proposed approach to specifically address co-development of EBM planning in the Vouga estuary for the mitigation of unintended impacts on biodiversity.

3.2 Policy Objectives, Synergies, and Gaps

The assessed EU water-related and Nature Directives showed that Natura 2000 network sites should be '*managed in a sustainable manner, both ecologically and economically*', involving local policy-makers and stakeholders. Priorities identified for improvement include:

- Harmonisation and integration of monitoring programmes of Water Framework and Habitats Directive in water-dependent Natura 2000 sites;
- Development of the Vouga estuary land use and management plan aiming to conserve and promote sustainable use of water resources, ecosystem functions, integrated management, and coordination between various territorial management instruments;
- Enhancement communication among entities and foster involvement and active participation of land users and landowners.

3.3 Understand Stakeholder Objectives

Stakeholders representing different sectoral interests or activities share key objectives to: foster sustainable development of economic activities and preservation of aquatic biodiversity; integrate territorial management instruments; enhance participatory management; and co-create adaptive management solutions. There were a number of aspects that stakeholders identified, both with respect to major beneficial effects and persisting concerns, regarding Ria de Aveiro and the Vouga estuary from the first workshop:

Ecosystems biodiversity—Stakeholders highlighted habitat richness as important and revealed concerns regarding impact of dredging on seagrasses, saltmarshes, and juvenile fauna due to changes in eco-hydrology and potential mobilization of contaminants due to dredging.

Water management—The need for targeted dredging (e.g., oriented for habitats, housing, and infrastructure) was acknowledged but concerns were expressed regarding changes at the system hydrology, specifically increase in tidal prism due to dredging. Consequently, low navigability in inner channels during low tide and the increase of ocean water volume in the lagoon during high tide are of concern.

Agriculture—Concerns were related to the loss of traditional agricultural activities that enable ecosystems and biodiversity maintenance, which could benefit from incentives and compensations. The stimulus for agriculture was acknowledged. To this end the need for the completion of the flood bank and the increase in agricultural land area was set forward.

Fisheries and aquaculture—Stakeholders highpoint the relevance of this coastal system to migratory species with high socio-economic value, such as sea lamprey (*Petromyzon marinus*), European eel (*Anguilla anguilla*) and allis shad (*Alosa alosa*). Concerns regarding these activities could be overcome through incentives and compensations for maintenance of traditional aquaculture activities that maintained ecosystems and biodiversity.

Tourism and recreational activities—Tourism was seen as an opportunity, namely marked walking trails, supporting recreational activities and ecotourism. The increased navigability conditions inside the lagoon after planned dredging will promote recreational boating and touristic activities, although some concerns remain as increases in water current velocity is expected to alter habitats used for other touristic and recreational purposes (e.g., loss of lagoon inner mud/sand-flats used either as beaches or preferential bird watching sites).

Transversal—Other beneficial aspects and persisting concerns were considered transversal to the previous issues, specifically benefits from development of different sectoral economic activities; as well as the recognised scientific knowledge on Ria de Aveiro natural capital. Main concerns were due to lack of communication, which is paramount for integrated management, the need for information and awareness in the municipal councils, as well as landowner involvement, and lack of regulatory surveillance of activities within Ria de Aveiro.

Considering the management measures to be implemented, stakeholder knowledge and perceptions supported baseline scenario development, formulation of objectives related to the unintended impacts on biodiversity, and narratives of possible futures to support planning the EBM response.

3.4 Understand the Social-Ecological System

To understand the impact of the dredging in the Ria de Aveiro, we identified key human activities (Fig. 4), resulting pressures, habitats (see Fig. 5 for EUNIS habitat types), and how these support valuable ecosystem services. Relevant activities are related to boating, diving, shipping, coastal defence, port facilities, saltworks, fishing, bait digging, aquaculture, agriculture, livestock and forestry.

The linkage framework for Drivers and Pressures considering the Vouga coastal watershed is shown in Fig. 6. It can be seen that transitional waters realm, which includes the EUNIS habitats type code A, is affected by several pressures resulting from specific human activities, including capital dredging and maintenance dredging. The linkage chain associated with these activities reflects the complexity of linkages relating activities with associated pressures that determine functions and services provided by the habitats they impact. Figure 7 highlights the linkage framework for Drivers-Pressures-Ecosystem Component-Ecosystem Function and Ecosystem Services in the Vouga estuary. The management options 'behind' the baseline scenario were plotted considering the aggregated primary activities of dredging (representing the dredging programme) and flood and coastal defence (representing the extension of the flood-bank) and ecosystems components, functions and services. These management options are also relevant for other activities, namely cultivation of crops and livestock. The extension of the flood bank will disable surface saltwater intrusion into Baixo Vouga allowing recovery of arable land for agriculture and livestock. Furthermore, they will also affect input of organic matter and litter into the aquatic environment. As such, changes in the mechanical and physical structuring will affect biogeochemical cycles and production (primary

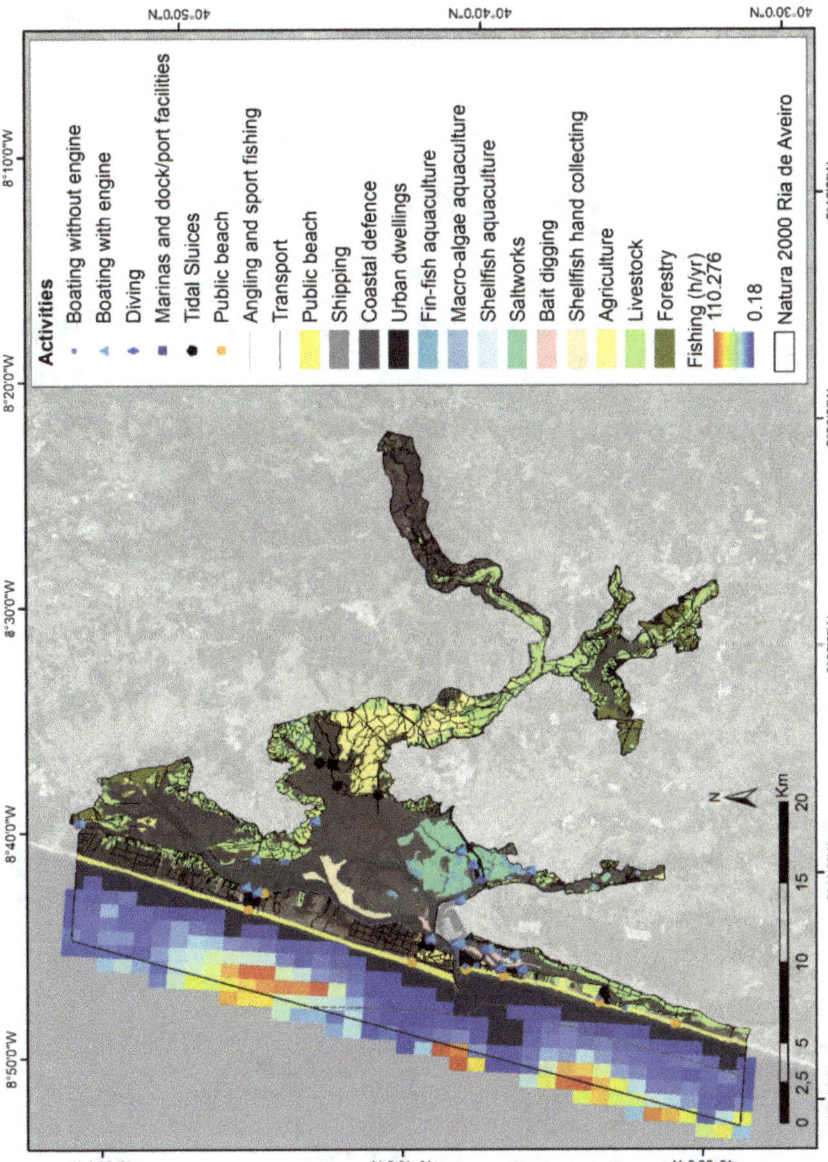

Fig. 4 The main activities identified at the Vouga river coastal watershed under classification of Natura 2000 network

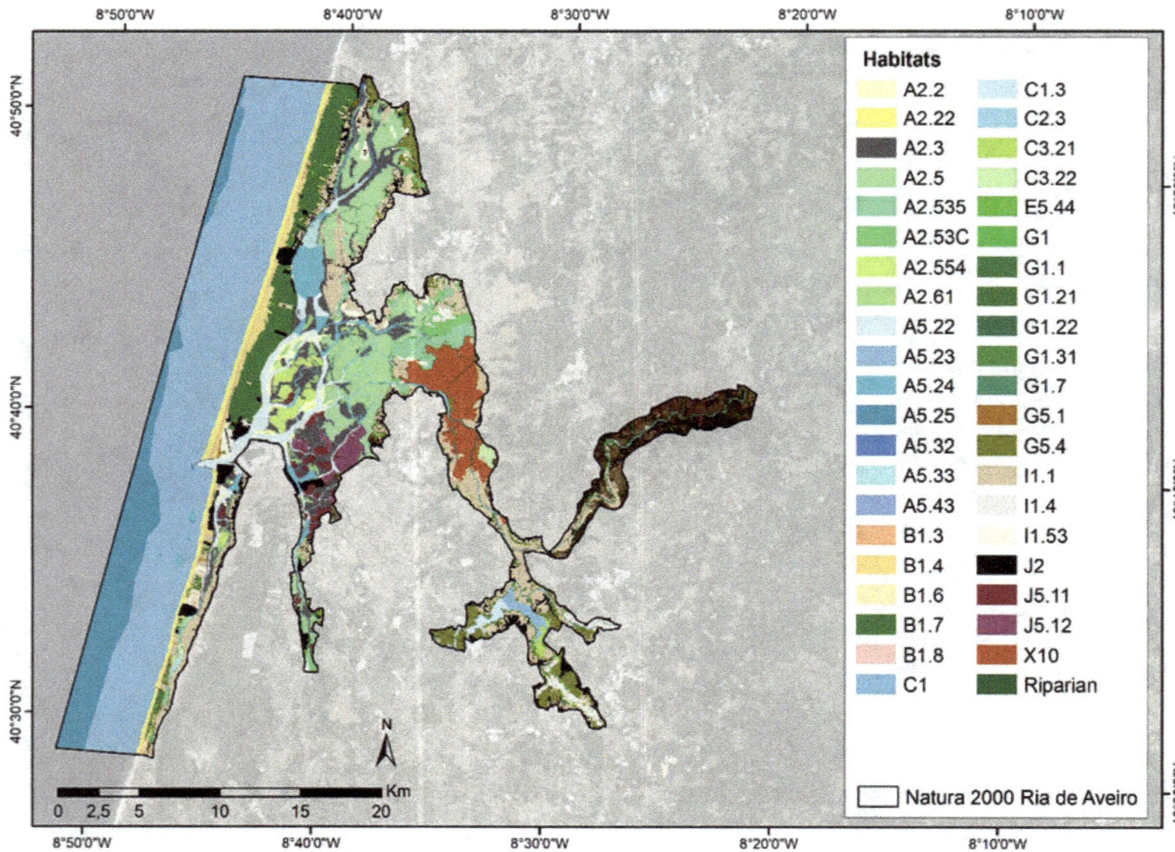

Fig. 5 The EUNIS habitats identified at the Vouga river coastal watershed under classification of Natura 2000 network. Code A relates to transitional waters domain, Code B to coastal/marine waters domain, Code C to freshwater domain, Code E to grasslands, Code G to riparian vegetation, Code I to terrestrial habitats, Code J to hypersaline habitats (e.g., salt pans), , Code X to habitat complexes, and Code X10 to Bocage

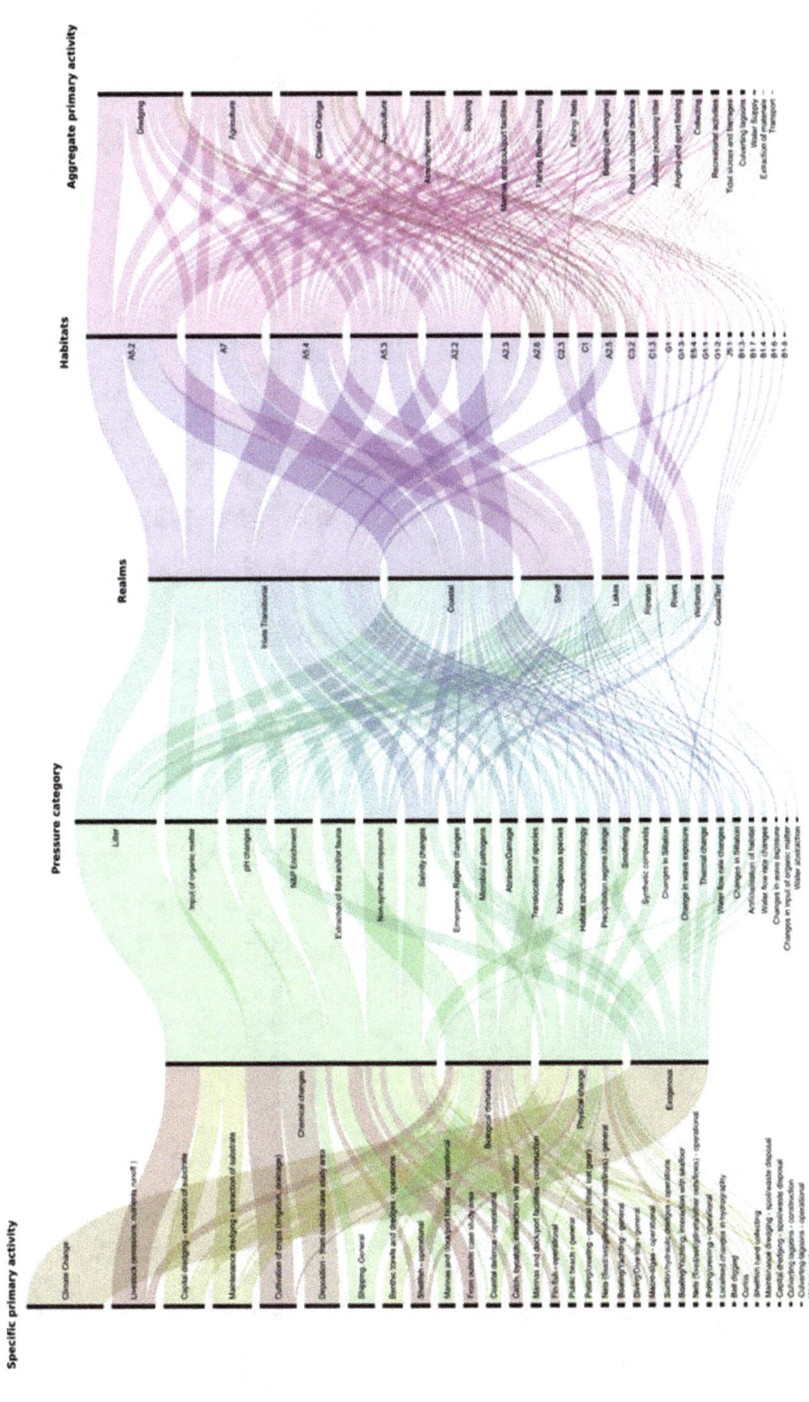

Fig. 6 Linking Drivers and Pressures in the Vouga river coastal watershed. (Plotted with Mauri et al. 2017)

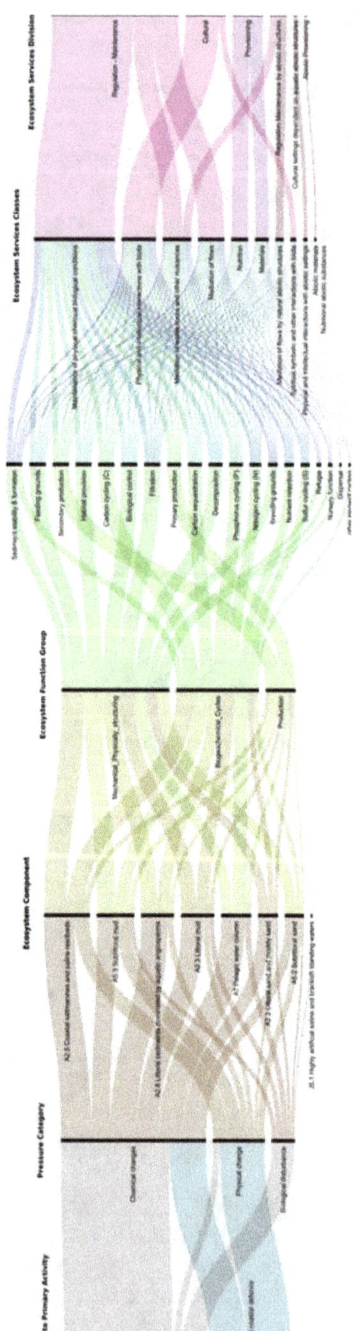

Fig. 7 Linkage framework for Drivers-Pressures-Ecosystem Component-Ecosystem Function and Ecosystem Services in the Vouga estuary, highlighting the management options, i.e., the dredging activities and the extension of the flood bank. (Plotted with Mauri et al. 2017)

Table 5 Description of proposed EBM responses supported by the prospective scenarios and considering the existing Sectoral Plan for Natura 2000 Network as well as the National Strategic Plan for climate change adaptation

	Seagrasses meadows restoration	Saltmarshes restoration
Main cause for mitigation measures	Compensate the loss due to changes in water current velocity and light availability.	Compensate the loss due to 'coastal squeeze' and the increase in tidal prism.
Ongoing science based knowledge to support mitigation measures	Intertidal *Zostera noltei* numerical modelling in Ria de Aveiro (e.g., Azevedo et al. 2013, 2017). Research projects BioPradaRia and Remoliço (PT MAR2020 funded) testing restoration techniques for *Z. noltei* populations *in situ* and under controlled laboratory conditions.	Running InVEST GIS-based modelling tool to support the selection of potential areas, as well as the restoration techniques. These might combine nature-based solutions to protect shorelines and actions to promote sediment accretion for salt marsh elevation.
Detailed restoration measures	(1) Protection of existing populations from fragmentation and increase resilience by enhancing sediment stability through application of coconuts fibber mats (e.g., Sousa et al. 2017a) (2) Transplantation of *Z. noltei* plots from selected donor sites within Ria de Aveiro (e.g., Suykerbuyk et al. 2016)	(1) Restoration of salt-marsh communities (Sousa et al., 2017b), namely *Juncus maritimus*, through revegetation of sheltered mudflats, considering submersion time. (2) Foster saltmarshes elevation through accretion.

Data source: Lillebø et al. (2019)

and secondary) as well as regulation and maintenance, cultural and provisioning ecosystem services and abiotic outputs.

Although dredging activities and extension of the flood bank are acknowledged as important for shipping and agriculture, concerns remain on the impact of dredging on seagrasses, saltmarshes and juvenile fauna due to changes in the ecosystem eco-hydrology.

Overall, stakeholders' sectoral activities (including public and private sectors), or spheres of interest (including direct and indirect users), generate conflicting interests that need to be considered in the context of co-creation of adaptive management solutions that consider better coordination among policies.

The specific policy plans and programmes aiming at mitigating negative unintended impacts on biodiversity in the Natura 2000 Vouga estuary will focus on restoration of tidal wetlands, namely seagrasses and saltmarshes (Table 5), development of the Vouga Estuary Management Plan, engagement of local users and landowners in restoration actions, and the promotion of the value of ecosystems services provided by tidal wetlands. Both measures, to compensate for the loss of seagrasses and saltmarshes, have as policy target the Water Frame Work Directive and the Birds and Habitats Directives. As well, the target policy instruments already in place are River Basin Management Plan and National Water Plan.

3.5 *Evaluate the EBM Solutions*

The EBM plan, co-created with stakeholders, is shown in Table 6. During the evaluation processes, special attention was given to seagrasses and saltmarshes restoration measures. These measures aim at recovering the ecological processes and services of these valuable coastal wetlands, being in this way ecologically sustainable, socially desirable, ethically defensible, and culturally inclusive.

Table 6 EBM plan alignment with EBM principles

EBM principles	EBM scenario
1. EBM considers ecological integrity, biodiversity, resilience and ecosystem services	The harmonised WFD and HD monitoring programmes will together with the proposed tidal wetlands restoration measures, and stakeholder participation, increase resilience and ecosystem services.
2. EBM is carried out at appropriate spatial scales	The EBM Plan considers the boundaries of the Vouga Estuary Management Plan and the interconnections with the Ria de Aveiro watershed.
3. EBM develops and uses multi-disciplinary knowledge	The Vouga Estuary Management Plan foresees the coordination between various territorial management instruments as well as stakeholders' sectoral activities, with support of science-based knowledge. The Vouga Estuary Management Plan should therefore involve complementary expertise between and within natural and social sciences, in a trans-disciplinary approach.
4. EBM builds on social–ecological interactions, stakeholder participation and transparency	The EBM plan was co-created with input from local stakeholders and policy-makers, and considers their perceptions, namely their concerns regarding the unintended pressures from the base-line scenario, their valuation of ecosystem services through spatial multi-criteria analysis and their recommendations regarding opportunities and constraints regarding implementation of the plan.
5. EBM supports policy coordination	The proposed EBM approach is timely to the Portuguese spatial planning and water planning systems, framed for the protection and management of estuarine systems. The EBM plan also proposes to harmonise Water Framework Directive and Habitats Directive monitoring programmes.
6. EBM incorporates adaptive management	The proposed measures, namely the habitats restoration measures, follow principles of resilience thinking and adaptive management, by considering ecology, management of natural capital and systems analysis.

Source: Lillebø et al. (2019)

Additionally, the relevance of these coastal wetlands as nursery areas, which support important economic activities in the region, is acknowledged by local populations (Dolbeth et al. 2016; Newton et al. 2018; Lillebø et al. 2019).

For the implementation, relevant EU funding instruments might be considered, namely R&I H2020 and the following Horizon Europe programmes, LIFE environmental programme, as well as Regional Development and/or Territorial Cooperation funds (Marino et al. 2014; UE 2016).

Both measures to restore tidal wetlands have the same policy target (i.e., Water Framework, Birds and Habitats Directives), are legally permissible, and are implementable using the same policy instrument (River Basin Management Plan; National Water Plan), therefore administratively achievable although it implies the commitment of several Institutions. This is foreseen with the proposed development of the Vouga Estuary Management Plan and is also effectively communicable and politically expedient for promoting the value of ecosystems services provided by tidal wetlands. In addition, effective implementation of proposed habitat restoration in the selected Natura 2000 area is consistent with the prevailing political climate and has explicit support of national political leaders. The main differences between the baseline and the proposed EBM solutions are shown in Table 7.

The performance of the proposed EBM measures is presented in Table 8. The baseline scenario corresponds to the unintended impacts on biodiversity, i.e., increase in tidal prism and water velocity; loss of coastal wetlands habitats (seagrasses and saltmarshes) and saltmarsh 'coastal squeeze' at the downstream area of the flood bank.

4 Vouga Estuary Natura 2000 Site Stakeholders' Evaluation and Feedback

Local stakeholders were supportive of the approach, "*ecosystem-based management allows for a 'correction' of less good results*" and appreciate that it is "*concerned with beneficiaries, as well as biodiversity*". Overall, stakeholders considered that:

- *The environment and biodiversity will be the main beneficiaries from tidal wetlands restoration;*
- *Some economic activities related to fisheries and ecotourism, which has a recognized potential, might benefit;*
- *Restoration actions need to ensure involvement of users due to conflicting activities and landowners, as most of the area is private property;*
- *Large interventions should include financing for implementation of the corresponding minimizing measures;*
- *There is a need for post-licensing supervision to ensure compliance with environmental protection obligations;*
- *There is a need for clear communication between institutions and enforcement of existing regulations;*

Table 7 Main differences between the baseline and proposed EBM solutions

Main differences	Baseline	EBM solutions
Environmental ambition/policy target	Protect biodiversity in line with Natura 2000 objectives. Whilst enabling economic and other activities in the area, aim to mitigate negative impacts of interventions and economic activity.	Same as in baseline.
Measures	Two measures are to be implemented in 2019/2020, which will have foreseen but unintended negatives impacts on biodiversity (dredging programme and the extension of a flood bank).	The same as baseline, but to minimise negative side effects, additional measures are proposed (see Table 5).
Policy instruments	Many policy instruments are implemented to achieve biodiversity goals, including protected areas.	Harmonise monitoring across water and environmental related Directives; Incorporate stakeholders into planning; Integrate territorial management institutions (and their multiple goals) into planning; Support development of Vouga Estuary Management Plan.
Sites	The boundaries of the Vouga Estuary Management Plan and the interconnections with the Ria de Aveiro watershed.	Same as in baseline; Seagrass and saltmarsh restoration sites will be selected considering multiple ecosystem services and with stakeholder input.
Governance/ Institutional context	Many separate; Limited, inconsistent stakeholder involvement in management.	Coordinated input from multiple institutions into integrated Vouga Estuary Management Plan. Ongoing, coordinated stakeholder engagement in management.

Source: Mattheiß et al. (2018)

- *There is a need for reinforcement of integrated management and development the Vouga Estuary Management Plan.*

As part of the co-creation process, stakeholders evaluated the produced maps, the ecosystem indicators' results, the proposed EBM solutions, and they discussed the benefits and constraints regarding its implementation. As a final remark stakeholders acknowledged that responses should be framed in the Sectoral Plan for Natura 2000 Network, and should consider climate change projections and the National Strategic Plan for climate change adaptation (Fig. 8 illustrates the spatial planning regulations to consider for climate change adaptation in the region).

At the Vouga river coastal watershed the Sectoral Plan for Natura 2000 Network establishes the strategic orientation and programme norms for the actions of central and local government, and the measures and guidelines provided therein should be transposed to the Municipal Planning of the territory and Special Plans. Thus, the management measures provided for the Sectoral Plan will only be binding measures when they are inserted in the Municipal and Special Plans. Within the considered

Table 8 Application of the pre-screening criteria: effectiveness, efficiency, and equity and fairness

Pre-screening criteria	EBM solutions
Effectiveness—hitting the environmental target	The proposed measures, i.e., saltmarshes restoration and seagrasses restoration clearly address the set of environmental targets defined under relevant policies.
Efficiency—making the most for human wellbeing	Although proposed measures were not supported by cost-effectiveness analysis or cost-benefit-analysis, they were supported by a spatial multi-criteria analysis performed by stakeholders in which they expressed preferences regarding the provided ecosystem services.
Equity and fairness—sharing the benefits	By integrating the proposed measures in the Vouga Estuary Management Plan, and by acknowledging the context of adaptation to climate change, strategy for Biodiversity and the Centro Portugal region strategy for smart specialisation (Portugal RIS3 Centro), stakeholders with very different interests participated actively and acknowledge the equity of benefits already achieved, or to be achieved, although they identified constraints regarding its implementation.

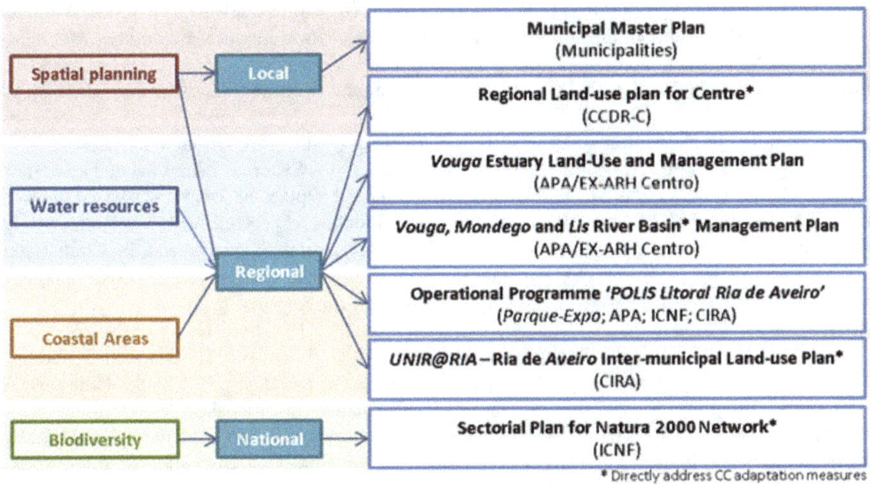

Fig. 8 Spatial planning regulations for climate change adaptation for Vouga river coastal watershed. (Source: ADAPT-MED 2015)

boundaries, this plan is implemented by the UNIR@RIA, which ensures articulation between the regional and municipal plans that are relevant for the Ria de Aveiro and associated protected areas.

The National Strategic Plan for climate change adaptation is framed on a Territorial Management System organized within a framework of coordinated interaction at three levels (Law no. 48/98, August 11): national, regional and local.

Acknowledgements The European Commission under the Horizon 2020 Programme for Research, Technological Development and Demonstration supported this study through the collaborative research project AQUACROSS (Grant Agreement no. 642317). Thanks, are also due to the Portuguese Foundation for Science and Technology (FCT) for the financial support to CESAM (UID/AMB/50017/2019; UIDP/50017/2020+UIDB/50017/2020). Ana Genua-Olmedo was funded by the project PORBIOTA—Portuguese E-Infrastructure for Information and Research on Biodiversity (POCI-01-0145-FEDER-022127), financed by FCT through PIDAC national funds and co-funded by the FEDER. Special thanks to Mariana Morgado for her work in the cartographic data collection, and to stakeholders participating at the workshops for their valuable contribution.

References

ADAPT-MED. (2015). *Baixo Vouga Lagunar Knowledge Database*. ADAPT-MED Report D2.1b. 92 pp.

APA. (2018). RECAPE da Transposição de Sedimentos para Otimização do Equilíbrio Hidrodinâmico na Ria de Aveiro – Canal de Ovar até ao Carregal, Canal de Ovar até ao Pardilhó e Canal da Murtosa Canal de Ílhavo, Canais do Lago Paraíso e Canais da Zona Central da Ria 35.

Azevedo, A., Sousa, A. I., Lencart e Silva, J. D., Dias, J. M., & Lillebø, A. I. (2013). Application of the generic DPSIR framework to seagrass communities of Ria de Aveiro: A better understanding of this coastal lagoon. *Journal of Coastal Research, 65*, 19–24.

Azevedo, A., Lillebø, A. I., Silva, J. L. E., & Dias, J. M. (2017). Intertidal seagrass models: Insights towards the development and implementation of a desiccation module. *Ecological Modelling, 354*, 20–25.

Curtin, C. G., & Parker, J. P. (2014). Foundations of resilience thinking. *Conservation Biology, 28*, 912–923.

Direção Regional de Agricultura e Desenvolvimento Rural. (2017). Aproveitamento Hidroagrić ola do Vouga Bloco do Baixo Vouga Lagunar. Intervenções nos Sistemas Primários de Drenagem e Defesa Contra Efeitos das Marés e Cheias a Candidatar à Operação 3.4.3 do PDR 2020.

Dolbeth, M., Stalnacke, P., Alves, F. L., Sousa, L. P., Gooch, G. D., Khokhlov, V., Tuchkovenko, Y., Lloret, J., Bielecka, M., Rozynski, G., Soares, J. A., Baggett, S., Margonski, P., Chubarenko, B. V., & Lillebø, A. I. (2016). An integrated Pan-European perspective on coastal lagoons management through a mosaic-DPSIR approach. *Scientific Reports, 6*, 19400.

Drakou, E. G., Kermagoret, C., Liquete, C., Ruiz-Frau, A., Burkhard, K., Lillebø, A. I., van Oudenhoven, A. P. E., Ballé-Béganton, J., Rodrigues, J. G., Nieminen, E., Oinonen, S., Ziemba, A., Gissi, E., Depellegrin, D., Veidemane, K., Ruskule, A., Delangue, J., Böhnke-Henrichs, A., Boon, A., Wenning, R., Martino, S., Hasler, B., Termansen, M., Rockel, M., Hummel, H., El Serafy, G., & Peev, P. (2017). Marine and coastal ecosystem services on the science–policy–practice nexus: Challenges and opportunities from 11 European case studies. *International Journal of Biodiversity Science, Ecosystem Services & Management, 13*(3), 51–67.

European Commission. (2011). *Our life insurance, our natural capital: An EU biodiversity strategy to 2020*. Brussels: European Commission.

Fidélis, T., & Carvalho, T. (2015). Estuary planning and management: The case of Vouga Estuary (Ria de Aveiro), Portugal. *Journal of Environmental Planning and Management, 58*(7), 1173–1195.

Fidélis, T., Teles, F., Roebeling, P., & Riazi, F. (2019). Governance for sustainability of estuarine areas—assessing alternative models using the case of Ria de Aveiro. *Portugal Water, 2019*(11), 846.

Gómez, C., Delacámara, G., Arévalo-torres, J., Barbière, J., Barbosa, A. L., Boteler, B., Culhane, F., Daam, M., Gosselin, M.-P., Hein, T., Iglesias-campos, A., Jähnig, S., Lago, M., Langhans, S., Martínez-López, J., Nogueira, A., Lillebø, A. I., O'Higgins, T., Piet, G., & Schlüter, M. (2016). The AQUACROSS innovative concept-deliverable 3.1.

Gómez, C., Delacámara, G., Jähnig, S., Langhans, S. D., Domisch, S., Hermoso, V., Piet, G., Martínez-López, J. B., Lago, M., Boteler, B., Rouillard, J., Abhold, K., Reichert, P., Schuwirth,

N., Hein, T., Pletterbauer, F., Funk, A., Nogueira, A., Lillebø, A. I., Daam, M., Teixeira, H., Robinson, L., Culhane, F., Schlüter, M., Martin, R., Iglesias-Campos, A., Luisa Barbosa, A., & Arévalo-Torres, J. (2017). Developing the AQUACROSS assessment framework deliverable 3.2.

Haines-Young, R., & Potschin, M. B. (2017). Common international classification of ecosystem services (CICES) V5.1 and guidance on the application of the revised structure. Retrieved from www.cices.eu.

Langhans, S. D., Domisch, S., Balbi, S., Delacámara, G., Hermoso, V., Kuemmerlen, M., Martin, R., Martínez-López, J., Vermeiren, P., Villa, F., & Jähnig, S. C. (2019). Combining eight research areas to foster the uptake of ecosystem-based management in fresh waters. *Aquatic Conservation: Marine and Freshwater Ecosystems, 29*(2), 1161–1173. https://doi.org/10.1002/aqc.3012.

Lillebø, A. I., Stålnacke, P., & Gooch, G. D. (Eds.). (2015). *Coastal lagoons in Europe: Integrated water resource strategies*. IWA Publishing; International Water Association (IWA), UK, 254pp. ISBN: 9781780406282; eISBN: 9781780406299. Retrieved from https://www.iwapublishing.com/sites/default/files/ebooks/9781780406299.full_.pdf.

Lillebø, A. I., Stålnacke, P., Gooc, G. D., Krysanova, V., & Bielecka, M. (2016). *Pan-European management of coastal lagoons: A science-policy-stakeholder interface perspective*. Coastal and Shelf Science: Estuarine.

Lillebø, A. I., Teixeira, H., Nogueira, A. J. A., Genua-Olmedo, A., Morgado, M., & Ferreira, V. (2018). AQUACROSS. 2018. Implementando uma gestão baseada nos ecossistemas aquáticos da Rede Natura 2000 da Ria de Aveiro: da Pateira até ao mar, passando pelo Baixo Vouga Lagunar (A. I. Lillebø, H. Teixeira, & A. J. A. Nogueira, Eds., 60 pp). ISBN: 978-989-99524-2-3.

Lillebø, A. I., Teixeira, H., Morgado, M., Martínez-López, J., Marhubi, A., Delacamara, G., Strosser, P., & Nogueira, A. J. A. (2019). Ecosystem-based management planning across aquatic realms at the Ria de Aveiro Natura 2000 territory. *Science of the Total Environment, 650*, 1898–1912.

Lopes, J. F., Ferreira, J. A., Cardoso, A. C., & Rocha, A. C. (2014). Variability of temperature and chlorophyll of the Iberian Peninsula near costal ecosystem during an upwelling event for the present climate and a future climate scenario. *Journal of Marine Systems, 129*, 271–288.

Lopes, M. L., Marques, B., Dias, J. M., Soares, A. M. V. M., & Lillebø, A. I. (2017). Challenges for the WFD second management cycle after the implementation of a regional multi-municipality sanitation system in a coastal lagoon (Ria de Aveiro, Portugal). *Science of the Total Environment, 586*, 215–225.

Luís, S., Lima, M. L., Roseta-Palma, C., Rodrigues, N., Sousa, L. P., Freitas, F., Alves, F. L., Lillebø, A. I., Parrod, C., Jolivet, V., Paramana, T., Alexandrakis, G., & Poulos, S. (2018). Psychosocial drivers for change: Understanding and promoting stakeholder engagement in local adaptation to climate change in three European Mediterranean case studies. *Journal of Environmental Management, 223*, 165–174.

Maes, J., Liquete, C., Teller, A., Erhard, M., Paracchini, M. L., Barredo, J. I., Grizzetti, B., Cardoso, A., et al. (2016). An indicator framework for assessing ecosystem services in support of the EU biodiversity strategy to 2020. *Ecosystem Services, 17*, 14–23.

Martin, R., Hellquist, F. K., Schlüter, M., Barbosa, A. L., Iglesias-Campos, A., Torres, J. A., Barbière, J., Martin, B., Delacámara, G., Gómez, C. M., Arenas, M., Domisch, S., Langhans, S., Martínez-López, J., Villa, F., Balbi, S., Schuwirth, N., & Rouillard, J. (2018). Scenario development. Deliverable 7.2, European Union's Horizon 2020 Framework Programme for Research and Innovation Grant Agreement No. 642317. 52 pp. [Online]. Retrieved from https://aquacross.eu.

Martínez-López, J., Bergillos, R. J., Bonet-García, F., & de Vente, J. (2019a). Connecting research infrastructures, scientific and sectorial networks to support integrated management of Mediterranean coastal and rural areas. *Environmental Research Letters*. https://doi.org/10.1088/1748-9326/ab4b22.

Martínez-López, J., Teixeira, H., Morgado, M., Almagro, M., Sousa, A. I., Villa, F., Balbi, S., Genua-Olmedo, A., Nogueira, A. J. A., & Lillebø, A. I. (2019b). Participatory coastal management through elicitation of ecosystem service preferences and modelling driven by "coastal squeeze". *Science of the Total Environment, 652*, 1113–1128.

Mattheiß, V., Strosser, P., Krautkraemer, A., Charbonnier, C., McDonald, H., Röschel, L., Hoffmann, H., Lago, M., Delacámara, G., Gómez, C.M., Piet, G., Schuwirth, N., Kuemmerlen, M., & Reichert, P. (2018). Evaluation of ecosystem-based management responses in case studies: AQUACROSS deliverable 8.2. European Union's Horizon 2020 Framework Programme For Research and Innovation Grant Agreement No. 642317.

Mauri, M., Elli, T., Caviglia, G., Uboldi, G., & Azzi, M., (2017). *RAWGraphs: A visualisation platform to create open outputs*. Proceedings of the 12th Biannual Conference on Italian SIGCHI Chapter. ACM, p. 28.

Marino, D., Gaglioppa, P., Schirpke, U., Guadagno, R., Marucci, A., Palmieri, M., Pellegrino, D., & Gusmerotti, N. (2014). Assessment and governance of ecosystem services for improving management effectiveness of natura 2000 sites. *Bio-based Applied Economics, 3*, 229–247.

Newton, A., Brito, A. C., Icely, J. D., Derolez, V., Clara, I., Angus, S., Schernewski, G., Inácio, M., Lillebø, A. I., & Sousa, A. I. (2018). Assessing, quantifying and valuing the ecosystem services of coastal lagoons. *Journal for Nature Conservation, 44*, 50–65.

O'Higgins, T., Nogueira, A. J. A., & Lillebø, A. I. (2019). A simple spatial typology for assessment of complex coastal ecosystem services across multiple scales. *Science of the Total Environment, 649*, 1452–1466.

Pereira, C., & Coelho, C. (2013). Mapping erosion risk under different scenarios of climate change for Aveiro coast, Portugal. *Natural Hazards, 69*, 1033–1050.

Piet, G., Delacámara, G., Gómez, C.M., Lago, M., Rouillard, J., Martin, R., & van Duinen, R. (2017). *Making ecosystem-based management operational*. Deliverable 8.1, European Union's Horizon 2020 Framework Programme for Research and Innovation Grant Agreement No. 642317. 49 pp. [Online]. Retrieved from https://aquacross.eu.

Rouillard, J., Lago, M., Abhold, K., Röschel, L., Kafyeke, T., Mattheiß, V., & Klimmek, H. (2018). Protecting aquatic biodiversity in Europe: How much do EU environmental policies support ecosystem-based management? *Ambio, 47*, 15–24.

Sousa, L. P. (2017). *Integration of ecosystem services and its value in an estuary governance model in the context of climate change: Application to Ria de Aveiro Coastal Lagoon*. PhD Thesis, University of Aveiro.

Sousa, L. P., Sousa, A. I., Alves, F. L., & Lillebø, A. I. (2016). Ecosystem services provided by a complex coastal region: Challenges of classification and mapping. *Scientific Reports, 6*, 22782.

Sousa, A. I., Valdemarsen, T., Lillebø, I., Jorgensen, L., & Flindt, M. R. (2017a). A new marine measure enhancing Zostera marina seed germination and seedling survival. *Ecological Engineering, 104*, 131–140.

Sousa, A. I., Santos, D. B., Ferreira da Silva, E., Sousa, L. P., Cleary, D. F. R., Soares, A. M. V. M., & Lillebø, A. I. (2017b). 'Blue carbon' and nutrient stocks of salt marshes at a temperate coastal lagoon (Ria De Aveiro, Portugal). *Scientific Reports, 7*, 41225.

Stefanova, A., Krysanova, V., Hesse, C., & Lillebø, A. I. (2015). Climate change impact assessment on water inflow to a coastal lagoon – Ria de Aveiro watershed, Portugal. *Hydrological Sciences Journal—Journal des Sciences Hydrologiques* (Special Issue: Evaluation of Water Resources with SWAT, *60*(5), 929–948.

Suykerbuyk, W., Govers, L. L., Bouma, T. J., Giesen, W. B., de Jong, D. J., van de Voort, R., Giesen, K., Giesen, P. T., & van Katwijk, M. M. (2016). Unpredictability in seagrass restoration: Analysing the role of positive feedback and environmental stress on *Zostera noltii* transplants. *Journal of Applied Ecology, 53*(3), 774–784.

Teixeira, H., Lillebø, A., Culhane, F., Robinson, L., Trauner, D., Borgwardt, F., Kummerlen, M., Barbosa A., McDonald, H., Funk, A., O'Higgins, T., Van der Wal, T., Piet, G., Hein, T., Arévalo-Torres, J., Barbière, J., & Nogueira, A. J. A. (2018). *Assessment of causalities, highlighting results from the application of meta-ecosystem analysis in the case studies - Synthesis report*. Deliverable 5.2, European Union's Horizon 2020 Framework Programme for Research and Innovation Grant Agreement No. 642317. [Online]. Retrieved from https://aquacross.eu.

Teixeira, H., Lillebø, A. I., Culhane, F., Robinson, L., Trauner, D., Borgwardt, B., Kummerlen, M., Barbosa, A., McDonald, H., Funk, A., O'Higgins, T., Van der Wal, J. T., Piet, G., Hein, T., Arévalo-Torres, J., Iglesias-Campos, A., Barbière, J., & Nogueira, A. J. A. (2019). Flow linkages from biodiversity to ecosystem services supply: Integrating across aquatic ecosystems. *Science of the Total Environment, 657*, 517–534.

Teles, F., Fidélis, T., Roebeling, P., Lillebø, A. I., & Lucas Pires, M. A. (2014). *Ria de Aveiro Governance*. Policy Paper prepared for the Intermunicipal Community of the Ria de Aveiro; University of Aveiro.

Villa, F., Tunesi, L., & Agardy, T. (2002). Zoning marine protected areas through spatial multiple-criteria analysis: The case of the Asinara Island National Marine Reserve of Italy. *Conservation Biology, 16*(2), 515–526.

12

Ecosystem-Based Management and Natural Capital Accounting

Marc Russell, Charles Rhodes, George Van Houtven, Paramita Sinha, Katherine Warnell, and Matthew C. Harwell

Abstract Natural capital includes the physical and biophysical components of an ecosystem working together to produce a flow of services to the economy and to society that support human well-being. An ecosystem service thus represents a type of transaction between natural capital and humans and can be considered within tools to support Ecosystem-Based Management (EBM) decision making. A natural capital perspective can be a useful way to put the value of ecosystems on par with other socioeconomic values in an EBM decision context. Further, the application of structured classifications for ecosystem components, human beneficiaries (users), and a suite of flows of final ecosystem services helps EBM practitioners organize information for a given decision context. This chapter explores the utility of natural capital accounting as a tool for EBM, outlines a standardized framework for natural capital accounting, and summarizes an ecosystem services classification system for natural capital accounting that can be used as an EBM tool, especially relevant for the decision-making step of evaluating management options (e.g., scenarios).

M. Russell (✉)
US Environmental Protection Agency, Center for Computational Toxicology and Exposure, Gulf Breeze, FL, USA
e-mail: russell.marc@epa.gov

C. Rhodes
Washington, DC, USA

G. Van Houtven · P. Sinha
RTI International, Durham, NC, USA
e-mail: gvh@rti.org; psinha@rti.org

K. Warnell
Nicholas Institute for Environmental Policy Solutions, Duke University, Durham, NC, USA
e-mail: katie.warnell@duke.edu

M. C. Harwell
Gulf Ecological Modeling and Measurement Division, U.S. Environmental Protection Agency, Gulf Breeze, FL, USA
e-mail: harwell.matthew@epa.gov

Lessons Learned

- The organizational structure of the National Ecosystem Services Classification System (NESCS) allows for the determination of environmental and valuation measures and metrics using classes and sub-classes specific enough that they minimize the possibility of double counting specific ecosystem service flows, an important criterion for establishing the creditability of ecosystem services assessments.
- The format of a NESCS code is **WW.XX.YY.ZZZ**, where each set of digits refers to the Environment, Ecological End-product, Direct Use/Non-use, and Direct User classes and subclasses, a useful feature for aggregating items in natural capital accounting tables, representing a unique potential pathway through which changes in Natural Capital may affect human welfare.
- Natural capital accounting efforts are directly relevant to the structured decision-making step “Estimate Consequences” in answering the question, “What ecosystem services models or data are needed to estimate consequences?”
- Natural capital accounting helps EBM practitioners organize: information for a given decision context; ways to approach the identification and valuation of the final ecosystem services for EBM decision making, especially relevant for evaluating management options (e.g., scenarios); and the standardized tracking of specific final ecosystem services over time.

Needs to Advance EBM

- As a new potential tool in the EBM toolbox, EBM practitioners need to examine natural capital accounting principles as part of efforts to analyze different scenario options as part of a decision context.
- Several natural capital accounting approaches (e.g., NESCS and InVEST) could be applied to the same decision context to inform the value-added benefits of natural capital accounting in EBM decisions.

1 Introduction

Management of ecosystems is a complicated affair (Delacámara et al. 2020). Not only do managers have to consider the many interactions in the ecology of a system, they are also mandated to manage the system for human-established goals. These goals might be conservation or restoration focused and oriented for ecological health and integrity or designated uses by humans. In either case the manager needs a way to break down a system into its component parts, quantify them in some way, link those parts to management goals, and assess trends over time, so they can adapt their management accordingly (Arkema et al. 2006; Guerry et al. 2015). Since physical structures, plants, and animals are easier to track than functions or processes, managers often rely on quantification of habitats, species, and abiotic factors monitored on a reoccurring basis. These factors can be thought of as natural capital (Costanza et al. 1997; Millennium Ecosystem Assessment 2005).

Natural capital is the physical components of an ecosystem working together to produce something of value to people (Costanza and Daly 1992). Ecosystem-based managers are focused on natural capital and on what that capital produces in their jurisdiction. Natural capital, just like typical economic capital, can be thought of as the machinery or structures that function to produce goods. In the case of natural capital, these goods are biophysical components that, when used or appreciated by humans, produce a flow of services to the economy and to society more generally, thereby supporting human well-being. Natural capital might act alone or may interact with other natural capital in a series of production functions—relationships between one feature of the environment and the production of another—to produce that which is directly valued, used, or otherwise consumed by humans. That biophysical component of nature that is used directly by humans has been referred to as a final ecosystem good (Landers and Nahlik 2013; DeWitt et al. 2020) or an ecological end-product (Boyd and Banzhaf 2007; Finisdore et al. 2019). These terms denote that the useful biophysical component is a result of production by natural capital. The Ecosystem Service (sometimes referred to as an ES in the literature) then represents a type of transaction between natural capital and humans. At that point, the Final Ecosystem Service (sometimes referred to as FES in the literature) is a transaction that adds value to economic production processes (which generate economic goods and services), or that directly contributes to human well-being (such as from the inspiration provided by a natural landscape).

Value is placed on the flow of ecosystem services produced when humans interact with ecosystems, as natural capital, through their use or appreciation of a final ecosystem good or ecological end-product (Farber et al. 2002). A natural capital perspective of ecosystems may help managers align ecological production and resulting flows of ecosystem services in a way that best supports or enhances human well-being. This chapter explores elements of natural capital, a standardized framework for natural capital accounting, ecosystem services classification in natural capital accounting, and the use of natural capital accounting as a tool for Ecosystem-Based Management (EBM).

2 Elements of Natural Capital

A natural capital perspective lends itself to the application of already accepted socioeconomic frameworks, tools, and approaches, and thus can be a useful way to put the value of ecosystems on par with other socioeconomic values in the system being managed (Wackernagel et al. 1999; Hein et al. 2016). An accounting framework, for example, can be applied to natural capital production to help differentiate what and how things produced by ecosystems are used, by whom, and whether supply is being maintained at levels that satisfy demand. Natural capital accounting is practiced by several countries to quantify and track natural resources such as in water (e.g., Hoekstra 2009), minerals (e.g., Lange 2004), and land accounts (e.g., Weber 2007). Specific ecological production in ecosystems is also being tracked using timber (e.g., Gundimeda et al. 2007), fisheries (e.g., Lange 2004), and wildlife accounts (e.g., Anderson 2003). These accounts have been called for by countries and regions experiencing scarcities, such as in droughts, or in areas that rely heavily on natural resources as the base of their economies. They are becoming more commonplace as the World Bank and the United Nations call for more standardized practices of ecosystem reporting and management (United Nations 2014). A natural capital accounting framework also allows one to tie ecosystem production to the economy by providing the structure to quantify biophysical supply and use by specific ecosystems and users, thereby providing the information needed for valuation and possible translation of that biophysical use into monetary terms (Guerry et al. 2015).

Natural capital accounting requires three types of classifications to work together. First, a list of ecosystems, natural areas, or other geographic groupings that separate out the various production areas with no overlapping areas is needed to cross reference to ecosystem production or supply from natural capital located within the boundaries of the area. Second, the supply from natural capital in the system needs to be apportioned to a list of users. Common to both the supply and use tables within an account, and tying them together, are the various ecosystem services. The quantified ecosystem goods, or ecological end-products (sometimes referred to as EEP in the literature), and their use by users (beneficiaries) for a given time period of a given natural capital account serves as the measure or count of ecosystem service flows that will populate the table cells. Within an accounting framework, for the accounts to balance, the total supply from ecosystems must equal the use by users. It is the list of these ecosystem goods or ecological end-products that is the third type of classification that is needed to construct natural capital accounts.

3 Standard Framework for Natural Capital Accounting

A standard framework for natural capital accounting called the System of Environmental-Economic Accounting, or SEEA, has been developed by the United Nations and partner organizations (Hein et al. 2016). Tangible environmental assets including land, water, minerals, and several resources such as timber and fish are part of the Central Framework, which makes up the core of SEEA, and is designed to quantify natural resources and their contributions to the economy in biophysical and monetary terms (United Nations 2014). The SEEA also houses Experimental Ecosystem Accounts (SEEA-EEA) which are tasked with tracking the extent and condition of ecosystem assets (ecosystems represented by spatial areas; e.g., forests, wetlands, cropland) and the flows of various ecosystem services that these ecosystem assets provide to people and the economy (United Nations 2014). These Experimental Ecosystem Accounts are needed to account for flows from complex ecosystems to people, flows that have not been treated in national accounts as natural resources and traded as commodities in the market. Ecosystem service flows in the Experimental Ecosystem Accounts are by nature difficult to translate into monetary contributions to the economy, as is done with the environmental assets of the Central Framework.

The SEEA-EEA definitions are laborious, detailed, and undergoing updates through 2020, but were constructed with extreme care to meet national accounting needs. Ecosystem assets are a type of capital, similar to machines in a factory, and yield a flow of ecosystem services, just as the factory machines yield a flow of services (e.g., they might stamp thick metal sheets into useful shapes faster and more accurately than humans with hammers could). The SEEA-EEA ecosystem services are "final" as in the definition in Boyd and Banzhaf (2007), and can be distinct from "benefits," where benefits may be the FES contribution to economic production or the FES may go directly to the users without further economic processing or inputs.

Within the SEEA-EEA framework of accounts, figures proposed as frames for "ecosystem services supply and use tables" list "ecosystem services" as rows in the supply-and-use tables (Fig. 1). The supply tables connect ecosystem services and products (in rows) to the types of assets that produce them (in columns) while the use tables connect them to the types of users that use them (in columns). The dark shaded shells in these tables represent "null sets" meaning that no supply or use connection exists for that row-column combination.

Ecosystem services supply-and-use tables are required to balance in accounting. This means that supply of an ecological end-product cannot be larger or smaller than the amount of that ecological end-product used in ecosystem service transactions, nor can it be missing for an ecosystem service to exist. In accounting, a supply with no use cannot represent an ecosystem service because there is no balancing entry in the corresponding use table. A nation cannot have 1000 km of swimmable shoreline and use only 100 km of it in the same accounting year and be able to correctly say they had more than 100 km of shoreline supplied and used as an ecological end-product within an ecosystem service. By the strictures of accounting, the other 900 km cannot be counted as an ecological end-product or part of a flow of ecosystem services during that year, because they were not used. Rows with no entries thus cannot represent ecological end-products, because the existence of transactions (depicted with entries in the intersecting cells) verifies final use, linking supply and demand within the flow of an ecosystem service. The numbers in the cells are a way to quantify particular ecosystem services as a transaction or flow between the supplying ecosystem asset and the human user.

Without use, there is not an ecosystem service, and thus proposed ecological end-products that ultimately do not have direct human use or appreciation (e.g., a wolf or eagle that lives and dies hundreds of miles from any human, say in remote reaches of Alaska, and is not valued specifically for its existence or bequest value) should not be included in accounting supply-and-use tables. Nonuse values are harder to derive in a transaction-value-based framework, which the SEEA is, so are currently not allowed. However, those same wolves or eagles not directly used might still be counted in another type of EEA account other than the supply-and-use tables, perhaps reflecting the condition of a distant ecosystem asset in a condition table or providing a service only between ecosystem assets. Thus, it is critical to define ecosystem services with a use and user component, and not just by identifying potential supply. Ecosystem services classification systems before those developed by the USEPA may have presumed, but did not directly incorporate, a use/user component. Without actual use, things classified as part of an ecosystem service are just ecosystem characteristics and processes that cycle through the environment, and thus not ecological end-products or part of final ecosystem services.

ECOSYSTEM SERVICES SUPPLY TABLE

		Type of economic unit	Type of Ecosystem Unit
Ecosystem services Provisioning services Regulating services Cultural services		A	B
Products		C	D

ECOSYSTEM SERVICES USE TABLE

		Type of economic unit	Type of Ecosystem Unit
Ecosystem services Provisioning services Regulating services Cultural services		E	F
Products		G	H

Fig. 1 SEEA EEA ecosystem services supply-and-use table structure. (Adapted from Table 4.4, December 2015 draft of United Nations 2014)

The U.S. natural capital accounting workgroup (Warnell et al. 2020) recognized the narrowness of the accounting definition and the difficulty of measuring actual ecosystem services and put important ecological measures and measures of known precursors of ecosystem services into an ecosystem condition account. For example, eagles and beaches that humans do not interact with might be tracked in a condition account. Accounts that are balanced, totaled, and specifically slated for integration with Standard National Accounts (SNA, the international accounting structure that provides a common economic foundation before environmental accounting is applied) need to have safeguards against double-counting. The supply-and-use tables are intended to be monetized and to be integrated with the SNA accounts in common terms. Condition accounts are not currently proposed to be integrated into the SNA or to be translated into dollar values, so can represent characteristics, processes and stocks of things (e.g., eagles) in the environment that are not directly used as part of ecosystem services.

4 Classification of Flows of Ecosystem Services for Natural Capital Accounting

The US EPA's National Ecosystem Services Classification System (NESCS) offers all three of the classification elements (i.e., types of supplying ecosystems, types of ecological end-products supplied and used, and types of users with different use demands) that are needed to do natural capital accounting (USEPA 2015). The NESCS and associated approach were designed to help identify and reference flows of services from ecosystems to human beings in a mutually exclusive way, which is critical for natural capital accounting.

The NESCS has a four-part structure (Fig. 2), with each part populated by a hierarchical set of classes and subclasses, which are intended to be as non-overlapping and comprehensive as possible for identifying distinct ecosystem service flows.

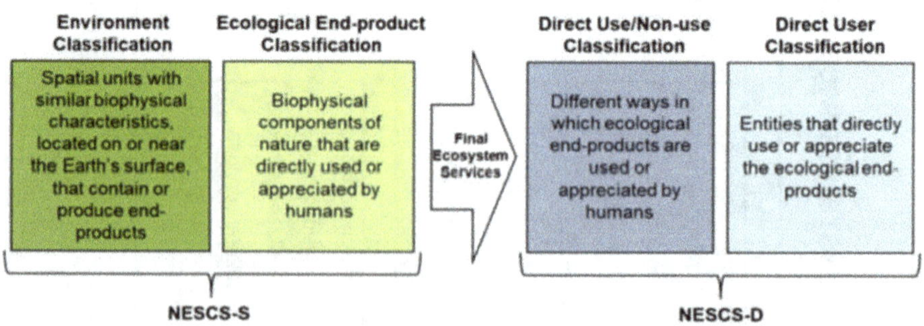

Fig. 2 NESCS four-part structure (from USEPA 2015)

The four parts of NESCS can be divided into two sides that are relevant for accounting: the supply-side (NESCS-S; Fig. 2 left) and demand-side (NESCS-D; Fig. 2 right) of the system. The NESCS-S side represents the source of ecosystem service flows. It characterizes different types of natural capital as environmental classes and represents the output of their ecological production as classes of ecological end-products. The NESCS-D side represents the recipients of the ecosystem service flows, by classifying the human users of the ecological end-products and how they are used. Taken together, the four parts hold the information necessary to designate individual flows of ecosystem services from natural capital to humans.

The NESCS adopts a nested hierarchical structure so that each part can be represented at multiple levels of aggregation or detail. The initial sets of classes presented by NESCS are meant to provide a high-level classification that provides a mutually exclusive partitioning of classes, each with its own set of subclasses. These initial classes and hierarchical structure were developed to meet the design requirements not found in other classification systems. The NESCS is intended to provide a broad and comprehensive structure capable of covering all the different ecosystem services that humans value from nature. In this regard, it is consistent with the total economic valuation (TEV) approach (e.g., Jewhurst and Mazzotta 2016) that is often used as a framework for valuing natural resources and environmental benefits. The NESCS is also expandable enough to identify specific classes and subclasses for determining environmental and valuation measures and metrics. It has classes that are intuitively separate from each other and thus specific enough that, when combined to represent flows of ecosystem services, they minimize the possibility of double counting specific ecosystem service flows, an important criterion for establishing the creditability of ecosystem services assessments (Fu et al. 2011). Moreover, by focusing on ecosystem services and the ecological end-products produced by natural systems, the NESCS structure also minimizes the possibility of double counting the contribution of "intermediate" ecological production processes in estimates of ecosystem service values (Landers et al. 2016). This is because the value of intermediate processes should be embedded within, and thus fully captured by, the value of final ecosystem services.

Each unique ecosystem service can also be easily referenced and identified by a NESCS code. The general format of the code is **WW.XX.YY.ZZZ**, where each set of digits (e.g., WW, XX, etc.) refer to the Environment, Ecological End-product, Direct Use/Non-use and Direct User classes and subclasses, respectively, as described below (Fig. 3). One example from Fig. 3, the WW.XX.YY.ZZZ NESCS code "41.12.11.2ZZ" would thus represent household extractive use of liquid water from a deciduous forest, as for a hiker taking a drink of water from a mountain stream below the tree line. Digits can be added or removed from each part of the four-part code to represent any further breakdown into more detailed sub-classes or to represent the roll up of subclasses into larger classes. While the ability to roll up classes into fewer classes is a useful feature for aggregating items in natural capital accounting tables, it may be less so for other uses of NESCS not covered here, such as the identification of metrics for each Flow of Ecosystem Service, mapping of areas of supply, or setting up scenario variables, which would primarily take advantage of

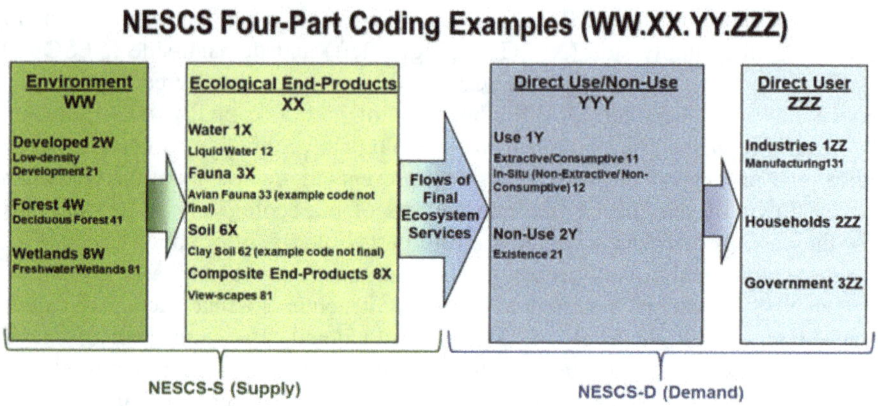

Fig. 3 NESCS coding examples (WW.XX.YY.ZZZ)

the flexible yet comprehensive nature of the hierarchical classification system. Each unique combination of one class or subclass from each of the four parts of NESCS defines a separate Flow of Ecosystem Service. In other words, it represents a unique potential pathway through which changes in Natural Capital may affect human welfare. The ability to define different combinations of classes allows the NESCS structure to be flexible and comprehensive. For example, the same Ecological End-Product category may be used in multiple ways or a single use category can be linked to multiple different user categories.

The first part of NESCS is the Environment classification (Fig. 3; far left box; "**WW**" part of the code), which is currently based on the system specified by Landers and Nahlik (2013). It spatially divides the Earth into non-overlapping areas with similar biophysical characteristics that, when taken together, can completely cover the surface of the Earth. The NESCS currently designates Environment classes down to a second-level of hierarchy, including a two-digit numeric coding structure, which provides a short-hand notation for the hierarchy and a numeric identifier for each element within each part. Example subclasses for the environment might include deciduous forests, freshwater wetlands, or low-density developments. When used within a four-part code designating a final ecosystem service, the environment classes and subclasses specifically refer to the environment in which the relevant ecological end-product is used or appreciated.

The second part of NESCS is the Ecological End-products classification (Fig. 3; second to left box; "XX" part of the code). End-products represent the components in nature that humans most directly use or appreciate (Farber et al. 2002). In its most aggregate form, classes of ecological end-products include Fauna, Flora, Water, Soil, Air, etc. As with Environment classes, these are subsequently broken down into a second-level hierarchy of subclasses, with codes designated for each one. Examples might include liquid water, avian fauna, or clay soils. One of the challenges in constructing this end-product classification is defining mutually exclusive categories while also recognizing that there can be substantial complexity and diversity in what people care about in nature. In addition to individual end-products, people often care about combinations of them. For example, people may value an entire landscape as more than the sum of individual value for the flora, fauna, water, etc., that are parts of the landscape. The NESCS includes a category called "Composite End-products" with sub-classes representing different types of natural features or phenomena that directly matter to humans but can be thought of as combinations of the other end-products. An example might be view-scapes. It is important to recognize that the joint nature of composite end-products may be problematic when applied to an accounting framework where there can be no overlaps between entries in the supply-and-use tables.

The third part of NESCS is the Direct Use/Non-Use classification (Fig. 3; second to right box; "**YY**" part of the code). The classes and sub-classes in this component describe distinct ways in which end-products can be directly used or appreciated by humans, again with the objective of providing categories that are non-overlapping and as comprehensive as possible. Examples include extraction of natural resources, such as wood, for transformation into economic products, such as timber, or non-extractive in-situ use associated with outdoor recreation, such as birdwatching. Consistent with the TEV framework, it includes separate "use" and "non-use" categories which make up the first hierarchical level. These use and non-use classes are then further subdivided into mutually exclusive extractive and in-situ use classes at the second-level of the hierarchy. Non-use use classes include existence, bequest, or other uses where humans do not have direct contact or physical use of the ecological end-product but might have a value associated with knowing that something exists or that something will be around for their descendants to enjoy.

The fourth and last NESCS part is the Direct User classification (Fig. 3; far right box; "**ZZZ**" part of the code). This component defines the separate economic sectors though which people directly use or appreciate end-products. Following established classification structures adopted by the U.S. Census Bureau and United Nations, the first level includes broad sectors of the economy, here: Industry, Households, and Government. To further subdivide the industry class, it adopts the existing North American Industrial Classification System (NAICS) and coding system, which is the standard used by U.S. federal statistical agencies in classifying business establishments (United States 2017). An example is the Manufacturing Industry sector which has a three-digit code, one digit for industry and two additional digits for the sector. Unlike commercial establishments, which tend to specialize in certain productive activities and can therefore be assigned to individual NAICS categories, households and governments do not specialize in the same way. For this reason, they are not yet divided into sub-classes as are NAICS categories. They currently are presented as first-level hierarchical classes, with further designation to subclasses remaining an option. One way to differentiate the many ways households and governments interact with nature is through the combination of the household or government user class with the different previously described use/non-use classes.

The NESCS approach helps guide the construction of supply-and-use tables within SEEA-EEA in that it clearly separates ecological end-products from ecosystem services by defining ecological end-products (i.e., components in nature that humans most directly use or appreciate) as part of a flow of ecosystem services (i.e., "transactions" that include human use of those components by a specific user). Without this specificity in terminology, confusion of, and mismatches between, what ecologists and other natural scientists measure as that which is supplied by an ecosystem asset and what social scientists measure as that which is used by a user could slow or stall efforts to generate useful estimates of ES for accounting.

The NESCS identifies and classifies components of final ecosystem services according to both the environmental supplier and human user of the service in order to identify where certain metrics best fit within the natural capital accounting structure (Fig. 4a). The NESCS can help distinguish between ecosystem services eligible for inclusion in supply-and-use accounts in standard statistical frameworks, such as SEEA-EEA. By viewing the standard statistical frameworks through the NESCS lens, one can separate ecosystem characteristics and processes that some have considered to be ecosystem services from those ecosystem services that are allowable by the structure of supply-and-use accounting (finality in accounting is proven by direct use; Boyd and Banzhaf 2007; Landers and Nahlik 2013; Landers et al. 2016; USEPA 2015; United Nations et al. 2017). The NESCS offers a practical and flexible structure and set of rules for naming ecosystem service flows as the central object of measure in the supply-and-use account.

Using the four component NESCS classification, "ecological end-products" may be a more functional label for types of SEES-EEA supply and use table row names than the label "ecosystem services." (Fig. 4b) Whether the rows in a SEEA-EEA supply-and-use table are named ecological end-products or ecosystem services however, cells in the rows sit at the intersection with columns that either designate supply of an ecological end-product by an ecosystem asset, or designate use of an ecological end-product by a particular user. Since an ecosystem service can be thought of as a transaction between nature and humans, row names must match in both the supply-and-use tables since an individual cell represents the common interaction point, or transaction (i.e., ecosystem service), between the two tables. For any one ecosystem service type defined by ecosystem assets, ecological end-products and types of (use and) users, supply must match demand for the account to balance. Quantities in cells within EEA supply and use tables should, thus, be measures of ecosystem services in that they simultaneously represent both the supply and use of a particular ecological end-product (i.e., the row).

5 Natural Capital Accounts as Tools for Ecosystem-Based Management (EBM)

Frameworks and approaches to develop ecosystem services classification and accounting systems focus on developing common, shared language and consistent approaches for identifying, assessing, and accounting of ecosystem goods and services for specific human benefits (DeWitt et al. 2020). These approaches are

(a)

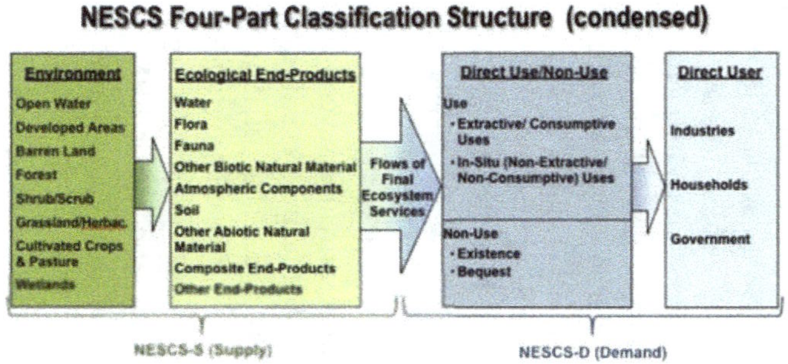

(b)

ECOSYSTEM SERVICE SUPPLY TABLE

		SEEA-EEA: Type of Economic Unit (columns)	SEEA-EEA: Type of Ecosystem Unit (columns)
		NESCS: Direct User*	NESCS: Environment
SEEA-EEA: Ecosystem Services (each type or class is a row)	NESCS: Ecological End-Product	A: no data here, Economic Units cannot supply ES	B: cells indicate supply of final ES by Ecosystem Unit; per NESCS, supply of Ecological End-Product by Environment to Direct User = supply of Final Ecosystem Service
SEEA-EEA: Economic Products (rows)	no NESCS equivalent	C: cells indicate supply of economic products by Economic Units; no ES supply or use here	D: no data here, Ecosystem Units cannot supply economic products

ECOSYSTEM SERVICE USE TABLE

		SEEA-EEA: Type of Economic Unit (columns)	SEEA-EEA: Type of Ecosystem Unit (columns)
		NESCS: Direct User*	NESCS: Environment
SEEA-EEA: Ecosystem Services (each type or class is a row)	NESCS: Ecological End-Product	E: cells indicate use of final ES by Economic Unit; per NESCS, use of Ecological End-Product by Direct User = use of Final Ecosystem Service	F: cells indicate use of ES by Ecosystem Units, not people or institutions, which the UNSD et al. 2017 defines as "intermediate ecosystem services"
SEEA-EEA: Economic Products (rows)	no NESCS equivalent	G: cells indicate use of economic products by Economic Units; no ES supply or use here	H: no data here, Ecosystem Units cannot use Economic Products

Fig. 4 Alignment of SEEA EEA ecosystem services supply and use tables with the National Ecosystem Services Classification System. (Adapted from Warnell et al. 2020). (**a**) National Ecosystem Services Classification System structure (adapted from USEPA 2015). (**b**) NESCS structure (**a**) superimposed on SEEA EEA ecosystem services supply and use table structure (Fig. 1) to show alignment. ∗ Direct Users could be associated with Direct Uses by nesting or dividing columns under Direct User (Economic Unit), so an Industry, Household, or Government might have Extractive or In-Situ uses of an Ecological End-Product

designed to be used primarily by scientists but developed to be useful for resource managers and ecosystem-based management practitioners. DeWitt et al. (2020) presents a crosswalk examining ecosystem services from a structured decision-making perspective with assorted EBM frameworks, including AQUACROSS (Piet et al. 2017; Delacámara et al. 2020). Natural capital accounting efforts are directly relevant to the structured decision-making step "Estimate Consequences" in answering the question: "What ecosystem services models or data are needed to estimate consequences?" This is translatable to Step 4 in the AQUACROSS framework, "Implementation, Monitoring and Evaluation," which focuses on either assessment of current state, or the application of forecasting and scenario tools to examine/predict consequences among alternative management decisions. This potential application of natural capital accounting also maps onto the step of "Evaluating Management Measures" (Cormier et al. 2017) and "Scenarios" (Foley et al. 2013) in other EBM frameworks. The InVEST (Integrated Valuation of Ecosystem Services and Tradeoffs) is one example of a modeling approach to utilize natural capital accounting for decision-making purposes (Daily et al. 2009); for further discussion on InVEST in this volume, the reader is directed to Fulford et al. (2020) and Lewis et al. (2020).

Overall, the natural capital accounting framework provides several important tools for EBM practitioners. First, the application of structured classifications for ecosystem components, human beneficiaries (users), and a suite of flows of final ecosystem services helps EBM practitioners organize information for a given decision context. Second, a natural capital accounting framework perspective informs the effort to identify and value the final ecosystem services for EBM decision making and to track them over time in a standardized way. Finally, the structured nature of natural capital accounting frameworks lends itself to supporting important EBM steps focused on evaluating management options among a suite of EBM alternatives for a given decision context (sensu DeWitt et al. 2020). With this perspective, EBM practitioners are encouraged to learn more about natural capital accounting in general, the meticulously detailed organization of the United Nation's SEEA-EEA (United Nations 2014), and the U.S. EPA's approach to natural capital accounting using NESCS (USEPA 2015), so they might add these tools to their EBM toolbox.

Disclaimer This chapter has been subjected to Agency review and has been approved for publication. The views expressed in this paper are those of the author(s) and do not necessarily reflect the views or policies of the U.S. Environmental Protection Agency.

References

Anderson, H. J. (2003). *An econometric analysis of the wildlife market in South Africa*. Dissertation, University of Cape Town.

Arkema, K. K., Abramson, S. C., & Dewsbury, B. M. (2006). Marine ecosystem-based management: From characterization to implementation. *Frontiers in Ecology and the Environment, 4* (10), 525–532.

Boyd, J. W., & Banzhaf, S. (2007). What are ecosystem services? The need for standardized environmental accounting units. *Ecological Economics, 63*, 616–626.

Cormier, R., Kelble, C. R., Anderson, M. R., Allen, J. I., Grehan, A., & Gregersen, O. (2017). Moving from ecosystem-based policy objectives to operational implementation of ecosystem-based management measures. *ICES Journal of Marine Science, 74*, 406–413.

Costanza, R., & Daly, H. E. (1992). Natural capital and sustainable development. *Conservation Biology, 6*(1), 37–46.

Costanza, C., d'Arge, R., de Groot, R., Farber, S., Grasso, M., Hannon, B., Limburg, K., Naeem, S., O'Neill, R. V., Paruelo, J., Raskin, R. G., Sutton, P., & van den Belt, M. (1997). The value of the world's ecosystem services and natural capital. *Nature, 387*, 253–260.

Daily, G. C., Polasky, S., Goldstein, J., Kareiva, P. M., Mooney, H. A., Pejchar, L., Ricketts, T. H., Salzman, J., & Shallenberger, R. (2009). Ecosystem services in decision making: Time to deliver. *Frontiers in Ecology and the Environment, 7*(1), 21–28.

Delacámara, G., O'Higgins, T., Lago, M., & Langhans, S. (2020). Ecosystem-based management: moving from concept to practice. In T. O'Higgins, M. Lago, & T. H. DeWitt (Eds.), *Ecosystem-based management, ecosystem services and aquatic biodiversity: Theory, tools and applications* (pp. 39–60). Amsterdam: Springer.

DeWitt, T. H., Berry, W. J., Canfield, T. J., Fulford, R. S., Harwell, M. C., Hoffman, J. C., Johnston, J. M., Newcomer-Johnson, T. A., Ringold, P. L., Russel, M. J., Sharpe, L. A., & Yee, S. J. H. (2020). The final ecosystem goods and services (FEGS) approach: A beneficiary-centric method to support ecosystem-based management. In T. O'Higgins, M. Lago, & T. H. DeWitt (Eds.), *Ecosystem-based management, ecosystem services and aquatic biodiversity: Theory, tools and applications* (pp. 127–148). Amsterdam: Springer.

Farber, S. C., Costanza, R., & Wilson, M. A. (2002). Economic and ecological concepts for valuing ecosystem services. *Ecological Economics, 41*(3), 4375–4392.

Finisdore, J., Rhodes, C., Haines-Young, R., Maynard, S., Wielgus, J., Dvarskas, A., Houdet, J., Quétier, F., Ding, H., Soulard, F., & Van Houtven, G. (2019). Expanding the field of ecosystem services practitioners—18 benefits from using classification systems. Sustainable Flows—Improving Financial & Ecosystem Services Flows. Retrieved October 9, 2019, from https://www.researchgate.net/profile/John_Finisdore/publication/329402275_Expanding_the_field_of_ecosystem_services_practitioners-18_benefits_from_using_classification_systems_Version_11_March_2019/links/5c794e60458515831f7b2bec/Expanding-the-field-of-ecosystem-services-practitioners-18-benefits-from-using-classification-systems-Version-11-March-2019.pdf.

Foley, M. M., Armsby, M. H., Prahler, E. E., Caldwell, M. R., Erickson, A. L., Kittinger, J. N., Crowder, L. B., & Levin, P. S. (2013). Improving ocean management through the use of ecological principles and integrated ecosystem assessments. *BioScience, 63*(8), 619–631.

Fu, B. J., Su, C. H., Wei, Y. P., Willett, I. R., Lü, Y. H., & Liu, G. H. (2011). Double counting in ecosystem services valuation: Causes and countermeasures. *Ecological Research, 26*(1), 1–14.

Fulford, R. S., Heymans, S. J. J., & Wu, W. (2020). Mathematical modelling for ecosystem-based management (EBM) and ecosystem goods and services (EGS) assessment. In T. O'Higgins, M. Lago, & T. H. DeWitt (Eds.), *Ecosystem-based management ecosystem services and aquatic biodiversity: Theory, tools and applications* (pp. 275–290). Amsterdam: Springer.

Guerry, A. D., Polasky, S., Lubchenco, J., Chaplin-Kramer, R., Daily, G. C., Griffin, R., Ruckelshaus, M., Bateman, I. J., Duraiappah, A., Elmqvist, T., & Feldman, M. W. (2015). Natural capital and ecosystem services informing decisions: From promise to practice. *PNAS, 112*(24), 7348–7355.

Gundimeda, H., Sukhdev, P., Sinha, R. K., & Sanyal, S. (2007). Natural resource accounting for Indian states—Illustrating the case of forest resources. *Ecological Economics, 61*(4), 635–649.

Hein, L., Bagstad, K., Edens, B., Obst, C., de Jong, R., & Lesschen, J. P. (2016). Defining ecosystem assets for natural capital accounting. *PLoS One, 11*(11), e0164460.

Hoekstra, A. Y. (2009). Human appropriation of natural capital: A comparison of ecological footprint and water footprint analysis. *Ecological Economics, 68*(7), 1963–1974.

Jewhurst, S., & Mazzotta, M. (2016). Economic tools for managing nitrogen in coastal watersheds. *EPA/600/R-16/036*. U.S. Environmental Protection Agency.

Landers, D. H., & Nahlik, A. M. (2013). Final Ecosystem Goods and Services Classification System (FEGS-CS). *EPA/600/R-13/ORD-004914*. U.S. Environmental Protection Agency.

Landers, D., Nahlik, A., & Rhodes, C. R. (2016). The beneficiary perspective—Benefits and beyond. In M. Potchin, R. Haines-Young, R. Fish, & K. Turner (Eds.), *Routledge handbook of ecosystem services* (pp. 74–87). New York: Routledge.

Lange, G. M. (2004). Wealth, natural capital, and sustainable development: Contrasting examples from Botswana and Namibia. *Environmental and Resource Economics, 29*(3), 257–283.

Lewis, N. S., Marois, D. E., Littles, C. J., & Fulford, R. S. (2020). Projecting changes to coastal and estuarine ecosystem goods and services - models and tools. In T. O'Higgins, M. Lago, & T. H. DeWitt (Eds.), *Ecosystem-based management, ecosystem services and aquatic biodiversity: Theory, tools and applications* (pp. 235–254). Amsterdam: Springer.

Millennium Ecosystem Assessment. (2005). *Ecosystems and human well-being: Synthesis*. Washington, DC: Island Press.

Piet, et al. (2017). Making ecosystem-based management operational: Deliverable 8.1 Executive Summary. Report as part of the Horizon 2020 project AQUACROSS (Knowledge, Assessment, and Management for AQUAtic Biodiversity and Ecosystem Services aCROSS EU policies). Retrieved October 9, 2019, from https://aquacross.eu/sites/default/files/AQUACROSS%20Executive%20Summary%20D8.1_v2_18062018.pdf.

United Nations. (2014). *System of environmental-economic accounting 2012—Experimental ecosystem accounting*. Retrieved October 9, 2019, from https://seea.un.org/sites/seea.un.org/files/seea_eea_final_en_1.pdf.

United States. (2017). North American Industry Classification System manual. Retrieved October 9, 2019, from https://www.census.gov/eos/www/naics/2017NAICS/2017_NAICS_Manual.pdf.

United States Environmental Protection Agency (USEPA). (2015). National Ecosystem Services Classification System (NESCS): Framework design and policy application. *EPA-800-R-15-002*. U.S. Environmental Protection Agency.

Wackernagel, M., Onisto, L., Bello, P., Linares, A. C., López Falfán, I. S., Garcıa, J. M., Guerrero, A. I. S., & Guerrero, M. G. S. (1999). National natural capital accounting with the ecological footprint concept. *Ecological Economics, 29*(3), 375–390.

Warnell, K., Russell, M., Rhodes, C., Bagstad, K., Olander, L., Nowak, D, Poudel, R., Glynn, P., Hass, J., Hirabayashi, S., Carter Ingram, J., Matuszak, J., Oleson, K., Posner, S., and Villa, F. (2020). Testing ecosystem accounting in the United States: A case study for the Southeast. *Ecosystem Services, 43,* https://doi.org/10.1016/j.ecoser.2020.101099.

Weber, J. L. (2007). Implementation of land and ecosystem accounts at the European Environment Agency. *Ecological Economics, 61*(4), 695–707.

PERMISSIONS

All chapters in this book were first published by Springer; hereby published with permission under the Creative Commons Attribution License or equivalent. Every chapter published in this book has been scrutinized by our experts. Their significance has been extensively debated. The topics covered herein carry significant findings which will fuel the growth of the discipline. They may even be implemented as practical applications or may be referred to as a beginning point for another development.

The contributors of this book come from diverse backgrounds, making this book a truly international effort. This book will bring forth new frontiers with its revolutionizing research information and detailed analysis of the nascent developments around the world.

We would like to thank all the contributing authors for lending their expertise to make the book truly unique. They have played a crucial role in the development of this book. Without their invaluable contributions this book wouldn't have been possible. They have made vital efforts to compile up to date information on the varied aspects of this subject to make this book a valuable addition to the collection of many professionals and students.

This book was conceptualized with the vision of imparting up-to-date information and advanced data in this field. To ensure the same, a matchless editorial board was set up. Every individual on the board went through rigorous rounds of assessment to prove their worth. After which they invested a large part of their time researching and compiling the most relevant data for our readers.

The editorial board has been involved in producing this book since its inception. They have spent rigorous hours researching and exploring the diverse topics which have resulted in the successful publishing of this book. They have passed on their knowledge of decades through this book. To expedite this challenging task, the publisher supported the team at every step. A small team of assistant editors was also appointed to further simplify the editing procedure and attain best results for the readers.

Apart from the editorial board, the designing team has also invested a significant amount of their time in understanding the subject and creating the most relevant covers. They scrutinized every image to scout for the most suitable representation of the subject and create an appropriate cover for the book.

The publishing team has been an ardent support to the editorial, designing and production team. Their endless efforts to recruit the best for this project, has resulted in the accomplishment of this book. They are a veteran in the field of academics and their pool of knowledge is as vast as their experience in printing. Their expertise and guidance has proved useful at every step. Their uncompromising quality standards have made this book an exceptional effort. Their encouragement from time to time has been an inspiration for everyone.

The publisher and the editorial board hope that this book will prove to be a valuable piece of knowledge for researchers, students, practitioners and scholars across the globe.

LIST OF CONTRIBUTORS

Fiona E. Culhane and Leonie A. Robinson
School of Environmental Sciences, University of Liverpool, Liverpool, UK

Ana I. Lillebø
Department of Biology & CESAM, Campus Universitário de Santiago, University of Aveiro, Aveiro, Portugal

Leah M. Sharpe
U.S. EPA, Gulf Ecosystem Measurement and Modeling Division, Gulf Breeze, FL, USA

Connie L. Hernandez and Chloe A. Jackson
Oak Ridge Institute for Science and Education (ORISE) at U.S. EPA, Pacific Ecological Systems Division, Newport, OR, USA

Anne Marie O'Hagan
MaREI: The SFI Research Centre for Energy, Climate and Marine, Environmental Research Institute: Beaufort Building, University College Cork, Cork, Ireland

Andrea Funk, Daniel Trauner and Thomas Hein
University of Natural Resources & Life Sciences, Vienna, Austria

Florian Borgwardt
University of Natural Resources & Life Sciences, Vienna, Austria

Rebekah Gibble
U.S. Fish and Wildlife Service, Boynton Beach, FL, USA

Lori Miller
U.S. Fish and Wildlife Service, Vero Beach, FL, USA

Michael Elliott
Department of Biological & Marine Sciences, University of Hull, Hull, UK
International Estuarine & Coastal Specialists Ltd., Leven, UK

Timothy G. O'Higgins
MaREI Center for Marine and Renewable Energy Ireland, Environmental Research Institute, University College Cork, Cork, Ireland

Richard S. Fulford
US Environmental Protection Agency, Gulf Ecosystem Measurement and Modeling Division, Gulf Breeze, FL, USA

Sheila J. J. Heymans
European Marine Board, Oostende, Belgium

Wei Wu
University of Southern Mississippi, Hattiesburg, MS, USA

John M. Johnston
US EPA Office Research and Development, Center for Environmental Measurement and Modeling, Athens, GA, USA

Robert J. Zomer and Ming-cheng Wang
World Agroforestry Center, Kunming, China

Gerjan Piet
Wageningen Marine Research, IJmuiden, The Netherlands

Gonzalo Delacámara
IMDEA Water Institute, Alcalá de Henares, Madrid, Spain

Marloes Kraan
Wageningen Marine Research, IJmuiden, The Netherlands
Environmental Policy Group, Wageningen University, Wageningen, The Netherlands

Christine Röckmann
Wageningen Marine Research, IJmuiden, The Netherlands
Wageningen Economic Research, The Hague, The Netherlands

Manuel Lago
Ecologic Institute, Berlin, Germany

Martin Le Tissier
MaREI Center for Marine and Renewable Energy, Beaufort Building, Environmental Research Institute, University College Cork, Co. Cork, Ireland

Heliana Teixeira, Ana Genua-Olmedo and António A. J. Nogueira
CESAM & Department of Biology, University of Aveiro, Aveiro, Portugal

Javier Martínez-López
BC3 - Basque Centre for Climate Change, Leioa, Spain

Asya Marhubi and Gonzalo Delacámara
IMDEA Water Institute, Parque Científico Tecnológico de la Universidad de Alcalá, Alcalá de Henares, Madrid, Spain

Verena Mattheiß and Pierre Strosser
AcTeon - 5 Place Sainte-Catherine, Colmar, France

Marc Russell
US Environmental Protection Agency, Center for Computational Toxicology and Exposure, Gulf Breeze, FL, USA

Charles Rhodes
Washington, DC, USA

George Van Houtven and Paramita Sinha
RTI International, Durham, NC, USA

Katherine Warnell
Nicholas Institute for Environmental Policy Solutions, Duke University, Durham, NC, USA

Matthew C. Harwell
Gulf Ecological Modeling and Measurement Division, U.S. Environmental Protection Agency, Gulf Breeze, FL, USA

Index

A

Adaptive Management, 3, 18, 46, 52-53, 58, 65, 73-74, 79, 92, 95, 101, 115, 122, 144, 165-167, 171, 174, 176, 179, 188, 196, 198, 207, 217, 223-224

Aquatic Ecosystems, 19, 21, 52, 56, 75, 77, 180, 182, 197, 231

B

Beneficiaries, 16, 22-23, 26, 28-29, 34, 36, 40-41, 59, 130, 139-140, 158, 206, 225, 232, 235, 244

Beneficiary Group, 29, 33-34, 38-39

Bioclimatic Zones, 143, 148-149, 151-157, 162

Biodiversity, 3, 5-6, 9, 15-17, 19-21, 28, 42-43, 45-53, 104-105, 116, 122-127, 131-132, 171, 179-180, 182, 184, 190-193, 195-199, 223, 225-231, 245-246

Biological Diversity, 5, 46, 48, 61-62, 119, 185, 191, 197, 203

Biomass, 209-210

Biotic Groups, 7, 69

C

Capital Accounting, 43, 232-235, 238-239, 242, 244-246

Climate Change, 3, 5-6, 11, 17, 35, 56-57, 63, 68, 110, 143-144, 146-151, 156, 158-162, 183-184, 195-196, 200, 202-204, 210, 223, 226-227, 229-230

Coastal Change, 184

Coastal Zones, 122, 127, 184, 186

Commonalities, 23, 33, 38

Conceptual Framework, 103, 105, 115-116, 119-120, 122, 165

Cost-effectiveness, 72, 74

D

Decision Makers, 6, 9, 18, 22, 24-36, 38-39, 41, 92, 132, 138-139, 146, 161, 170

Double Counting, 26, 233, 239, 245

E

Ecological Process, 103, 122

Economic Value, 17, 147, 189, 217

Ecosystem, 1-7, 9-23, 26-28, 33-35, 38-59, 61-65, 68-70, 72-73, 75-80, 83, 86, 88, 90-95, 158-164, 166-173, 175-176, 178-198, 200, 207-209, 222-242, 244-246

Ecosystem Approach, 46-53, 55-59, 61-63, 104, 116, 119, 123-124, 142, 180, 185, 188, 192-193

Ecosystem Asset, 236, 242

Ecosystem Condition, 16, 238

Ecosystem Restoration, 91-92, 95, 100

Ecosystem Services, 1-7, 9-12, 15-22, 26-28, 33, 35, 42-43, 68-70, 75-77, 79, 90-91, 99-100, 103-104, 207-209, 214-215, 218, 222-224, 226-239, 241-242, 244-246

Ecosystem State, 12-14, 16, 117

Ecosystem Types, 73

Ecosystem-based Management, 1, 3, 6, 15, 18-23, 35, 106, 120-127, 129, 138, 140, 144, 146, 163-164, 167, 175, 179-182, 184-185, 229-230, 232, 242, 244-246

Effective Management, 61, 99, 104, 148

Endangered Species, 48, 83, 87, 91, 189

Environmental Impacts, 54, 116, 216

Environmental Law, 55, 100, 104, 126, 188, 190, 192

Environmental Management, 18, 43, 45, 66, 76, 101, 104, 124, 126, 142, 160, 162, 166, 180, 182, 193, 229

Environmental Movement, 83
Environmental Problems, 49, 64, 103, 105, 127, 185
Environmental Quality, 106, 176

F

Fisheries, 2, 43, 46, 48, 50, 54-56, 59-60, 62-63, 66, 141-142, 146, 158, 170, 175, 179-182, 187, 191-192, 197, 202, 204-206, 211, 217, 225, 234
Fragmentation, 53, 64, 187, 223

H

Habitat, 2, 4, 11-12, 16, 18, 21, 40-41, 48, 51-52, 60, 64, 69, 71-72, 76, 79-80, 83, 87-88, 91, 136-137, 140-141, 144, 150, 158, 176, 189, 208, 211, 215, 217-218, 220, 225
Human Activities, 11, 17-18, 48-49, 58, 63-65, 69, 72, 75, 103-105, 111, 113-114, 117, 120, 166, 169, 173, 180-181, 184-187, 190, 193, 208, 215, 218
Human Capital, 111, 119, 177
Human Pressures, 109, 117, 181
Human User, 236, 242
Human Wellbeing, 2, 9, 16-18, 28, 51, 113, 177, 216, 234
Hydrodynamics, 92, 199

M

Management Actions, 2, 39, 58, 86, 104, 110, 207
Management Decisions, 22, 25, 27, 41, 83, 91-92, 95, 138, 244
Management Measures, 43, 51, 54, 77, 109, 111, 114-115, 117, 121, 126, 141, 164, 166, 170, 172, 176, 178, 180, 215, 218, 226, 244-245
Management Tool, 121, 202-203
Marine Ecosystems, 2, 19, 48, 52, 54-55, 57, 109, 116-117, 123, 125, 139, 141, 161, 181, 186-187, 189
Marine Environments, 1-3, 6, 11, 16, 21, 42, 104-105, 123, 127, 185
Marine Protected Areas, 54, 62, 123, 125, 137, 175, 231

N

Natural Capital, 5, 19, 43, 62, 76, 123, 146-147, 158-159, 165, 198, 200, 215, 218, 224, 228, 232-235, 238-240, 242, 244-246
Natural Environment, 113, 117, 188
Nonuse Values, 236

O

Ownership, 130

P

Policy Decisions, 104
Policy Objectives, 14-15, 75, 138, 164-165, 168, 172, 179-180, 195-196, 200, 216, 245
Prioritization, 7, 9, 22, 28-41, 43, 68, 70, 75, 207, 214-215

R

Regional Seas, 1, 12, 19, 21, 48-49, 53-54, 126, 194
Resource Management, 9, 39, 43, 63, 98, 116, 161, 185-186, 216
Risk Assessment, 7, 11, 51, 88, 104-105, 115, 117, 119, 135, 140, 164, 168, 170, 180-182, 215
Risk Management, 67, 104, 115, 121, 138, 172, 180

S

Scoping Tool, 22-23, 27-30, 34-35, 43
Sediment Retention, 147, 149
Social Sciences, 26, 224
Social-ecological System, 7, 21, 73-74, 126, 163, 166, 169, 171, 195, 208, 218
Stakeholder, 2, 7, 18, 22, 25-43, 58-60, 65, 74, 78-80, 84, 97, 99-100, 129, 131-132, 137, 167-169, 171, 173-176, 179, 182, 188, 196, 198, 205, 214-218, 224, 226, 229
Stakeholder Input, 74, 226

State Information, 12-14

State-service Relationship, 12-14

Sustainability Goals, 57, 165

Sustainable Development, 3, 5-6, 17, 45, 49, 58-59, 116, 146, 149, 158, 177, 185, 187-192, 194, 205, 208, 217, 245-246

T

Terminology, 184, 187-188, 192, 242

Territorial Management, 196, 200, 202-203, 217, 224, 226-227

Timber Production, 147

Typologies, 3, 16, 168

V

Vegetation, 83, 143-144, 148-150, 156, 220

www.ingramcontent.com/pod-product-compliance
Lightning Source LLC
LaVergne TN
LVHW051047220926
848485LV00001BA/40
* 9 7 8 1 6 4 7 4 0 3 7 7 5 *